AutoCAD® 2015
Beginning
and
Intermediate

LICENSE, DISCLAIMER OF LIABILITY, AND LIMITED WARRANTY

By purchasing or using this book (the "Work"), you agree that this license grants permission to use the contents contained herein, but does not give you the right of ownership to any of the textual content in the book or ownership to any of the information or products contained in it. *This license does not permit uploading of the Work onto the Internet or on a network (of any kind) without the written consent of the Publisher.* Duplication or dissemination of any text, code, simulations, images, etc. contained herein is limited to and subject to licensing terms for the respective products, and permission must be obtained from the Publisher or the owner of the content, etc., in order to reproduce or network any portion of the textual material (in any media) that is contained in the Work.

MERCURY LEARNING AND INFORMATION ("MLI" or "the Publisher") and anyone involved in the creation, writing, or production of the companion disc, accompanying algorithms, code, or computer programs ("the software"), and any accompanying Website or software of the Work, cannot and do not warrant the performance or results that might be obtained by using the contents of the Work. The author, developers, and the Publisher have used their best efforts to insure the accuracy and functionality of the textual material and/or programs contained in this package; we, however, make no warranty of any kind, express or implied, regarding the performance of these contents or programs. The Work is sold "as is" without warranty (except for defective materials used in manufacturing the book or due to faulty workmanship).

The author, developers, and the publisher of any accompanying content, and anyone involved in the composition, production, and manufacturing of this work will not be liable for damages of any kind arising out of the use of (or the inability to use) the algorithms, source code, computer programs, or textual material contained in this publication. This includes, but is not limited to, loss of revenue or profit, or other incidental, physical, or consequential damages arising out of the use of this Work.

The sole remedy in the event of a claim of any kind is expressly limited to replacement of the book, and only at the discretion of the Publisher. The use of "implied warranty" and certain "exclusions" vary from state to state, and might not apply to the purchaser of this product.

AutoCAD® 2015
Beginning
and
Intermediate

By
Munir M. Hamad

Autodesk™ Approved Instructor

Mercury Learning and Information
*Dulles, Virginia
Boston, Massachusetts
New Delhi*

Copyright ©2015 by MERCURY LEARNING AND INFORMATION LLC. All rights reserved.

This publication, portions of it, or any accompanying software may not be reproduced in any way, stored in a retrieval system of any type, or transmitted by any means, media, electronic display or mechanical display, including, but not limited to, photocopy, recording, Internet postings, or scanning, without prior permission in writing from the publisher.

Publisher: David Pallai

MERCURY LEARNING AND INFORMATION
22841 Quicksilver Drive
Dulles, VA 20166
info@merclearning.com
www.merclearning.com
(800) 232-0223

This book is printed on acid-free paper.

Munir M. Hamad, AUTOCAD® 2015 BEGINNING AND INTERMEDIATE.
ISBN: 978-1-937585-36-5

The publisher recognizes and respects all marks used by companies, manufacturers, and developers as a means to distinguish their products. All brand names and product names mentioned in this book are trademarks or service marks of their respective companies. Any omission or misuse (of any kind) of service marks or trademarks, etc. is not an attempt to infringe on the property of others.

Library of Congress Control Number: 2014950127
15161732

Our titles are available for adoption, license, or bulk purchase by institutions, corporations, etc. For additional information, please contact the Customer Service Dept. at (800) 232-0223 (toll free).

All of our titles are available in digital format at authorcloudware.com and other digital vendors. Companion disc files for this title are available by contacting info@merclearning.com. The sole obligation of MERCURY LEARNING AND INFORMATION to the purchaser is to replace the disc, based on defective materials or faulty workmanship, but not based on the operation or functionality of the product.

Contents

About the Book   xxiii

Preface   xxv

About the DVD   xxvii

Chapter 1: AutoCAD 2015 Basics   1
- 1.1 How to Start AutoCAD   1
- 1.2 AutoCAD 2015 Interface   3
  - 1.2.1 Application Menu   4
  - 1.2.2 Quick Access Toolbar   5
  - 1.2.3 Ribbons   6
  - 1.2.4 InfoCenter   9
  - 1.2.5 Command Window   10
  - 1.2.6 Graphical Area   10
  - 1.2.7 Status Bar   10
- 1.3 AutoCAD Defaults   12
- 1.4 Drawing Units   13
- 1.5 Creating a New AutoCAD Drawing   14
- 1.6 Opening an Existing AutoCAD Drawing   16
  - 1.6.1 File Tab   17
- 1.7 Closing Drawing File(s)   18
- 1.8 Undo and Redo Commands   19
  - 1.8.1 Undo Command   20
  - 1.8.2 Redo Command   20
- Practice 1-1 AutoCAD Basics   22
- Chapter Review   23
- Chapter Review Answers   24

Chapter 2: Precise Drafting in AutoCAD 2015 — 25

- 2.1 Drafting Priorities — 25
- 2.2 Drawing Lines Using Line Command — 26
- 2.3 Dynamic Input in AutoCAD — 27
  - 2.3.1 Example for Showing Prompts — 27
  - 2.3.2 Example for Specifying Lengths and Angles — 28
  - Practice 2-1 Drawing Lines Using Dynamic Input — 29
- 2.4 Exact Angles (Ortho Versus Polar Tracking) — 29
  - 2.4.1 Increment Angle — 31
  - 2.4.2 Additional Angles — 31
  - 2.4.3 Polar Angle Measurement — 32
  - Practice 2-2 Exact Angles — 32
- 2.5 Precise Drafting Using Object Snap — 33
  - 2.5.1 Activating Running OSNAPs — 34
  - 2.5.2 OSNAP Override — 36
  - Practice 2-3 Object Snap (OSNAP) — 37
- 2.6 Drawing Circles Using Circle Command — 37
- 2.7 Drawing Circular Arcs Using Arc Command — 39
- 2.8 Object Snaps Related to Circle and Arc — 41
- 2.9 Using Object Snap Tracking with OSNAP — 41
  - Practice 2-4a Drawing Using OSNAP and OTRACK — 43
  - Practice 2-4b Drawing Using OSNAP and OTRACK — 46
- 2.10 Drawing Lines and Arcs Using Polyline Command — 47
- 2.11 Converting Polylines to Lines and Arcs, and Vice-Versa — 49
  - 2.11.1 Converting Polylines to Lines and Arcs — 49
  - 2.11.2 Joining Lines and Arcs to Form a Polyline — 50
  - Practice 2-5 Drawing Polylines and Converting — 50
- 2.12 Using Snap and Grid to Specify Points Accurately — 51
- 2.13 Using Polar Snap — 53
  - Practice 2-6 Snap and Grid — 54
  - Practice 2-7 Small Project — 55
  - Chapter Review — 57
  - Chapter Review Answers — 58

Chapter 3: Modifying Commands Part I — 59

- 3.1 How to Select Objects in AutoCAD — 59
  - 3.1.1 Window Mode (W) — 61
  - 3.1.2 Crossing Mode (C) — 61
  - 3.1.3 Window Polygon Mode (WP) — 61
  - 3.1.4 Crossing Polygon Mode (CP) — 62
  - 3.1.5 Lasso Selection — 62
  - 3.1.6 Fence Mode (F) — 63

		3.1.7 Last (L), Previous (P), and All Modes	63
		3.1.8 Other Methods to Select Objects	64
	3.2	Selection Cycling	65
	3.3	Erase Command	66
		Practice 3-1 Selecting Objects and Erase Command	67
	3.4	Move Command	70
		3.4.1 Nudge Functionality	71
		Practice 3-2 Moving Objects	71
	3.5	Copy Command	72
		Practice 3-3 Copying Objects	73
	3.6	Rotate Command	74
		3.6.1 Reference Option	75
		Practice 3-4 Rotating Objects	75
	3.7	Scale Command	76
		3.7.1 Reference Option	77
		Practice 3-5 Scaling Objects	78
	3.8	Mirror Command	78
		Practice 3-6 Mirroring Objects	80
	3.9	Stretch Command	80
		Practice 3-7 Stretching Objects	82
		Practice 3-8 Stretching Objects	83
	3.10	Lengthening Objects	84
	3.11	Joining Objects	85
		Practice 3-9 Lengthen and Joining Objects	86
	3.12	Using Grips to Edit Objects	87
	3.13	Grips and Dynamic Input	91
	3.14	Grips and Perpendicular and Tangent OSNAPs	93
		Practice 3-10 Using Grips to Edit Objects	94
		Chapter Review	98
		Chapter Review Answers	99
Chapter 4:	**Modifying Commands Part II**		**101**
	4.1	Introduction	101
	4.2	Offsetting Objects	102
		4.2.1 Offsetting Using Offset Distance Option	102
		4.2.2 Offsetting Using Through Option	103
		4.2.3 Using Multiple Option	103
		Practice 4-1 Offsetting Objects	104
		Practice 4-2 Offsetting Objects	105
	4.3	Filleting Objects	106
		Practice 4-3 Filleting Objects	108
	4.4	Chamfering Objects	108

		4.4.1 Chamfering Using Distance Option	109
		4.4.2 Chamfering Using Distance and Angle	109
		Practice 4-4 Chamfering Objects	111
	4.5	Trimming Objects	111
		Practice 4-5 Trimming Objects	113
	4.6	Extending Objects	114
		Practice 4-6 Extending Objects	115
	4.7	Arraying Objects – Rectangular Array	116
		4.7.1 The First Step	116
		4.7.2 Using Array Creation Context Tab	117
		4.7.3 Editing Rectangular Array Using Grips	118
		4.7.4 Editing Rectangular Array Using Context Tab	120
		4.7.5 Editing Rectangular Array Using Quick Properties	120
		Practice 4-7 Arraying Objects Using Rectangular Array	121
	4.8	Arraying Objects – Path Array	122
		Practice 4-8 Arraying Objects Using Path	128
	4.9	Arraying Objects – Polar Array	129
		Practice 4-9 Arraying Objects Using Polar Array	132
	4.10	Break Command	133
		Practice 4-10 Breaking Objects	135
		Chapter Review	138
		Chapter Review Answers	139
Chapter 5:	**Layers and Inquiry Commands**		**141**
	5.1	Layers Concept in AutoCAD	141
	5.2	Creating and Setting Layer Properties	143
		5.2.1 How to Create a New Layer	143
		5.2.2 How to Set a Color for a Layer(s)	144
		5.2.3 How to Set the Linetype for Layer(s)	145
		5.2.4 How to Set a Lineweight for Layer(s)	146
		5.2.5 How to Set the Current Layer	147
		Practice 5-1 Creating and Setting Layer Properties	147
	5.3	Layer Controls	148
		5.3.1 Controlling Layer Visibility, Locking, and Plotting	148
		5.3.2 Deleting and Renaming Layers	150
		5.3.3 How to Make an Object's Layer Current Layer	151
		5.3.4 How to Undo Only Layers Actions	152
		5.3.5 Moving Objects from One Layer to Another	152
	5.4	Using the Layer Properties Manager	153
		Practice 5-2 Layer Controls	155
	5.5	Changing Object's Layer, Quick Properties, and Properties	156
		5.5.1 Reading Instantaneous Information About an Object	156
		5.5.2 How to Move an Object from a Layer to Another Layer	156

		5.5.3 Quick Properties	156
		5.5.4 Properties	158
		Practice 5-3 Changing an Object's Layer, Quick Properties, and Properties	159
	5.6	Inquiry Commands – Introduction	159
	5.7	Measuring Distance	160
	5.8	Inquiring Radius	161
	5.9	Measuring Angle	162
	5.10	Measuring Area	163
		5.10.1 How to Calculate Simple Area	163
		5.10.2 How to Calculate Complex Area	164
		Practice 5-4 Inquiry Commands	166
		Practice 5-5 Inquiry Commands	167
		Chapter Review	169
		Chapter Review Answers	169
Chapter 6:	**Blocks and Hatch**		**171**
	6.1	What are Blocks?	171
	6.2	How to Create a Block	172
	6.3	How to Use (Insert) Blocks	174
		6.3.1 Block Insertion Point OSNAP	176
		Practice 6-1 Creating and Inserting Blocks	177
	6.4	Exploding Blocks and Converting Them to Files	178
		6.4.1 Exploding Blocks	178
		6.4.2 Converting Blocks to Files	178
		Practice 6-2 Exploding and Converting	179
	6.5	Hatching in AutoCAD	180
	6.6	Hatch Command: First Step	180
	6.7	Controlling Hatch Properties	181
		Practice 6-3 Inputting Hatch and Controlling Hatch Properties	183
	6.8	Specifying Hatch Origin	185
	6.9	Controlling Hatch Options	186
		6.9.1 How to Use Associative Hatching	187
		6.9.2 How to Make Your Hatch Annotative	187
		6.9.3 Using Match Properties to Create Identical Hatches	187
		6.9.4 Hatching an Open Area	188
		6.9.5 Creating Separate Hatches in the Same Command	189
		6.9.6 Island Detection	190
		6.9.7 Set Hatch Draw Order	191
		Practice 6-4 Hatch Origin and Options	192
	6.10	Hatch Boundary	192
	6.11	Editing Hatch	195

		Practice 6-5 Hatch Boundary and Hatch Editing	197
		Chapter Review	200
		Chapter Review Answers	200

Chapter 7: Writing Text — 201

	7.1	Writing Text Using Single Line Text	201
		Practice 7-1 Creating Text Style and Single Line Text	202
	7.2	Writing Text Using Multiline Text	203
		7.2.1 Style Panel	205
		7.2.2 Formatting Panel	206
		7.2.3 Paragraph Panel	206
		7.2.4 Insert Panel	208
		7.2.5 Spell Check Panel	210
		7.2.6 Tools Panel	211
		7.2.7 Options Panel	212
		7.2.8 Close Panel	213
		7.2.9 While You Are in the Text Editor	213
		Practice 7-2 Writing Using Multiline Text	214
	7.3	Text Editing	216
		7.3.1 Double-Click Text	216
		7.3.2 Quick Properties and Properties	217
		7.3.3 Editing Using Grips	218
	7.4	Spell Check and Find and Replace	219
		Practice 7-3 Editing Text	220
		Chapter Review	223
		Chapter Review Answers	223

Chapter 8: Dimensions — 225

	8.1	What is Dimensioning in AutoCAD?	225
	8.2	Dimension Types	226
	8.3	How to Insert a Linear Dimension	229
	8.4	How to Insert an Aligned Dimension	230
	8.5	How to Insert an Angular Dimension	231
		Practice 8-1 Inserting Linear, Aligned, and Angular Dimension	232
	8.6	How to Insert an Arc Length Dimension	233
	8.7	How to Insert a Radius Dimension	324
	8.8	How to Insert a Diameter Dimension	236
		Practice 8-2 Inserting Arc Length, Radius, and Diameter Dimension	237
	8.9	How to Insert a Jogged Dimension	238
	8.10	How to Insert Ordinate Dimension	239
		Practice 8-3 Inserting Dimensions	240

	8.11	Inserting Series of Dimensions Using Continue Command	241
	8.12	Inserting Series of Dimensions Using Baseline Command	242
		Practice 8-4 Continue Command	243
		Practice 8-5 Baseline Command	244
	8.13	Using the Quick Dimension Command	245
	8.14	Editing a Dimension Block Using Grips	248
	8.15	Editing a Dimension Block Using Right-Click Menu	252
	8.16	Editing a Dimension Block Using Quick Properties and Properties	253
		Practice 8-6 Quick Dimension and Editing	254
		Chapter Review	257
		Chapter Review Answers	257

Chapter 9: Plotting 259

	9.1	What are Model Space and Paper Space?	259
	9.2	Introduction to Layouts	260
	9.3	Steps to Create a New Layout from Scratch	260
	9.4	Steps to Create a New Layout Using a Template	264
	9.5	Creating Layouts Using Copying	265
		Practice 9-1 Creating New Layouts	268
	9.6	Creating Viewports	269
		9.6.1 Adding Single Rectangular Viewports	269
		9.6.2 Adding Multiple Rectangular Viewports	270
		9.6.3 Adding Polygonal Viewport	271
		9.6.4 Creating Viewports by Converting Existing Objects	272
		9.6.5 Creating Viewports by Clipping Existing Viewports	273
		9.6.6 Dealing with Viewports After Creation	274
	9.7	Scaling and Maximizing Viewports	275
	9.8	Freezing Layers in Viewport	278
	9.9	Layer Override in Viewport	279
		Practice 9-2 Creating and Controlling Viewports	280
	9.10	Plot Command	281
		Chapter Review	284
		Chapter Review Answers	284

Chapter 10: Projects 285

	10.1	How to Prepare Your Drawing for a New Project	285
	10.2	Architectural Project (Imperial)	288
	10.3	Architectural Project (Metric)	295
	10.4	Mechanical Project – I (Metric)	303
	10.5	Mechanical Project – I (Imperial)	306
	10.6	Mechanical Project – II (Metric)	310
	10.7	Mechanical Project – II (Imperial)	311

Chapter 11: More on 2D Objects   **313**

    11.1 Introduction   313
    11.2 Drawing Lines and Arcs Using the Polyline Command – Revision   314
    11.3 Converting Polylines to Lines and Arcs, and Vice-Versa   315
        11.3.1 Converting Polylines to Lines and Arcs   315
        11.3.2 Joining Lines and Arcs to Form a Polyline   316
        Practice 11-1 Drawing Polylines and Converting   317
    11.4 Drawing Using the Rectangle Command   317
        11.4.1 Chamfer Option   318
        11.4.2 Elevation Option   318
        11.4.3 Fillet Option   318
        11.4.4 Thickness Option   319
        11.4.5 Width Option   319
        11.4.6 Area Option   319
        11.4.7 Dimensions Option   320
        11.4.8 Rotation Option   320
    11.5 Drawing Using the Polygon Command   320
        11.5.1 Using an Imaginary Circle   321
        11.5.2 Using Length and Angle of One of the Edges   322
        Practice 11-2 Drawing Rectangles and Polygons   322
    11.6 Drawing Using the Donut Command   323
    11.7 Drawing Using the Revision Cloud Command   324
        Practice 11-3 Drawing Using Donut and Revision Cloud   325
    11.8 Using the Edit Polyline Command   326
        11.8.1 Open and Close Options   327
        11.8.2 Join Option   327
        11.8.3 Width Option   327
        11.8.4 Edit Vertex Option   327
        11.8.5 Fit, Spline, and Decurve Options   328
        11.8.6 Ltype gen Option   329
        11.8.7 Reverse Option   329
        11.8.8 Multiple Option   330
        Practice 11-4 Using the Polyline Edit Command   332
    11.9 Using Construction Lines and Rays   333
        11.9.1 Construction Lines   333
        11.9.2 Rays   334
        Practice 11-5 Using Construction Lines and Rays   335
    11.10 Using the Point Style and Point Commands   336
        11.10.1 Point Style Command   337
        11.10.2 Point Command   338
    11.11 Using the Divide and Measure Commands   338
        11.11.1 Divide Command   339
        11.11.2 Measure Command   339

	11.11.3 Divide and Measure Commands with Block Option	340
	Practice 11-6 Using Point Style, Point, Divide, and Measure	340
11.12	Using the Spline Command	342
	11.12.1 Fit Points Method	342
	11.12.2 Control Vertices Method	343
	11.12.3 Editing Spline	344
	Practice 11-7 Using the Spline Command	345
11.13	Using the Ellipse Command	346
	11.13.1 Drawing an Ellipse Using the Center Option	347
	11.13.2 Drawing an Ellipse Using Axis Points	347
	11.13.3 Drawing an Elliptical Arc	348
	Practice 11-8 Using the Ellipse Command	348
11.14	Boundary Command	349
	Practice 11-9 Using the Boundary Command	350
11.15	Using the Region Command	351
	11.15.1 Performing Boolean Operation on Regions	352
	Practice 11-10 Using the Region Command	353
	Chapter Review	355
	Chapter Review Answers	356

Chapter 12: Advanced Practices – Part I — 357

12.1	Offset Command – Advanced Options	357
	12.1.1 Erase Source Option	358
	12.1.2 Layer Option	358
	12.1.3 System Variable: offsetgaptype	358
12.2	Trim and Extend – Edge Option	358
	Practice 12-1 Using Advanced Options in Offset, Trim, and Extend	359
12.3	Using Match Properties	360
12.4	Copy/Paste Objects and Match Properties Across Files	361
	12.4.1 Copying Objects	362
	12.4.2 Pasting Objects	363
	12.4.3 Drag-and-Drop Method	363
	12.4.4 Match Properties Across Files	363
	Practice 12-2 Using Match Properties, Copy/Paste Across Files	363
12.5	Sharing Excel and Word Content in AutoCAD	364
	12.5.1 Sharing Data Coming from MS Word	364
	12.5.2 Sharing Data Coming from MS Excel	366
	12.5.3 Pasting a Linked Table from Excel	367
	Practice 12-3 Sharing Excel and Word Content in AutoCAD	371
12.6	Hyperlinking AutoCAD Objects	372
	Practice 12-4 Hyperlinking AutoCAD Objects	373

		12.7	Purging Items	373
			Practice 12-5 Purging Items	375
		12.8	Using Views and Viewports	376
			12.8.1 Creating Views	376
			12.8.2 Using Views in Viewports	378
			12.8.3 Creating Named Viewport Arrangement – Method (I)	379
			12.8.4 Creating Named Viewport Arrangement – Method (II)	382
			Practice 12-6 Using Views and Viewports	384
			Chapter Review	387
			Chapter Review Answers	388

Chapter 13: Advanced Practices – Part II — **389**

	13.1	Using Autodesk Content Explorer	389
		Practice 13-1 Using Autodesk Content Explorer	394
	13.2	Using Quick Select	394
		Practice 13-2 Using Quick Select	396
	13.3	Using Select Similar and Add Selected	397
		13.3.1 Select Similar Command	397
		13.3.2 Add Selected Command	398
	13.4	What is Object Visibility in AutoCAD?	399
		Practice 13-3 Using Select Similar and Add Selected	400
	13.5	Advanced Layer Commands	401
		13.5.1 Isolate and Unisolate Commands	401
		13.5.2 Using Freeze and Off Commands	401
		13.5.3 Using Turn All Layers On and Thaw All Layers Commands	402
		13.5.4 Lock and Unlock Commands	402
		13.5.5 Change to Current Layer Command	402
		13.5.6 Copy Objects to New Layer Command	402
		13.5.7 Layer Walk Command	403
		13.5.8 Isolate to Current Viewport Command	403
		13.5.9 Merge Command	403
		13.5.10 Delete Command	404
	13.6	Layer's Transparency	404
		Practice 13-4 Using Advanced Layer Commands	405
	13.7	Using Fields in AutoCAD	406
		Practice 13-5 Using Fields in AutoCAD	411
	13.8	Using Partially Opened Files	412
		13.8.1 How to Open a File Partially	412
		13.8.2 Using Partial Load	413
		Practice 13-6 Using Partially Opened Files	414
		Chapter Review	415
		Chapter Review Answers	416

Chapter 14: Using Block Tools and Block Editing — 417
- 14.1 Automatic Scaling Feature — 417
- 14.2 Design Center — 419
  - Practice 14-1 Using Design Center — 421
- 14.3 Tool Palettes — 422
  - 14.3.1 How to Create a Tool Palette from Scratch — 424
  - 14.3.2 How to Fill the New Palette with Content — 424
  - 14.3.3 How to Create a Palette from Design Center Blocks — 425
  - 14.3.4 How to Customize Tools Properties — 425
- 14.4 Hatch and Tool Palette — 427
  - Practice 14-2 Using Tool Palettes — 428
- 14.5 Customizing Tool Palettes — 429
  - 14.5.1 Allow Docking — 429
  - 14.5.2 Transparency — 430
  - 14.5.3 View Options — 431
  - 14.5.4 Add Text and Add Separator — 432
  - 14.5.5 New / Delete / Rename Palette — 433
  - 14.5.6 Customize Palettes — 433
  - Practice 14-3 Customizing Tool Palettes — 435
- 14.6 Editing Blocks — 436
  - Practice 14-4 Editing Blocks — 437
  - Chapter Review — 440
  - Chapter Review Answers — 441

Chapter 15: Creating Text, Table Styles, and Formulas in Tables — 443
- 15.1 Steps to Create Text and Tables — 443
- 15.2 How to Create a Text Style — 444
  - Practice 15-1 Creating Text Style and Single Line Text — 447
- 15.3 Creating Table Style — 448
  - 15.3.1 General Tab — 451
  - 15.3.2 Text Tab — 452
  - 15.3.3 Borders Tab — 453
- 15.4 Inserting a Table in the Drawing — 454
  - 15.4.1 Specify Insertion Point Option — 455
  - 15.4.2 Specify Window Option — 455
  - Practice 15-2 Creating Table Style and Inserting a Table in the Current Drawing — 456
- 15.5 Using Formulas in Table Cells — 457
- 15.6 Using Table Cell Functions — 458
  - 15.6.1 Using Rows Panel — 459
  - 15.6.2 Using Columns Panel — 459
  - 15.6.3 Using Merge Panel — 459

	15.6.4 Using Cell Styles Panel	460
	15.6.5 Using Cell Format Panel	462
	15.6.6 Using Insert Panel	464
	15.6.7 Using Data Panel	464
	Practice 15-3 Formulas and Table Cell Functions	465
	Chapter Review	468
	Chapter Review Answers	468

Chapter 16: Dimension and Multileader Styles — 469

- 16.1 What is Dimensioning? — 469
- 16.2 How to Create a New Dimension Style — 470
- 16.3 Dimension Style: Lines Tab — 472
- 16.4 Dimension Style: Symbols and Arrows Tab — 475
- 16.5 Dimension Style: Text Tab — 478
- 16.6 Dimension Style: Fit Tab — 484
- 16.7 Dimension Style: Primary Units Tab — 486
- 16.8 Dimension Style: Alternate Units Tab — 488
- 16.9 Dimension Style: Tolerances Tab — 489
- 16.10 Creating a Sub Dimension Style — 492
- Practice 16-1 Creating Dimension Style — 493
- 16.11 More Dimension Functions — 495
  - 16.11.1 Dimension Break — 495
  - 16.11.2 Dimension Adjust Space — 496
  - 16.11.3 Dimension Jog Line — 498
  - 16.11.4 Dimension Center Mark — 499
  - 16.11.5 Dimension Oblique — 499
  - 16.11.6 Dimension Text Angle — 500
  - 16.11.7 Dimension Justify — 501
  - 16.11.8 Dimension Override — 501
- Practice 16-2 More Dimension Functions — 502
- 16.12 How to Create a Multileader Style? — 504
  - 16.12.1 Leader Format Tab — 505
  - 16.12.2 Leader Structure Tab — 506
  - 16.12.3 Content Tab — 507
- 16.13 Inserting a Multileader Dimension — 509
- Practice 16-3 Creating the Multileader Style and Inserting Multileaders — 512
- Chapter Review — 516
- Chapter Review Answers — 516

Chapter 17: Plot Style, Annotative, and DWF — 517

- 17.1 Plot Style Tables – First Look — 517
- 17.2 Color-Dependent Plot Style Table — 517

	17.3	Named Plot Style Table	522
		Practice 17-1 Color-Dependent Plot Style Table	527
		Practice 17-2 Named Plot Style Table	528
	17.4	What is the Annotative Feature?	528
		Practice 17-3 Annotative Feature	532
	17.5	Design Web Format (DWF) File	534
	17.6	Exporting DWF, DWFx, and PDF Files	534
	17.7	Using the Batch Plot Command	537
	17.8	Viewing DWF and DWFx Files	539
		Practice 17-4 Creating and Viewing DWF File	540
		Chapter Review	542
		Chapter Review Answers	542

Chapter 18: How to Create a Template File and Interface Customization 543

	18.1	What is a Template File and How Do You Create One?	543
	18.2	Editing a Template File	546
		Practice 18-1 Creating and Editing a Template File	546
	18.3	Customizing the Interface – Introduction	547
	18.4	How to Create a New Panel	548
	18.5	How to Create a New Tab	553
	18.6	How to Create a Quick Access Toolbar	553
	18.7	How to Create a New Workspace	554
		Practice 18-2 Customizing Interface	555
	18.8	How to Create Your Own Command	557
		Practice 18-3 Creating New Commands	559
		Chapter Review	561
		Chapter Review Answers	562

Chapter 19: Parametric Constraints 563

	19.1	What are Parametric Constraints?	563
	19.2	Using Geometric Constraints	564
		19.2.1 Using the Coincident Constraint	564
		19.2.2 Using the Collinear Constraint	566
		19.2.3 Using the Concentric Constraint	567
		19.2.4 Using the Fix Constraint	568
		19.2.5 Using the Parallel Constraint	568
		19.2.6 Using the Perpendicular Constraint	569
		19.2.7 Using the Horizontal Constraint	570
		19.2.8 Using the Vertical Constraint	571
		19.2.9 Using the Tangent Constraint	572
		19.2.10 Using the Smooth Constraint	572
		19.2.11 Using the Symmetric Constraint	573
		19.2.12 Using the Equal Constraint	574

	19.3	Geometric Constraints Settings	575
	19.4	What is the Infer Constraint?	576
	19.5	What is Auto Constrain?	578
	19.6	Constraint Bar and Showing and Hiding	580
		19.6.1 Constraint Bar	580
		19.6.2 Showing and Hiding	582
	19.7	Relaxing and Over Constraining Objects	582
		19.7.1 Relaxing Constraints	582
		19.7.2 Over Constraining an Object	583
		Practice 19-1 Applying Geometric Constraint	583
	19.8	Using Dimensional Constraints	584
		19.8.1 Using Linear, Horizontal, and Vertical Constraints	585
		19.8.2 Using Aligned Constraint	586
		19.8.3 Using Radial and Diameter Constraints	587
		19.8.4 Using Angular Constraints	588
		19.8.5 Using the Convert Command	589
	19.9	Controlling Dimensional Constraints	590
		19.9.1 Constraint Settings Dialog Box	590
		19.9.2 Deleting Constraints	592
		19.9.3 Showing and Hiding Dimensional Constraints	592
	19.10	Using the Parameters Manager	593
	19.11	What is Annotational Constraint Mode?	596
		19.11.1 Annotational Constraint Mode	596
		19.11.2 Converting Dimensional Constraints to Annotational	597
	19.12	Using Dimensional Grips	598
		Practice 19-2 Applying Dimensional Constraint	599
		Chapter Review	602
		Chapter Review Answers	603

Chapter 20: Dynamic Blocks — 605

	20.1	Introduction to Dynamic Blocks	605
	20.2	Methods to Create a Dynamic Block	606
		20.2.1 Using the Block Definition Dialog Box	607
		20.2.2 Double-Clicking an Existing Block	607
		20.2.3 Block Editor Command	608
	20.3	Inside Block Editor	608
	20.4	What Are Parameters and Actions?	609
	20.5	Controlling Parameter Properties	614
	20.6	Controlling the Visibility Parameter	616
	20.7	Using Lookup Parameter and Action	618
	20.8	Final Steps	621
		Practice 20-1 Dynamic Blocks – Creating a Chest of Drawers	622
		Practice 20-2 Dynamic Blocks – Door Control	623

Contents   xix

	Practice 20-3 Dynamic Blocks – Wide Flange Beams	624
20.9	Using Constraints for Dynamic Blocks	626
	20.9.1 Block Table Button	626
	20.9.2 Construction Button	631
	20.9.3 Constraint Status Button	631
	Practice 20-4 Dynamic Blocks Using Constraints – Wide Flange Beams	632
	Practice 20-5 Dynamic Blocks Using Constraints – Creating a Window	632
	Chapter Review	635
	Chapter Review Answers	636

Chapter 21: Block Attributes   **637**

21.1	What are Block Attributes?	637
21.2	How to Define Attributes?	638
	21.2.1 Attribute Part	639
	21.2.2 Mode Part	639
	21.2.3 Text Settings Part	639
	21.2.4 Insertion Point Part	640
21.3	Inserting Blocks with Attributes	640
21.4	How to Control Attribute Visibility?	641
	Practice 21-1 Defining and Inserting Blocks with Attributes	641
21.5	How to Edit Individual Attribute Values?	642
21.6	How to Edit Attribute Values Globally?	644
	Practice 21-2 Editing Attribute Values Individually and Globally	645
21.7	How to Redefine and Sync Attribute Definitions?	646
	Practice 21-3 Redefining Attribute Definitions	649
21.8	How to Extract Attributes from File(s)?	650
	Practice 21-4 Extracting Attributes	658
	Chapter Review	659
	Chapter Review Answers	660

Chapter 22: External Referencing (XREF)   **661**

22.1	Introduction to External Reference	661
22.2	Inserting External Reference Different File Formats	662
	22.2.1 Attach a DWG File	663
	22.2.2 Attach an Image File	665
	22.2.3 Attach a DWF File	666
	22.2.4 Attach a DGN File	667
	22.2.5 Attach a PDF File	668
22.3	External Reference Palette Contents	669
22.4	Using the Attach Command	670

22.5	Reference File and Layers	671
22.6	Controlling Fading of a Reference File	672
	Practice 22-1 Attaching and Controlling Reference Files	673
22.7	Editing an External Reference DWG File	674
	22.7.1 Using the Edit Reference Command	675
	22.7.2 Using the Open Command	676
22.8	External Reference File Related Functions	677
	22.8.1 Unload Command	678
	22.8.2 Reload Command	678
	22.8.3 Detach Command	678
	22.8.4 Bind Command	678
	22.8.5 Xref Type	679
	22.8.6 Path	679
	Practice 22-2 External Reference Editing	679
22.9	Clipping an External Reference File	680
22.10	Clicking and Right-Clicking a Reference File	682
	22.10.1 Clicking and Right-Clicking a DWG File	683
	22.10.2 Clicking and Right-Clicking an Image File	683
	22.10.3 Clicking and Right-Clicking a DWF File	684
	Practice 22-3 External Reference File Clipping and Controlling	685
22.11	Using the eTransmit Command with External Reference Files	686
	Practice 22-4 Using the eTransmit Command	690
	Chapter Review	692
	Chapter Review Answers	693

Chapter 23: Sheet Sets — 695

23.1	Introduction to Sheet Sets	695
23.2	Dealing with Sheet Set Manager Palette	696
	23.2.1 How to Open and Close an Existing Sheet Set	696
	23.2.2 Working with the Sheet Set Manager Palette	699
23.3	Sheet Set File Setup	701
	Practice 23-1 Opening, Manipulating, and Closing Sheet Set	702
23.4	Sheet Set Using an Example	702
	23.4.1 Adding Sheets in Subsets	707
	23.4.2 Sheet Control	708
	Practice 23-2 Creating a Sheet Set Using an Example Sheet Set	709
23.5	Adding and Scaling Model Views	710
	Practice 23-3 Adding and Scaling Model Views	713
23.6	Sheet Sets Using Existing Drawings	714
	Practice 23-4 Creating a Sheet Set Using Existing Drawings	719
23.7	Publishing Sheet Sets	720
23.8	Using eTransmit and Archive Commands	722
	23.8.1 Using the eTransmit Command	722

	23.8.2 Using the Archive Command	723
	Practice 23-5 Publishing and eTransmitting a Sheet Set	724
	Chapter Review	726
	Chapter Review Answers	727

Chapter 24: CAD Standards and Advanced Layers — 729

24.1	Why Do We Need CAD Standards?	729
24.2	How to Create a CAD Standard File	730
24.3	How to Link DWS to DWG Files and Check Them	731
	24.3.1 Configuring (Linking) DWS to DWG	731
	24.3.2 Checking DWG Files	734
	Practice 24-1 Using CAD Standards Commands	736
24.4	Using the Layer Translator	737
	Practice 24-2 Using the Layer Translator	740
24.5	Dealing with Layer Properties Manager	741
24.6	Creating Property Filters	743
24.7	Creating Group Filters	745
24.8	Things You Can Do with Filters	746
	24.8.1 Property Filter Menu	746
	24.8.2 Group Filter Menu	747
	Practice 24-3 Layer Advanced Features and Filters	747
24.9	Creating Layer States	748
	Practice 24-4 Using Layer States	752
24.10	Settings Dialog Box	752
	24.10.1 New Layer Notification	754
	24.10.2 Isolate Layer Settings	754
	24.10.3 Dialog Settings	755
	Practice 24-5 Using Settings Dialog Box	756
	Chapter Review	759
	Chapter Review Answers	759

Chapter 25: Drawing Review — 761

25.1	Introduction to Drawing Review	761
25.2	First Step – Creating a DWF File	762
25.3	Second Step – Using Autodesk Design Review	763
25.4	Markup and Measure Tab	764
	25.4.1 Clipboard Panel	765
	25.4.2 Formatting Panel	765
	25.4.3 Callouts Panel	766
	25.4.4 Draw Panel	767
	25.4.5 Measure Panel	768
	25.4.6 Stamps and Symbols Panel	769
25.5	How to Edit Markup Objects	770

	25.6	Controlling Markups	771
		25.6.1 Markup Window	771
		25.6.2 Markup Properties Window	772
		Practice 25-1 Creating DWF and Using Markup Tools	772
	25.7	Using Markup Set Manager in AutoCAD	773
		Practice 25-2 Creating a DWF and Using Markup Tools	776
	25.8	Comparing DWF Files	777
		Practice 25-3 Comparing DWF Files	780
		Chapter Review	781
		Chapter Review Answers	781

Index   **783**

ABOUT THE BOOK

This book is the most comprehensive book you will find on AutoCAD 2015 – 2D Drafting. It is divided into three major parts. The first, Essentials, covers ten chapters and three projects (one architectural and two mechanical using both imperial and metric units). The second part, Intermediate, contains eight chapters. It contains topics covered in the first section, is more depth, or additional knowledge needed to fill in the gaps left in the first part. The final part of the book, Advanced, contains seven chapters discussing the most advanced features of AutoCAD 2015.

If you don't have any prior experience using AutoCAD, you can start with any chapter of the book. But if you want to be an advanced AutoCAD user, you should go through all 25 chapters and complete all the projects and practices.

This book can also help you prepare for the *AutoCAD Certified Professional* exam, given by Autodesk, Inc.

This book's chapters are divided as follows:

- Chapter 1 covers the basics of AutoCAD along with the interface.
- Chapter 2 covers AutoCAD techniques for drawing with accuracy.
- Chapters 3 & 4 cover the modifying commands used to modify and construct drawings.
- Chapter 5 covers layers and inquiry commands.
- Chapter 6 covers creating and editing blocks and inserting and editing hatches.
- Chapter 7 covers AutoCAD methods for writing text.
- Chapter 8 covers how to create and edit dimensions in AutoCAD.

- Chapter 9 covers how to plot your drawing.
- Chapter 10 includes three projects, one architect and two mechanical covering both metric and imperial units.
- Chapter 11 covers more 2D objects creation.
- Chapters 12 & 13 cover advanced practices and techniques.
- Chapter 14 covers block tools and block editing.
- Chapter 15 covers text styles and table styles along with formulas in tables.
- Chapter 16 covers dimension styles and multileaders.
- Chapter 17 covers plot styles, the meaning of Annotative, and creating DWF files.
- Chapter 18 covers how to create a template file and customizing the AutoCAD interface.
- Chapter 19 covers parametric constraints.
- Chapter 20 covers dynamic blocks.
- Chapter 21 covers block attributes.
- Chapter 22 covers Xref.
- Chapter 23 covers sheets sets.
- Chapter 24 covers CAD standards and advanced layer commands.
- Chapter 25 covers the drawing review.

PREFACE

- Since its inception, AutoCAD has enjoyed a very wide user base and has been the most widely used CAD software since the 1980s. This popularity is due to its logic and simplicity, which makes it very easy to learn.
- This book addresses all levels of AutoCAD 2D drafting: Essentials, Intermediate, and Advanced AutoCAD techniques. It is not a replacement for the manual(s) that comes with the software, but is considered complementary with its practices and projects, meant to strengthen the knowledge gained and solidify the techniques discussed.
- Solving all practices is essential because AutoCAD is a practical tool and not theoretical.
- At the end of each chapter you will find "Chapter Review Questions." These are the same sort of questions you might see in an Autodesk exam. The answers to the odd questions are included at the end of each chapter to check your work.
- Chapter 10 contains three projects: One for an architectural plan and the other two for mechanical engineering. (One of these is explained in detail, so you can follow the steps by themselves without any help.) Solving these will allow you to master the knowledge needed to land a job in today's market. These projects are presented in metric and imperial units.

ABOUT THE DVD

- The DVD included with this book contains:
  - A link to the AutoCAD 2015 Trial version, which will last for 30 days starting from the day of installation. This version will help you solve all the exercises and workshops in this book. *The student trial version can be extended beyond the 30 day period.*
  - The practices files which will be your starting point to solve all exercises and workshops in the book.
  - Copy the folder named "Practices and Projects" onto the hard drive of your computer. In the Project folder, you will find two folders; the first is called "Metric" for metric units projects and the second one is called "Imperial" for imperial units projects.

CHAPTER 1

AUTOCAD 2015 BASICS

In This Chapter

- How to start AutoCAD
- How to deal with the AutoCAD interface
- AutoCAD defaults and drawing units
- How to deal with file oriented commands
- Undo and Redo commands

1.1 HOW TO START AUTOCAD

- AutoCAD was released in 1982 by Autodesk, Inc. – a small company at that time – and designed for PCs only. Since then, AutoCAD has grown the biggest user base in the world in the CAD business. Users can use AutoCAD for both 2D and 3D drafting and designing. AutoCAD can also be used for architectural, structural, mechanical, electrical, road and highway designs, environmental, and manufacturing drawings.
- Though the theme these days is BIM (Building Information Modeling), AutoCAD remains the most profitable software for Autodesk, Inc. because of its ease-of-use and totality. Another version of AutoCAD, called AutoCAD LT, is used for 2D drafting only.
- To start AutoCAD 2015, double-click the shortcut that appears on your desktop and is created during the installation process. AutoCAD shows the following **Welcome** window:

2 • AutoCAD® 2015 Beginning and Intermediate

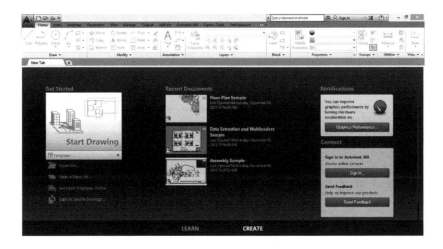

- As you can see in the bottom toolbar, there are two choices: CREATE and LEARN. This figure shows CREATE, which offers the following options:
    - Start a new drawing
    - Open an existing file
    - Open a Sheet Set
    - Download more templates from online
    - Explore the sample files that come with the software
    - See recently opened files
    - Check if AutoCAD has any notifications concerning software or hardware
    - Connect to Autodesk 360 (Autodesk Cloud)
    - Send your feedback to Autodesk
- Click the LEARN option to see the following window:

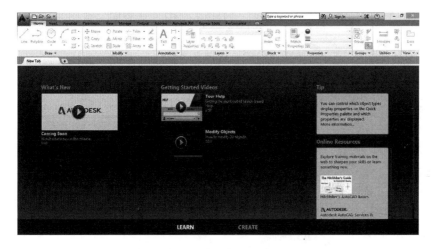

- LEARN offers the following choices:
  - Videos of the new features of AutoCAD 2015
  - Other videos discussing features of AutoCAD 2015 such as how to use modifying commands
  - Tips from Autodesk (normally different tips are shown every time AutoCAD starts)
  - Online sources to help and train such as Hitchhiker videos and Lynda.com
- Starting a new file or opening an existing file launches the interface of AutoCAD 2015, which looks like the following:

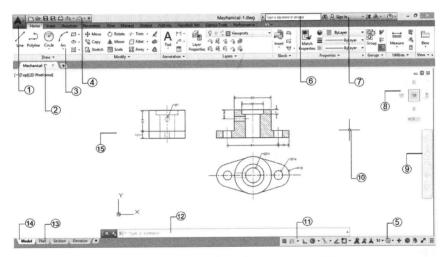

1. Application Menu
2. File Tab
3. Ribbon
4. Quick Access Toolbar
5. Workspace
6. Info Center
7. Autodesk 360
8. ViewCube
9. Navigation Bar
10. Cross Hairs
11. Status Bar
12. Command Window
13. Layout tab
14. Model tab
15. Graphical Area

1.2 AUTOCAD 2015 INTERFACE

- AutoCAD interface is based on the Ribbons and Application Menu. The most important feature of this interface is **Graphical Area**, which allows more area.

1.2.1 Application Menu

- The Application Menu contains the File related commands.

- These commands include creating a new file, opening an existing file, saving the current file, saving as the current file under a new name and different folder, exporting the current file to a different file format, printing and publishing the current file, etc. We discuss almost all of these commands within this book. By default, you see the recent files. You can choose how to display the recent files in the Application Menu, using this control:

- Also, you can choose how to sort the recent files, using this control:

1.2.2 Quick Access Toolbar

- This toolbar contains all the File commands along with Workspace and Undo/Redo.

- You can customize this toolbar by clicking the arrow at the end. You will see the following menu:

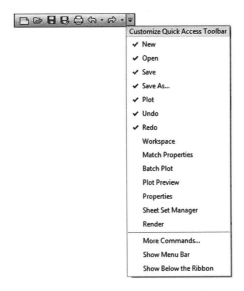

- As shown, you can add or remove commands and choose **Show Menu Bar**, which is useful because Ribbon does not include all the AutoCAD commands.

1.2.3 Ribbons

- As you can see, Ribbons consist of two parts, tabs and panels:

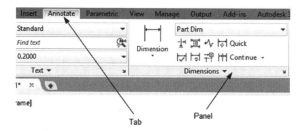

- Some panels have more buttons than shown. The following is the **Modify** panel:

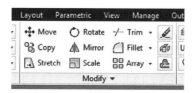

- Click the small triangle near the title to view the following buttons:

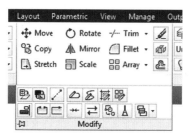

- If you move away from the panel, these buttons disappear. To make them visible again, click the push pin and the new view looks like the following:

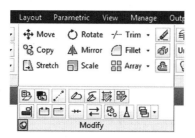

- For some commands, there are many options. To make your life easier, AutoCAD put all the corresponding options in the same button. See the following illustration:

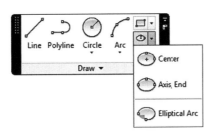

- Ribbons have a very simple help feature. If you move the mouse over a button, a small help screen appears:

- If you hover the mouse over the button, AutoCAD shows more detailed help:

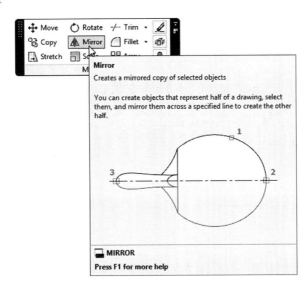

- Other buttons may have video help, similar to the following:

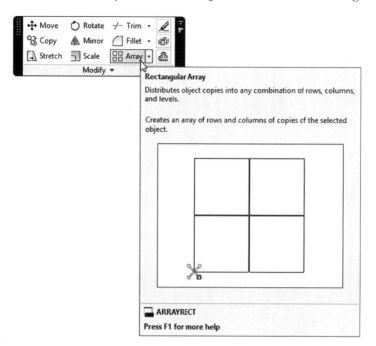

- Panels have two states, either *docked* or *floating*. By default all panels are docked in their respective tab. Drag and drop the panel in the graphical area to make it floating. One important feature of making the panel floating is you can see it while other tabs are active.
- You can also send the panel back to its respective tab, by clicking the small button at the top-right side:

- While the panel is floating, you can toggle the orientation:

- It will either extend to the right:

- Or down:

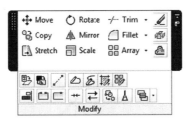

- The small arrow at the end of each tab name allows you to cycle through the different states of the ribbons. The main objective of this feature is to give you more graphical area. Clicking the small arrow offers the following options:

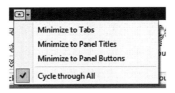

1.2.4 InfoCenter

- The InfoCenter is where you can review help topics online and offline, along with other helpful tools:

- For example, if you type a word or phrase in the field shown, AutoCAD opens the Autodesk Exchange window with all the related topics online and offline. (Online means the search feature encompasses all Autodesk websites along with some popular blogs.)

- Sign In grants access to the Autodesk Online Services. The X at the right activates the Autodesk Exchange Apps website. The last button at the right with the question mark then shows the following:

1.2.5 Command Window

- By default, the Command Window is floating but you can dock it at the bottom or top of the screen. Reading the command window saves wasted time trying to figure out what AutoCAD wants from you. AutoCAD's command window shows two things: your commands and the AutoCAD prompts asking you to do something such as specify a point or input an angle, etc. See the following illustration:

```
Arc creation direction: Counter-clockwise (hold Ctrl to switch direction).
ARC Specify start point of arc or [Center]:
```

1.2.6 Graphical Area

- Graphical area is your drafting area. This is where you draw all your lines, arcs, and circles. It is the precise environment with X,Y,Z space for 3D and X,Y plane for 2D. You can monitor coordinates in the left part of the status bar.

1.2.7 Status Bar

- The status bar in AutoCAD contains coordinates along with important functions for precise drafting in 2D and 3D.

- All of the available buttons may not be showing. To customize this menu, click on the last button at the right (the one with three horizontal lines) and you will see the following list:

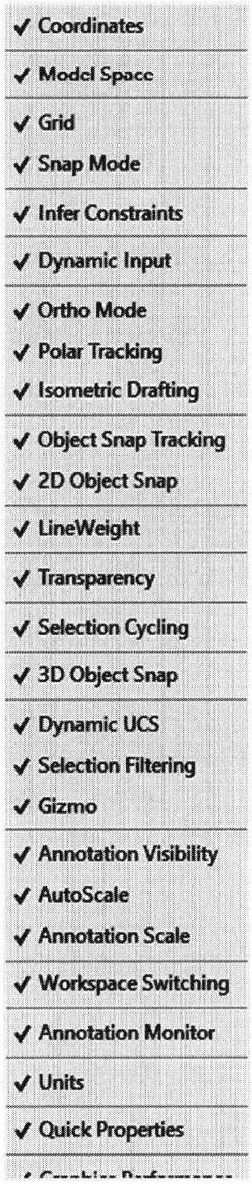

  - ✓ Coordinates
  - ✓ Model Space
  - ✓ Grid
  - ✓ Snap Mode
  - ✓ Infer Constraints
  - ✓ Dynamic Input
  - ✓ Ortho Mode
  - ✓ Polar Tracking
  - ✓ Isometric Drafting
  - ✓ Object Snap Tracking
  - ✓ 2D Object Snap
  - ✓ LineWeight
  - ✓ Transparency
  - ✓ Selection Cycling
  - ✓ 3D Object Snap
  - ✓ Dynamic UCS
  - ✓ Selection Filtering
  - ✓ Gizmo
  - ✓ Annotation Visibility
  - ✓ AutoScale
  - ✓ Annotation Scale
  - ✓ Workspace Switching
  - ✓ Annotation Monitor
  - ✓ Units
  - ✓ Quick Properties

1.3 AUTOCAD DEFAULTS

- There are some settings in AutoCAD you should be familiar with before using the AutoCAD environment.
  - AutoCAD saves points as cartesian coordinates (X,Y), for both metric and imperial numbers. This is the first method of precise input in AutoCAD; therefore, type the coordinates using the keyboard.

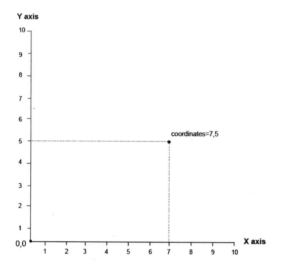

  - To specify angles in AutoCAD, assume East (to your right) is 0° and then go counterclockwise.

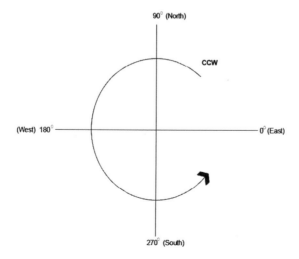

- The wheel in the mouse has four zoom functions: Zoom In (move wheel forward), Zoom out (move wheel backward), Panning (press and hold the wheel), and Zoom Extents (double-click the wheel).
- Pressing [Enter] or [Spacebar] is equal in AutoCAD.
- Pressing [Enter] without typing any command in AutoCAD repeats the last command. If it is the first thing you do in a current session, it starts Help.
- Pressing [Esc] cancels the current command.
- Pressing [F2] shows the following Text Window:

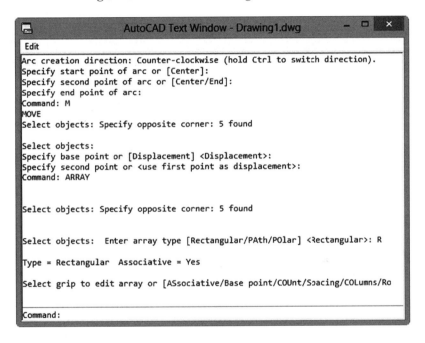

1.4 DRAWING UNITS

- If you draw a 6-unit line in AutoCAD, what units AutoCAD will use? Will the line be 6 m or 6 ft, or neither? AutoCAD deals with any unit you want; if you mean 6 m, AutoCAD will use this unit. If you mean 6 ft, AutoCAD will use this one as well. Overall, just be sure your measurement units are consistent throughout your entire file.

This is relevant in the Model Space where you do your drafting, but also when printing because you have to set your drawing scale accordingly (Chapter 9 discusses printing). At the bottom-left of the screen you can see the model tab and the layouts, as in the following:

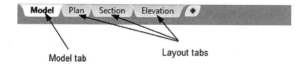

1.5 CREATING A NEW AUTOCAD DRAWING

- This command creates a new drawing based on a pre-made template. Use the **Quick Access Toolbar** and click the **New** button:

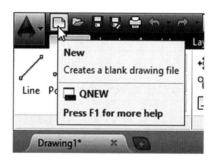

- Or you can click the (+) sign in the File tab:

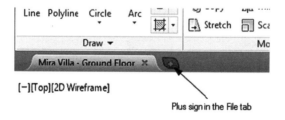

- You then see the following dialog box:

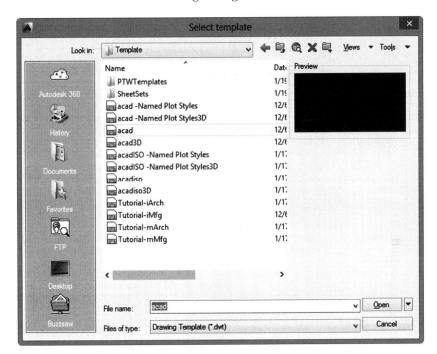

- Complete the following steps:
  - Select the desired template file (AutoCAD template files has an extension *.dwt). AutoCAD 2015 comes with many pre-made templates that you can use (however, it is often preferable to create your own template files).
  - Once you choose a template, click the **Open** button.
  - AutoCAD drawing files have a file extension of *.dwg.
  - When AutoCAD starts a new file, it has a temporary name such as Drawing1.dwg, so rename it to something more meaningful.

NOTE *In AutoCAD 2015, if no files are open, you can use the Template drop down list under Start Drawing to show the list of existing templates. Pick one of them to start a new file, like in the following window:*

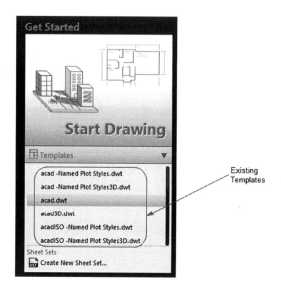

1.6 OPENING AN EXISTING AUTOCAD DRAWING

- This command allows you to open an existing drawing file for additional modifications. From the **Quick Access Toolbar**, click the **Open** button:

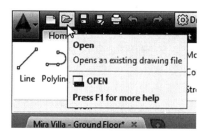

- You then see the following dialog box:

- Complete the following steps:
  - Specify your desired drive and folder.
  - You can open a single file by selecting its name from the list and clicking the **Open** button, or you can double-click on the file's name. You can also open more than one file by selecting the first file name, holding the [Ctrl] key and clicking the other file names in the list (a common MS Windows skill), and then clicking the **Open** button.

1.6.1 File Tab

- Using the File tab beneath the ribbon, you can view a tab for each opened file, just like the following:

- The color used is gray. The current file (tab) is a lighter gray whereas the other unused tabs are darker gray. If you hover over one of the tabs, two things take place:
  - The path of this file will appear above the tab.
  - The model space and the layouts of this file will appear beneath.

- See the following window:

- There is a blue frame around the model space view. Moving your mouse to the right shows the graphical area and its layout. When you find what you are looking for, click the layout view to move to it.

NOTE

A star beside the name of the file in the file tab means this file has changed and you need to save changes.

Click (x) beside the name to close the file, hence closing the tab.

You can customize the File tabs and the Layout tabs, by switching them off. Go to the View tab and locate the Interface panel; the two buttons will be blue if they are on, but if you want to turn them off click once on each button:

1.7 CLOSING DRAWING FILE(S)

- This command allows you to close the current opened file(s) or all opened files, depending on the command you choose. Use the **Application Menu** and move your mouse to the **Close** button, then select either **Current Drawing** to close the current file or **All Drawings** to close all the opened files in a single command:

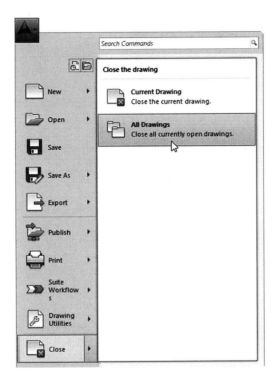

- If any of the open files were modified, AutoCAD will ask if you want to save or close without saving. See the following dialog box:

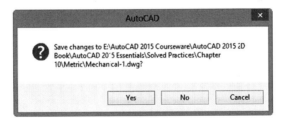

1.8 UNDO AND REDO COMMANDS

- Undo and Redo are commands to help you correct mistakes. They can be used in the current session only.

1.8.1 Undo Command

- This command allows you to undo the effects of the last command. You can reach this command by going to the **Quick Access toolbar** and clicking the **Undo** button. If you want to undo several commands, click the small arrow at the right. You then see a list of the commands; select the group and undo them:

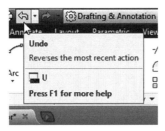

- Also, you can type **u** in the command window (do not type **undo**, because it has a different meaning) or press Ctrl + Z on the keyboard.

1.8.2 Redo Command

- This command allows you to undo the Undo command. You can reach this command from the **Quick Access toolbar**, by clicking the **Redo** button. If you want to redo several commands, click the small arrow at the right. You then see a list of the commands; select the group and redo them:

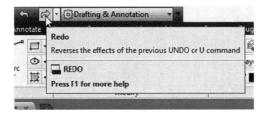

- You can also type **redo** in the command window or press **Ctrl + Y** on the keyboard.

NOTES:

PRACTICE 1-1

AutoCAD Basics

1. Start AutoCAD 2015.
2. Open the following files:
    - a. Ground Floor
    - b. Mechanical-1
    - c. Mechanical-2
3. Using the File tab, view the three files and their layouts.
4. Use the different zoom techniques with the mouse wheel.
5. Using the Application menu, close all the files without saving.

CHAPTER REVIEW

1. The File tab allows you to view the Model space and layouts of the opened files.

    a. True

    b. False

2. AutoCAD template files have the _____ extension.

    a. *.dwt

    b. *.dwg

    c. *.tmp

    d. *.temp

3. AutoCAD units can be a meter or foot, whichever your preference.

    a. True

    b. False

4. Moving the mouse wheel forward will _____

5. To undo any command in AutoCAD, you can:

    a. Click the Undo icon in the Quick Access Toolbar.

    b. Type **u** in the command window.

    c. Use [Ctrl] + Z.

    d. All of the above

6. Ribbons consist of _____ and _____.

7. The menu bar is not shown by default, but you can make it visible.

    a. True

    b. False

8. The AutoCAD drawing file extension is _____.

9. Positive angles in AutoCAD are _____.

CHAPTER REVIEW ANSWERS

1. a
3. a
5. d
7. a
9. CCW

CHAPTER 2

PRECISE DRAFTING IN AUTOCAD 2015

In This Chapter
- Drafting priorities
- How to draw lines, circles, and arcs using precise methods
- How to draw polylines using precise methods
- How to convert lines and arcs to polylines and vice versa
- Object Snap and Object Track

2.1 DRAFTING PRIORITIES

- When drafting, there are two main priorities: accuracy and speed. Most people want to finish their drawings fast, but without compromising accuracy. Experts tend to put accuracy first and sacrifice speed.
- This tension reflects the "life cycle" of drafting. If your drawing is accurate, all the other people who modify it will accomplish their mission without hassle. On the other hand, their work will be difficult if you finished your original drawing fast, but none of the objects are accurate.
- In this chapter, you learn how to use the most important four drafting commands. They are:
  - Line command, used to draw line segments.
  - Arc command, used to draw circular arcs.
  - Circle command, used to draw circles.
  - Polyline command, used to draw lines and arcs jointly.
- While discussing the four drafting commands, I also introduce accuracy tools, which will help you speed up the drafting process.

2.2 DRAWING LINES USING LINE COMMAND

- This command enables you to draw straight lines; each line segment presents a single object. To issue this command, go to the **Home** tab, locate the **Draw** panel, then select the **Line** button:

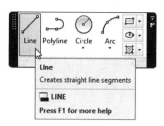

- You then see the following AutoCAD prompts:

```
Specify first point:
Specify next point or [Undo]:
Specify next point or [Undo]:
Specify next point or [Close/Undo]:
```

- Using the first prompt specify the coordinates of your first point. Keep specifying points until you are done, but keep in mind the following:
  - If you want to stop without closing the shape, simply press [Enter]. ([Esc] will do the job as well, but don't make it a habit, as [Esc] generally means abort.)
  - If you want to close the shape and finish the command, press **C** on the keyboard, or right-click and select **Close** option.
  - If you made any mistake, you can undo the last point, by typing **U** on the keyboard or right-clicking and selecting the **Undo** option.
- This is what the right-click menu looks like:

2.3 DYNAMIC INPUT IN AUTOCAD

- Dynamic Input has two functions:
  - It shows all the prompts at command window in the graphical area.
  - It shows the lengths and angles of the lines before drafting, which allows you to accurately specify them.
- In order to turn on/off the **Dynamic Input**, click the following button in the Status bar:

2.3.1 Example for Showing Prompts

- By default, if you type any command using the command window, AutoCAD helps you by showing all the commands starting with the same letter(s). See the following example:

- Type the letter **m** and AutoCAD gives all the commands starting with this letter. While Dynamic Input is on, this is also applicable to the crosshairs, as seen in the following:

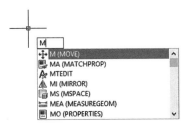

- Select command Line from the list then press [Enter]. The following prompt appears:

- Type in the X and Y coordinates, using the [Tab] key to move between the two fields:

2.3.2 Example for Specifying Lengths and Angles

- Once you specify the starting point, AutoCAD uses Dynamic Input to show the length and the angle of the line using rubberband mode:

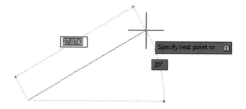

NOTE *Angles are measured CCW starting from the east, but only for 180°, unlike the angle system in AutoCAD, which uses the whole 360°.*

- Type the length of the line, then using [Tab] input the angle (it can increase by 1° increment). Once you are done, press [Enter] to specify the first line then continue doing the same for the other segments:

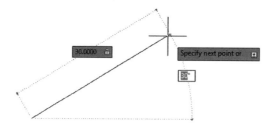

PRACTICE 2-1

Drawing Lines Using Dynamic Input

1. Start AutoCAD 2015.
2. Open the file, Practice 2-1.dwg.
3. Using status bar, click off, Polar Tracking, Ortho, Object Snap, and make sure Dynamic Input is on.
4. Draw the following shape, using 0,0 as your start point, bearing in mind all sides = 4, and all angles are multiples of 45°:

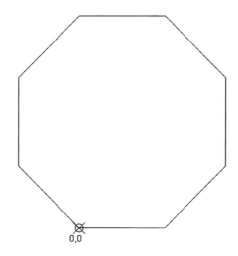

5. Save the file and close it.

2.4 EXACT ANGLES (ORTHO VERSUS POLAR TRACKING)

- Using Dynamic Input angles are incremented by 1°, but you cannot depend on AutoCAD to specify angles precisely.
- **Ortho** function will force the lines to be at right angles (orthogonal) using the following angles: 0, 90, 180, and 270.

- In order to turn on/off the **Ortho**, use the following button in the status bar:

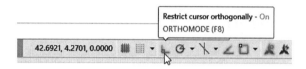

- What if we want to use other angles such as 30, 45, 60, etc.? For this reason, AutoCAD introduced another function called Polar Tracking, which shows in the graphical area rays starting from current point pointing toward angles like 30, 45, etc., and based on the settings, you can specify angles. Since Ortho and Polar Tracking are contradicting each other, if you switch one, the other is automatically turned off.
- To turn on/off the **Polar Tracking**, use the following button in the status bar:

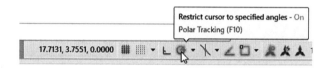

- If you click the small triangle at the right of the button, you see the following menu:

- You can select the desired angle or select **Tracking Settings** to change some of the default settings of Polar Tracking. You then see the following dialog box:

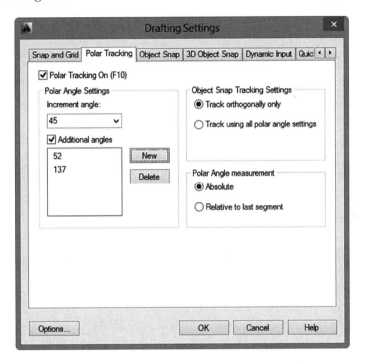

2.4.1 Increment Angle

- Increment Angle is the angle to be used along with its multiples. Select one from the list or type your own.

2.4.2 Additional Angles

- If you are using 30 as your increment angles, then 45 is not among the angles that Polar Tracking allow you to use. Therefore, you need to specify it as an additional angle. Be aware you do not use its multiples.

2.4.3 Polar Angle Measurement

- When you are using Polar Tracking, you have the ability to specify angles as an absolute angle (based on 0° at the East) or using the last line segment to be your 0 angle. See the following illustration:

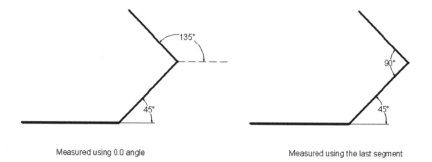

Measured using 0.0 angle   Measured using the last segment

NOTE *While you control the angle using either Ortho or Polar Tracking, you can type in the distance desired and then press [Enter]. This is called Direct Distance Entry and it allows you to draw accurate distances.*

PRACTICE 2-2

Exact Angles

1. Start AutoCAD 2015.

2. Open the file, Practice 2-2.dwg.

3. Draw the following shape (without dimension) using line command, starting from 0,0 as your starting point, keeping in mind you have to use Polar Tracking. Set the proper Increment angle and additional angles, using the Direct Distance Entry method to input the exact distances:

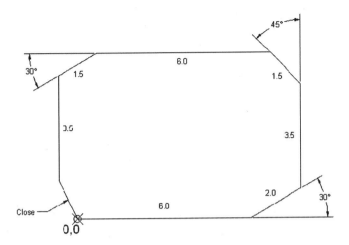

4. Save and close the file.

2.5 PRECISE DRAFTING USING OBJECT SNAP

- Object Snap, or OSNAP, is the most important accuracy tool to be used in AutoCAD for 2D and 3D. It is a way to specify points on objects precisely using the AutoCAD database stored in the drawing file.
- Some OSNAPs include:
  - **Endpoint**: To catch the Endpoint of a line.
  - **Midpoint**: To catch the Midpoint of a line.
  - **Intersection**: To catch the Intersection of two objects (any two objects).
  - **Perpendicular**: To catch the Perpendicular point on an object (any object).
  - **Nearest**: To catch a point on an object Nearest to your click point (any object).
- We learn more about object snaps in the discussion on drawing objects.

- Here are some graphical presentations of these OSNAPs:

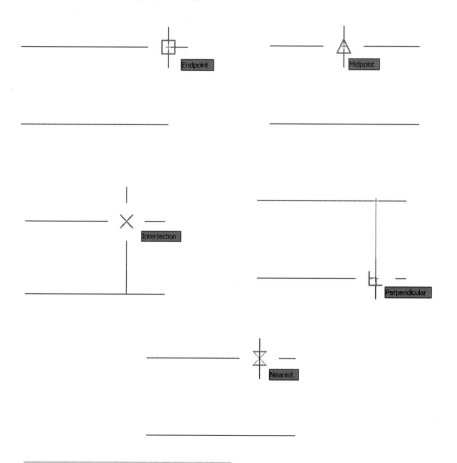

2.5.1 Activating Running OSNAPs

- To activate running OSNAPs in the drawing, click on the Object Snap button in the status bar:

- If you click the small triangle at the right of the button, you will see the following menu:

- You can switch on the desired OSNAPs one-by-one.
- If you want to make more convenient change, select the **Object Snap Settings** option and the following dialog box appears:

- There are two buttons at the right: **Select All** and **Clear All**. I recommend using Clear All first and then select the desired OSNAPs. When you are finished, click **OK**.

2.5.2 OSNAP Override

- While the Object Snap button is on, several OSNAPs are working and the other is not. Sometimes you you want to temporarily switch all of them off and only use a single one, whereas other times you may want to switch everything back to normal. This is called OSNAP Override.
- There are two ways to activate an override. They are:
  - Using the keyboard, type the first three letters of the desired OSNAP.
  - Using the keyboard, hold the [Shift] key, and then right-click. You then see the following pop-up menu:

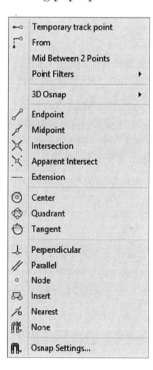

PRACTICE 2-3

Object Snap (OSNAP)

1. Start AutoCAD 2015.
2. Open the file, Practice 2-3.dwg.
3. Check the Object Snap at the status bar and make sure that Endpoint, and Midpoint, Intersection, and Perpendicular are the only OSNAPs switched on.
4. Using the Object Snap and Line command draw lines in the drawing to make it look like the following:

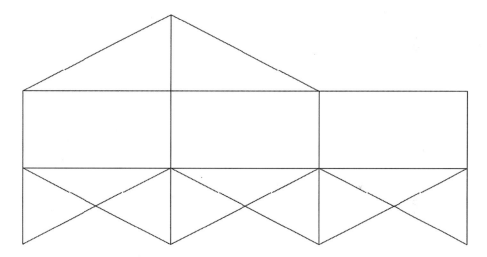

5. Save and close the file.

2.6 DRAWING CIRCLES USING CIRCLE COMMAND

- This command allows you to draw a circle using different methods based on the available data. If you know the coordinate of the center, there are two methods. If you know the coordinates of points at the parameter of the circle, there are another two methods. Finally, if there are drawn objects such as lines, arcs, or other circles, which can be used as tangents for the to-be-created circles, there are two more methods. There are six methods to draw a circle in AutoCAD:

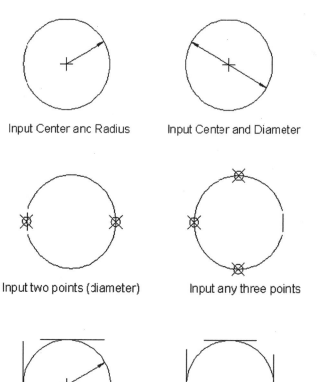

- To issue this command, go to the **Home** tab, locate the **Draw** panel, then select the arrow near the **Circle** button to see all the available methods:

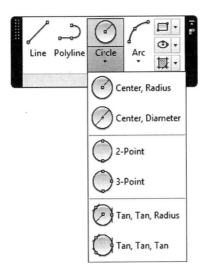

2.7 DRAWING CIRCULAR ARCS USING ARC COMMAND

- This command allows you to draw an arc part of a circle. To make our lives easier, AutoCAD deals with eight pieces of information related to a circular arc. These are:
    - The starting point of the arc.
    - Any point as a second point on the parameter of the arc.
    - The ending point of the arc.
    - The Direction of the arc, which is the tangent that passes through the Start point. You should input the angle of the tangent.
    - The distance between the starting point and the ending point, which is called the Length of Chord.
    - The center point of the arc.
    - The radius.
    - The angle between Start-Center-End, which is called the Included Angle.

- See the following illustration:

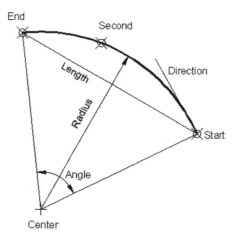

- If you provide three of these eight, AutoCAD can draw an arc, but you cannot just any three. The combination of the information needed can be found in the **Home** tab, using the **Draw** panel, while clicking the arrow near the **Arc** button to see all the available methods:

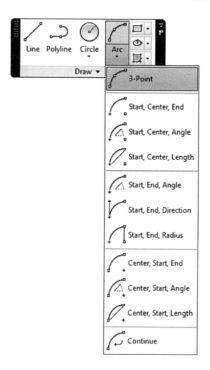

- As you can see, Start point is always required information. Normally you should think counterclockwise when specifying points, but if you want to work clockwise, simply hold the [Ctrl] key and it will change.

2.8 OBJECT SNAPS RELATED TO CIRCLE AND ARC

- Some of the Object Snaps related to circles and arcs are:
  - **Center**: To catch the Center of arc or circle.
  - **Quadrant**: To catch the Quadrant of arc or circle.
  - **Tangent**: To catch the Tangent of arc or circle.
- Here are some graphics showing each of these OSNAPs:

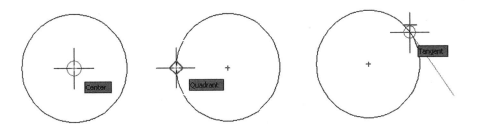

2.9 USING OBJECT SNAP TRACKING WITH OSNAP

- Sometimes OSNAP alone is not enough to specify desired points, especially if we need complex points. To solve this problem in the past, we used to draw dummy objects to help specify complex points, like the case when you want to specify the center of the circle at the center of a rectangle. We used to draw a line from the midpoints of the two vertical lines and then the same for the horizontal lines. But since the introduction of Object Snap Tracking, or OTRACK, in AutoCAD 2000, drawing dummy objects is no longer needed. OTRACK depends on active OSNAP modes, which means if you want to use midpoint with OTRACK, you have to first switch on midpoint. To activate OTRACK, go to the status bar, and turn the **Object Snap Tracking** button on:

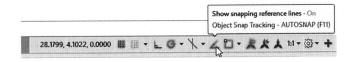

- The procedure is very simple:
  - Using OSNAP, go to the desired point and hover over it for couple of seconds (do not click), then move to the right or left (also up and down depending on the next point), and you will see an infinite line extending in both directions (this line will be horizontal or vertical depending on your movement).
  - If you want to use a single point to specify your desired point, move to the needed direction, type in the desired distance, and press [Enter].
  - If you need two points, go the next point and hover for a couple of seconds, then move toward the desired direction. Another infinite line will appear. Go to the intersection point of the two infinite lines and this will be your point.

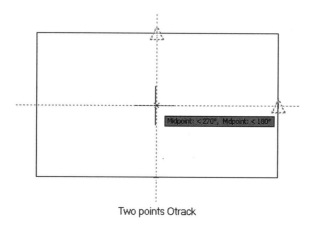

Two points Otrack

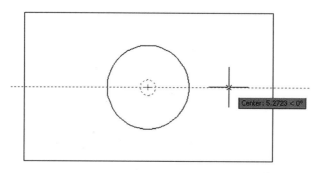

Single point Otrack

- The Polar command is a major help here as well. Let's go back to the same dialog box:

- Under **Object Snap Tracking Settings**, there are two choices:
  - Track orthogonally only (default option).
  - Track using all polar tracking settings.
- This means you can use the current Polar angles (increment and additional angles) to specify points using OTRACK.

NOTE *To deactivate an OTRACK point, stay at the same point again for a couple of seconds and it becomes deactivated.*

PRACTICE 2-4A

Drawing Using OSNAP and OTRACK

1. Start AutoCAD 2015.
2. Open the file, Practice 2-4a.dwg
3. Using the proper OSNAP and OTRACK, create the four arcs as shown.

4. Create the two circles as shown (Radius = 1).

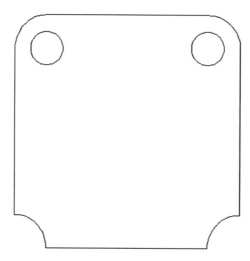

5. Using OSNAP and OTRACK (using two points), draw the circle at the center of the shape (Radius = 3.0).

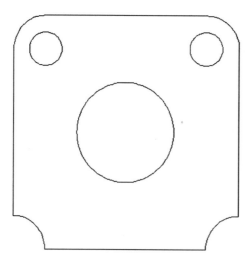

6. Using OSNAP and OTRACK (one point), draw the two circles at the right and left (distance center-to-center = 5.0 and Radius = 0.5).

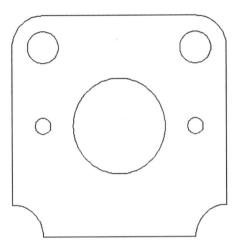

7. Change the increment angle in Polar Tracking dialog box to 45. Make sure that **Track using all polar angle settings** is on, draw a circle (Radius = 0.5) with its center specified using the OSNAP an OTRACK, and polar tracking as shown.

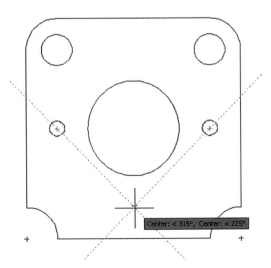

8. Do the same procedure to draw a circle at the top and you end up with the final shape as shown.

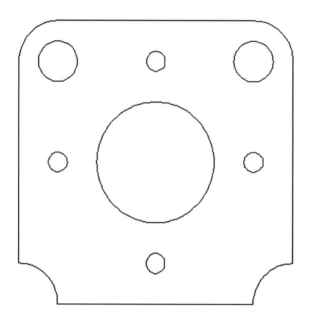

9. Save and close the file.

PRACTICE 2-4B

Drawing Using OSNAP and OTRACK

1. Start AutoCAD 2015.

2. Open the file, Practice 2-4b.dwg.

3. Using the proper OSNAP and OTRACK add lines and circles to make the shape look like the following:

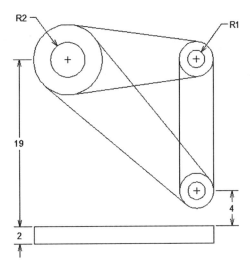

4. Save and close the file.

2.10 DRAWING LINES AND ARCS USING POLYLINE COMMAND

- Polyline command allows you to do any of the following:
  - Draw both line segments and arc segments.
  - Draw a single object in the same command rather than drawing segments of lines and arcs, like in Line and Arc commands.
  - Draw lines and arcs with starting and ending widths.
- To use the command go to the **Home** tab, locate the **Draw** panel, then select the **Polyline** button.

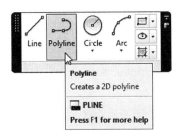

- The following prompt appears:

```
Specify start point:
Current line-width is 1.0000
Specify next point or [Arc/Halfwidth/Length/Undo/ Width]:
```

- AutoCAD then asks you to specify the first point. When you do, AutoCAD will report to you the current line-width. If you like it, continue specifying points using the same method we learned with the Line command; if not, change the width as a first step by typing the letter **W** or right-click and select the **Width** option. You will see the following prompts:

```
Specify starting width <1.0000>:
Specify ending width <1.0000>:
```

- Specify the starting width, press [Enter], and then specify the ending width. The next time you use the same file, AutoCAD will report these values for you when you issue the Polyline command. Halfwidth is the same, but instead of specifying the full width, you specify halfwidth.
- The Undo and Close options are identical to the ones in the Line command.
- Length allows you to specify the length of the line using the angle of the last segment.
- Arc allows you to draw an arc attached to the line segment. You then see the following prompt:

```
Specify endpoint of arc or [Angle/CEnter/CLose/ Direc-
tion/ Halfwidth/Line/Radius/Second pt/Undo/Width]:
```

- Arc is attached to the last segment of line or is the first object in a Polyline command. Using either method, the first point of the arc is already known, so we need two more pieces. AutoCAD will make an assumption (which you have the right to reject); AutoCAD will then assume that the angle of the last line segment will be considered the direction (tangent) of the arc. If you accept this assumption, you should specify the endpoint. If not, choose from the following to specify the second piece of information:
    - The Angle of the arc
    - The Center point of the arc

- Another Direction to the arc
- The Radius of the arc
- The Second point, which can be any point on the parameter of the arc
- Based on the information selected as the second point, AutoCAD then asks you to supply the third piece of information.

NOTE *Normally you should think counterclockwise when specifying points. On the other hand, if you want to work clockwise, simply hold the [Ctrl] key and it will change.*

2.11 CONVERTING POLYLINES TO LINES AND ARCS, AND VICE-VERSA

- This is a very essential technique that allows you to convert any polyline to lines and arcs, and convert lines and arcs to polylines.

2.11.1 Converting Polylines to Lines and Arcs

- The Command **Explode** allows you to explode a polyline to lines and arcs. To issue this command, go to the **Home** tab, locate **Modify** panel, then select the **Explode** button:

- AutoCAD then shows the following prompt:

```
Select objects:
```

- Select the desired polylines and press [Enter] when done. The new shape will have lines and arcs.

2.11.2 Joining Lines and Arcs to Form a Polyline

- This discussion focuses on an option called **Join** within a command called **Edit Polyline**. To issue this command, go to the **Home** tab, locate the **Modify** panel, then select the **Edit Polyline** button:

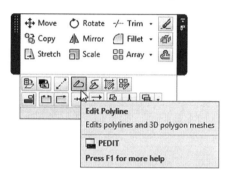

- You then see the following prompts:

  ```
  Select polyline or [Multiple]:
  Object selected is not a polyline
  Do you want to turn it into one? <Y>
  Enter an option [Close/Join/Width/Edit vertex/Fit/
  Spline/Decurve/Ltype gen/Reverse/Undo]: J
  ```

- Start by selecting one of the lines or arcs that you want to convert. AutoCAD will respond by telling you that the selected object is not a polyline and giving you the option to convert this specific line or arc to a polyline. If you accept, options will appear. One of these options is **Join**, so select the **Join** option then select the rest of the lines and arcs. At the end, press [Enter] twice. The objects were converted to a polyline.

PRACTICE 2-5

Drawing Polylines and Converting

1. Start AutoCAD 2015.
2. Open the file, Practice 2-5.dwg.

3. Draw the following polyline using a start point of 18.5 and width = 0.1.

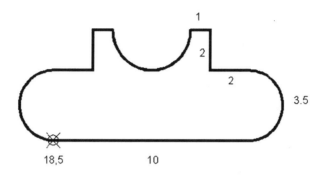

4. Then explode the polyline. As evidence, the width will disappear.

5. Check the objects after exploding, they are lines and arcs.

6. Save and close the file.

2.12 USING SNAP AND GRID TO SPECIFY POINTS ACCURATELY

- Snap and Grid is another method to help you specify points accurately in the XY plane.
- Using the mouse is not as accurate, so we cannot depend on it to specify points. We need to control its movement, which is the sole function of Snap. Snap can control the mouse to jump in the X and Y with exact distances.
- Grid by itself is *not* an accurate tool, but it does complement the Snap function. It shows horizontal and vertical lines replicating the drawing sheets. In order to turn on/off **Snap** mode, use the following button in the status bar:

- In order to turn on/off the **Grid** tool, use the following button in the status bar:

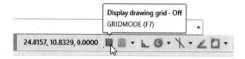

- Most likely, switching both on will not help because you need to modify the settings to set your own requirements. Using the small triangle at the right of the Snap button, you will see the following:

- Select the **Snap Settings** option and the Drafting Settings dialog box will pop up:

- Input the Snap X Spacing and Snap Y Spacing (by default, they are equal). Switch off the checkbox to make them unequal. Do the same for the Grid Spacing in X and Y. If you want Grid to follow Snap, set the Grid spacing to zero. In Grid, there are major and minor lines; set the major line frequency.

- Set if you want to see the Grid in dots (before AutoCAD 2011) and where (2D Model space, Block editor, or Sheet layout)?
- Grid behavior is for 3D only.
- Specify the type of Snap: is it Grid Snap or Polar Snap?
- You can use function keys to turn on/off both Snap and Grid:
  - F9 = Snap on/off
  - F7 = Grid on/off

2.13 USING POLAR SNAP

- While using the Snap command, the mouse will use the increment distance to specify exact distance only in horizontal and vertical directions. While going diagonal distances, Snap will not help you. To solve this problem, AutoCAD introduced Polar Snap, which specifies exact increments using all angles specified in the Polar Tracking dialog box. To activate the Polar Snap, AutoCAD switches off the normal Snap (which is called Grid Snap). To do this, go to the Snap button, click the triangle at the right of the button, and select the **Snap Settings** option. You will see the following dialog box:

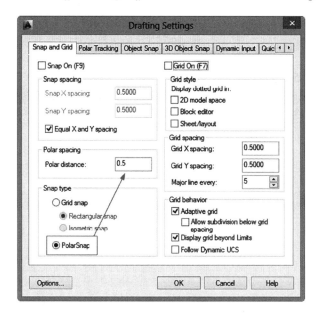

- Select the Polar Snap button, then specify the Polar distance as shown in the preceding illustration.

PRACTICE 2-6

Snap and Grid

1. Start AutoCAD 2015.
2. Open the file, Practice 2-6.dwg.
3. Change the Polar Tracking to use angle = 45.
4. Change the Snap to Polar Snap and set the distance to 0.5.
5. Draw the following shape, starting from point 22,5, with all segment lengths as 6.5, and the angle used as 45:

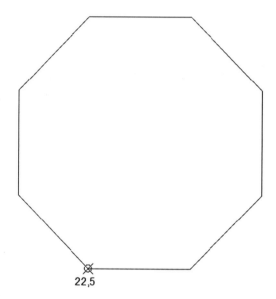

6. Save and close the file.

PRACTICE 2-7

Small Project

1. Start AutoCAD 2015.
2. Start a new drawing using acad.dwt.
3. Using all the commands and techniques you learned in Chapter 2, draw the following drawing (without dimension) starting from any point you wish:

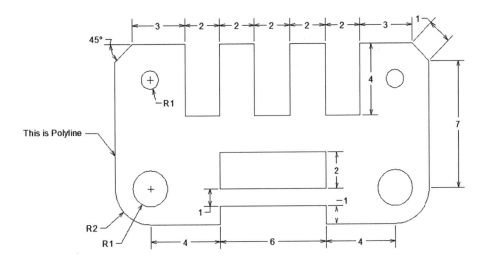

4. Save and close the file.

NOTES:

CHAPTER REVIEW

1. Polyline command is different in comparison to Line command because:

    a. It will produce lines and arcs.

    b. All segments drawn using the same command are considered a single object.

    c. You can specify a starting and ending width.

    d. All of the above.

2. Perpendicular, Tangent, and Endpoint are some _____ available in AutoCAD.

3. OTRACK can work by itself.

    a. True

    b. False

4. One of the following is not part of the eight pieces of information AutoCAD needs to draw an arc:

    a. End point of the arc

    b. Midpoint of the arc

    c. Center point of the arc

    d. Angle

5. To convert lines and arcs from a polyline, use the _____ command.

6. If you want Grid to follow the Snap settings, set it up to be _____ in both X and Y.

7. Using the Polar Tracking dialog box, you can only track the orthogonal angles.

    a. True

    b. False

8. _____ is always required to draw an arc.

9. To convert lines and arcs to polylines use:

   a. Polyline Edit, Convert option

   b. Polyline Edit, Union option

   c. Polyline Edit, Join option

   d. None of the above.

CHAPTER REVIEW ANSWERS

1. d

3. b

5. Explode

7. b

9. c

CHAPTER 3

MODIFYING COMMANDS PART I

In This Chapter
- Different methods to select objects and selection cycling
- How to erase objects
- How to move and copy objects
- How to rotate and scale objects
- How to mirror and stretch
- How to lengthen and join objects
- How to use grips for editing

3.1 HOW TO SELECT OBJECTS IN AUTOCAD

- To use any of the modification commands discussed in this chapter, you have to select the desired objects. Once you issue any of the modifying commands, the following prompt will appear:

    ```
    Select objects:
    ```

- The cursor will change to Pick Box. At this prompt, you can work without typing anything or can type a few letters to activate a certain mode.
- Without typing a letter, you can do the following things:
  - Select objects by clicking them using pick box one-by-one.

- Without selecting any object, click and move to the right to start Window mode, which will select all objects contained fully inside the Window.
- Without selecting any object, click and move to the left to start Crossing mode, which will select all objects contained fully inside Crossing or touched (crossed) by it.

- See the following two examples.
- Window Example:

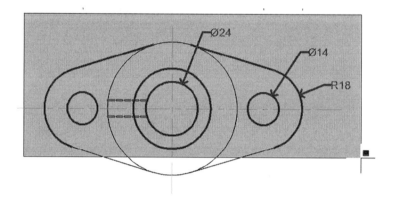

- At the cursor, you will see a small rectangle with a blue circle inside it. All of the objects will be selected except the big circle and the bottom diagonal lines. Why? Because they are not fully contained inside the window.
- Crossing Example:

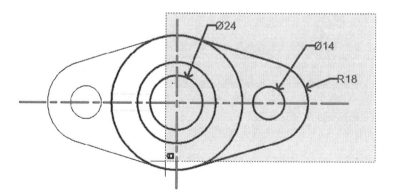

- At the cursor, you will see a small rectangle crossing a green circle. All the objects will be selected except the two diagonal lines at the left along with the arc and circle. Why? Because they were neither contained nor crossed.

- Another way to use Select objects prompt is to type a few letters to activate a certain mode. These letters are:

3.1.1 Window Mode (W)

- At the command prompt, typing **W** switches the selecting mode to **Window**, which is available whether you go to the right or left.

3.1.2 Crossing Mode (C)

- At the command prompt, typing **C** switches the selecting mode to **Crossing**, which is available whether you go to the right or left.

3.1.3 Window Polygon Mode (WP)

- WP mode will specify a non-rectangular window, by specifying points in any way you like. After typing **WP** and pressing [Enter], the following prompts will appear:

```
First polygon point:
Specify endpoint of line or [Undo]:
Specify endpoint of line or [Undo]:
```

- Press [Enter] to end WP mode. WP is just like W; it needs to contain the object fully in order to select it.
- See the following example:

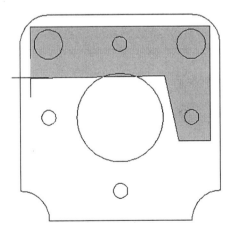

- Only four circles will be selected and they are fully contained inside the WP.

3.1.4 Crossing Polygon Mode (CP)

- Because there is W and WP, there is also C and CP. Crossing Polygon is just like WP, in that it will specify non-rectangular shapes to contain and cross objects. See the following illustration:

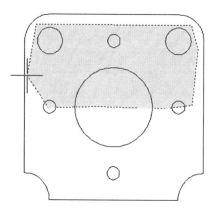

- Three circles will be selected because they are fully contained inside the CP and three more will be selected because they are crossed by CP.

3.1.5 Lasso Selection

- You can combine the two modes WP and CP without typing a letter. While the pick box is displayed *click and hold*; if you go to the right, you have WP and if you go to the left you have CP, but with an irregular shape. See the following example:

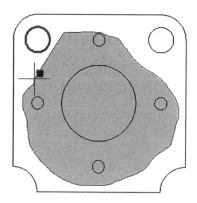

3.1.6 Fence Mode (F)

- Fence mode allows you to select multiple objects by crossing (touching) them. The Lines of Fence mode can be crossing itself, unlike when using WP and CP. See the following example:

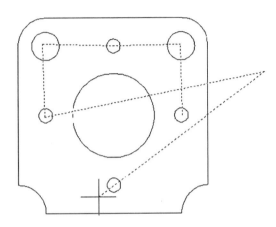

- All the circles are selected because they are touched by the fence. The vertical line at the right is selected as well. You can see that fence lines are crossed, which is 100% acceptable in AutoCAD.

3.1.7 Last (L), Previous (P), and All Modes

- You can also select objects using the following modes:
  - Last (L): To select the last object drawn.
  - Previous (P): To select the last selected objects.
  - All: To select all objects in the drawing.

NOTE
To deselect objects, hold the [Shift] key on the keyboard and then click the objects. While you are in this mode, you can use the window and crossing modes.

3.1.8 Other Methods to Select Objects

- There are two ways to use Modifying commands: either by issuing the command then selecting the objects or selecting the objects then issuing the command. This technique is called the **Noun/Verb** technique. Without issuing any command, you can:
    - Select a single object by clicking it.
    - Click an empty space and go to the right for **Window** mode.
    - Click an empty space and go to the left for **Crossing** mode.
    - Click an empty space, then type W for **Window polygon** mode.
    - Click an empty space, then type C for **Crossing polygon** mode.
    - Click an empty space, then type F for **Fence** mode.
- When you select objects, you can go to the **Home** tab, locate the **Modify** panel, and issue the desired command. You can also right-click and get a shortcut menu, which contains five modifying commands: Erase, Move, Copy Selection, Scale, and Rotate, as shown.

- By default, Noun/Verb is working, but if you want to know from where to control it, do the following steps:
    - Go to the Application Menu and select the **Options** button.
    - Select the **Selection** tab.
    - Under **Selection modes**, make sure that the Noun/Verb selection is on.

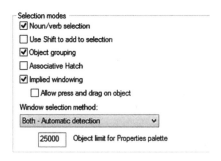

- Use this dialog box as well to do the following:
  - Turn on **Allow press and drag on object**, which allows you to create a window or crossing even if the cursor is not over a clean spot for picking.
  - Under **Window selection method**, make sure that **Both-Automatic detection** is selected so if you click then release the mouse or you click and drag, both methods are accepted.

3.2 SELECTION CYCLING

- While you are drafting using AutoCAD, we may unintentionally draft objects over each other or you may click on an object using a point that is sharable by other objects. This issue used to be a problem in the past but not anymore. AutoCAD introduced with **Selection Cycling,** which notifies you if your click touched more than one object and gives you the ability to pick the desired one. To activate Selection Cycling (by default it is active), go to the **Drafting Settings** dialog box and select the last tab titled Selection Cycling as shown:

- After activating and clicking on one object, you may see the following:

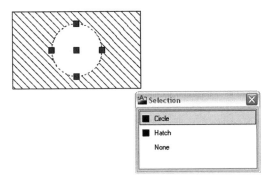

- The small window will tell you there are two possible objects at the click position, Circle and Hatch, and that it is selecting the Circle. Using this window, if you move the mouse to Hatch, you get the following picture:

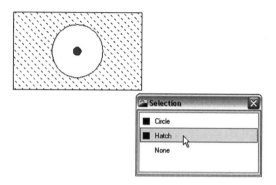

3.3 ERASE COMMAND

- The **Erase** command will delete any object you select. To issue the command, go to the **Home** tab, locate the **Modify** panel, and then select the **Erase** button:

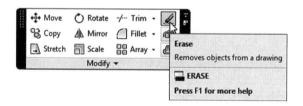

- You then see the following prompt in the command window:

  ```
  Select Objects:
  ```

- If you use the one-by-one method, you will see the cursor change to the following shape:

- Select the desired objects. Select Objects prompt is repetitive so you always need to end it by pressing [Enter] or by right-clicking.
- You can erase using other methods:
  - Click on the desired object(s) and then press the [Del] key on the keyboard.
  - Click on the desired object(s) and then right-click, a shortcut menu will appear; select **Erase**:

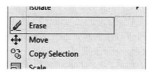

PRACTICE 3-1

Selecting Objects and Erase Command

1. Start AutoCAD 2015.

2. Open the file, Practice 3-1.dwg.

3. Start Erase command (for all steps, after you finish selecting, press [Enter] then Undo).

4. Without typing a letter, find an empty space at the left, click (release your finger but do not hold the mouse button down) and go to the right. Try to get the following result:

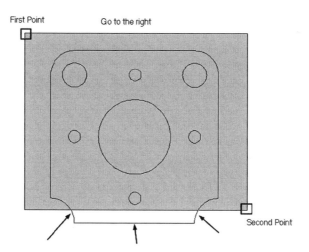

5. What was selected? Did you expect this result?

6. Do the same steps to simulate the following:

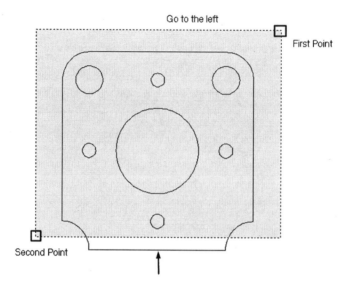

7. What was selected and why? _____

8. Try to simulate the following using WP:

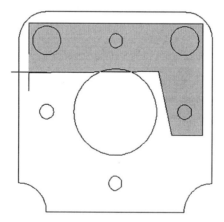

9. Try to simulate the following using CP:

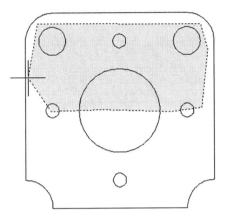

10. Try to simulate the following using Fence:

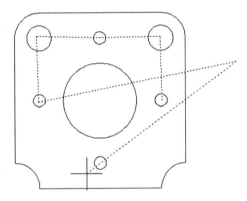

11. Without issuing the Erase command, select the objects then press [Del] on the keyboard to delete them. Undo what you did.

12. Without issuing the Erase command, select the objects, then right-click and select the Erase command from the menu. Undo what you did.

13. Start the Erase command and select objects, then hold down the [Shift] key to deselect some of the selected objects; press [Enter]. Undo what you did.

14. Close the file without saving.

3.4 MOVE COMMAND

- This command allows you to move objects from one place to another in the drawing. To issue this command, go to the **Home** tab, locate the **Modify** panel, then select the **Move** button:

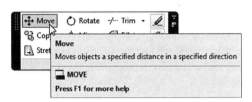

- The following prompts are shown:

```
Select objects:
Specify base point or [Displacement] <Displacement>:
Specify second point or <use first point as displacement>:
```

- The first step is to select the objects and then you should select the base point. Base point is the point that represent the objects. It will move a distance at an angle and objects will follow. The main objective of base point is accuracy.
- After you select the base point, the cursor shape will change to the following:

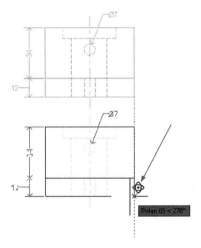

- The last prompt will ask you to specify the second point or the destination of your movement.

3.4.1 Nudge Functionality

- This function is very simple and enables you to make an orthogonal move for selected objects. All you have to do is to select objects, hold [Ctrl] key at the keyboard, then use the four arrows on the keyboard. You will see objects move toward the desired direction.

PRACTICE 3-2

Moving Objects

1. Start AutoCAD 2015.
2. Open the file, Practice 3-2.dwg.

   Using the Move command move the three circles and the rectangle to the right, so you get the following result (you need OSNAP and OTRACK to move the rectangle accurately).

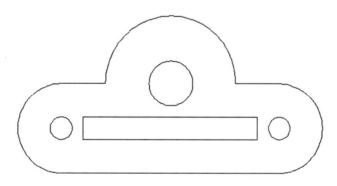

3. Save and close the file.

3.5 COPY COMMAND

- This command enables you to copy objects. To issue this command, go to the **Home** tab, locate the **Modify** panel, then select the **Copy** button:

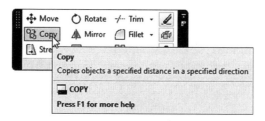

- You then see the following prompts:

```
Select objects:
Current settings: Copy mode = Multiple
Specify base point or [Displacement/mOde] <Displacement>:
Specify second point or [Array] <use first point as dis-
placement>:
Specify second point or [Array/Exit/Undo] <Exit>:
Specify second point or [Array/Exit/Undo] <Exit>:
```

- After you select the desired objects, AutoCAD reports to you the current mode, which is in your case Multiple. This mode means you can create several copies in the same command. The other mode is Single copy. The first prompt will ask you to specify the base point. AutoCAD then asks you to specify the second point to complete a single copy process; it repeats this prompt to create another one, and so on. There are three options you can use:
    - Undo, to undo the last copy.
    - Exit, to end the command.
    - Array, which creates an array of the same object using distance and angle. When you select the array option, you will see the following prompts:

```
Enter number of items to array:
Specify second point or [Fit]:
```

- The first prompt is to input the number of array (including the original object), then to specify the distance between the objects. You can use the **Fit** option to specify the total distance and AutoCAD will equally divide the distance over the number of objects.

- After you select the base point, the cursor shape changes to the following:

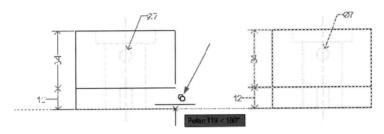

PRACTICE 3-3

Copying Objects

1. Start AutoCAD 2015.
2. Open the file, Practice 3-3.dwg.
3. Copy the door using multiple copying. Next, copy the toilet using the Array option (use Midpoint OSNAP for the toilet) to achieve the following:

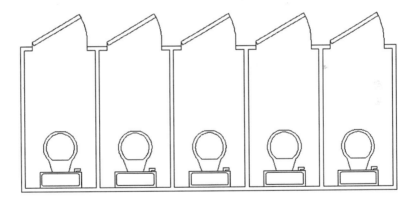

4. Save and close the file.

3.6 ROTATE COMMAND

- This command allows you to rotate objects around the base point, using rotation angle or reference. To issue the command, go to the **Home** tab, locate the **Modify** panel, and select the **Rotate** button:

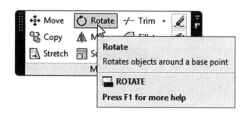

- You then see the following prompts:

```
Current positive angle in UCS: ANGDIR=counterclockwise
ANGBASE=0
Select objects:
Select objects:
Specify base point:
Specify rotation angle or [Copy/Reference] <0>:
```

- The first message tells you the current angle direction and Angle Base value. The base point here is the rotation point because all the selected objects rotate around it.
- Use the **Copy** option to make a copy of the selected objects and then rotate them.
- After you select the base point, the cursor shape changes to the following:

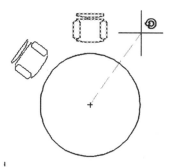

3.6.1 Reference Option

- This option is very helpful if you do not know the rotation angle, instead you can specify two points to indicate the current angle and two points to input the new angle. You then see the following prompts:

```
Specify the reference angle <0>:
Specify second point:
Specify the new angle or [Points] <0>:
```

- You can input the angle by typing it. If you want to input angles using two points, the first point you pick will be for both angles; the current and the new. See the following illustration:

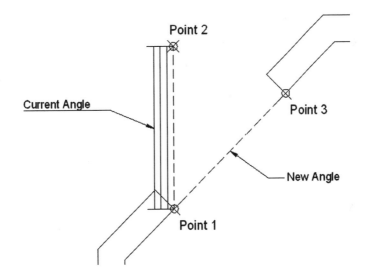

PRACTICE 3-4

Rotating Objects

1. Start AutoCAD 2015.
2. Open the file, Practice 3-4.dwg.
3. Rotate the lower window using Angle.
4. Rotate the upper window using Reference.

5. Rotate the chair with Copy mode for the following result:

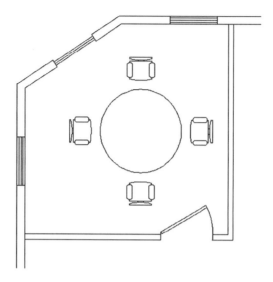

6. Save and close the file.

3.7 SCALE COMMAND

- This command allows you to create bigger or smaller objects using scale factor or reference. To issue this command, go to the **Home** tab, locate the **Modify** panel, then select the **Scale** button:

- You will see the following prompts:

```
Select objects:
Specify base point:
Specify scale factor or [Copy/Reference] <1.0000>:
```

- The base point here is the scaling point, meaning all the selected objects will be bigger or smaller in relation to it. Use the **Copy** option to make a copy of the selected objects and then scale them.
- After you select the base point, the cursor shape changes to the following:

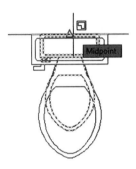

3.7.1 Reference Option

- This option is very handy if you do not know the scaling factor as a number. Instead, you can specify two points to indicate the current length and two points to input the new length. You then see the following prompts:

```
Specify reference length <0'-1">:
Specify second point:
Specify new length or [Points] <0'-1">:
```

- You can input length by typing it. If you want to input lengths using two points, the first point you pick will be for both lengths; the current and the new. See the following illustration:

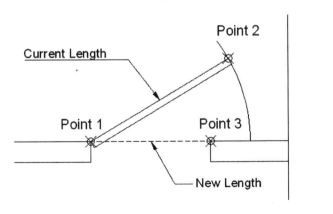

PRACTICE 3-5

Scaling Objects

1. Start AutoCAD 2015.
2. Open the file, Practice 3-5.dwg.
3. Scale the toilet by scale factor = 0.9 using the midpoint of the wall.
4. Scale the sink by scale factor = 1.2 using the quadrant of the sink.
5. Scale the door using the Reference option to fit in the door opening.
6. You get the following:

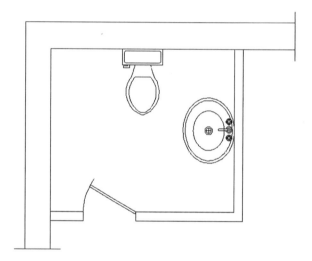

7. Save and close the file.

3.8 MIRROR COMMAND

- This command enables you to create a mirror image of the selected objects using mirror line. To issue this command, go to the **Home** tab, locate the **Modify** panel, then select the **Mirror** button:

- You then see the following prompt:

```
Select objects:
Specify first point of mirror line:
Specify second point of mirror line:
Erase source objects? [Yes/No] <N>:
```

- After selecting the objects, specify mirror line by specifying two points (you do not need to draw a line to specify these two points). The last prompt will ask you to keep or delete the original objects.
- Text can be part of the selection set to be mirrored. You can tell AutoCAD what to do with it (copying or mirroring) by using the system variable MIRRTEXT. To issue this command, type it in the command window and you will see the following:

```
Enter new value for MIRRTEXT <0>:
```

- You can input either 1 (which means mirror the text just like the other objects) or 0 (which means copy the text).
- See the following example:

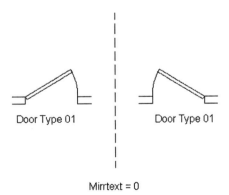

PRACTICE 3-6

Mirroring Objects

1. Start AutoCAD 2015.
2. Open the file, Practice 3-6.dwg.
3. Mirror the entrance door to open inside rather than outside, keeping the text without mirroring.
4. Mirror the furniture of the room at the right along with the two windows to the room at the left.
5. You should get the following result:

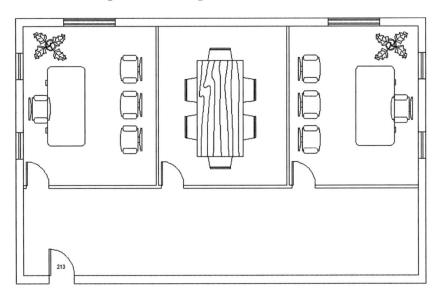

6. Save and close the file.

3.9 STRETCH COMMAND

- This command allows you to change the length of selected objects by stretching and using distance and angle. To issue this command, go to the **Home** tab, locate the **Modify** panel, and select the **Stretch** button:

- You will see the following prompts:

```
Select objects to stretch by crossing-window or crossing-
polygon...
Select objects:
Specify base point or [Displacement] <Displacement>:
Specify second point or <use first point as displacement>:
```

- The stretch command is different than the other modifying commands we have learned about because it asks you to select the objects desired using C or CP modes. Why? Because Stretch will utilize both features of C and CP, containing and crossing. All objects, contained fully inside the C or CP, are moving, whereas objects crossed are stretching, either by increasing or decreasing the length. You should specify the base point (the same principle of both the Move and Copy commands), then, finally, specify the second point. See the following example:

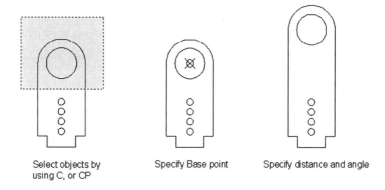

Select objects by using C, or CP   Specify Base point   Specify distance and angle

PRACTICE 3-7

Stretching Objects

1. Start AutoCAD 2015.

2. Open the file, Practice 3-7.dwg.

3. The three rooms vertical distance is not correct because it should be 1'-0" more. Use the Stretch command to fix it, keeping the entrance vertical distance as is.

4. The door of the room at the right is positioned incorrectly. Stretch it to the right, for 2'-5".

5. You should have the following picture:

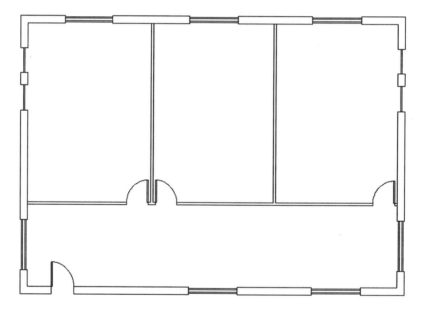

6. Save and close the file.

PRACTICE 3-8

Stretching Objects

1. Start AutoCAD 2015.
2. Open the file, Practice 3-8.dwg.
3. Stretch the upper part to look like the lower part by using distance = 1.00 and crossing polygon.
4. You should have the following picture:

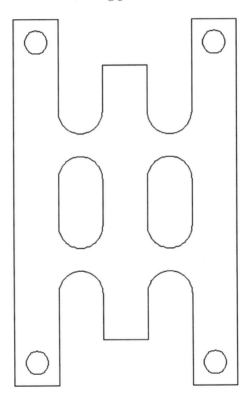

5. Save and close the file.

3.10 LENGTHENING OBJECTS

- Using this command, you can add more length or subtract length from objects using different methods. To issue this command, go to the **Home** tab, locate the **Modify** panel, and select the **Lengthen** button:

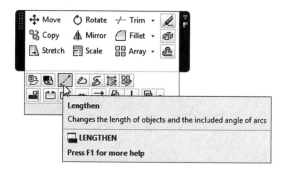

- You then see the following prompt:

```
Select an object or [DElta/Percent/Total/DYnamic]:
```

- If you click an object, it will report the current length using the current length units. There are multiple methods to lengthen (or shorten) objects in AutoCAD.
- The first method is called **Delta**, which allows you to add (remove) to (from) the current length. A positive value means adding and a negative value means subtracting.
- If you choose Delta option, you see the following prompt:

```
Enter delta length or [Angle] <0.0000>:
```

- The second method is **Percentage**, which allows you to add (remove) to (from) the length, by specifying a percentage of the current length. To add more length, input a value more than 100; to remove length input a value less than 100.
- If you choose Percentage option, you will see the following prompt:

```
Enter percentage length <100.0000>:
```

- The third method is **Total**, which allows you to input a new total length of the object. This means the object will add more length, if the new value is greater than the current value and vice-versa.

- If you choose Total option, you will see the following prompt:

  ```
  Specify total length or [Angle] <1.0000)>:
  ```

- The fourth method is **Dynamic**, which allows you to increase/decrease the length dynamically using the mouse. You will see the following prompt:

  ```
  Specify new end point:
  ```

NOTE *You lengthen/shorten a single object using a single method per command.*

3.11 JOINING OBJECTS

- This Join command is a helpful command because it enables joining lines to lines, arcs to arcs, and polylines to polylines.
- To issue this command, go to the **Home** tab, locate the **Modify** panel, and select the **Join** button:

- AutoCAD then shows the following prompts:

  ```
  Select source object or multiple objects to join at
  once:
  Select lines to join to source:
  Select lines to join to source:
  1 line joined to source
  ```

- The preceding prompts are for joining lines and they may differ for arcs and polylines. There are some conditions for the joining to succeed:
  - You can join lines, arcs, and polylines to form a single polyline but they should be connected to the ends of each other.
  - If the lines, arc, and polylines are not connected, each connected group will be considered a single polyline.
  - If you want to join lines to form a line they should be always collinear.
  - If you want to join arcs to form a single arc, they should have the same center point.
  - While joining arcs, there is a special prompt that asks if you are interested in creating a circle (this is the only command which creates a circle from an arc).

PRACTICE 3-9

Lengthen and Joining Objects

1. Start AutoCAD 2015.
2. Open the file, Practice 3-9.dwg.
3. Using the Lengthen command, make the total length of the line at the upper left = 1.75".
4. Using the Lengthen command, make the arc length at the right side of the shape 200% of the current length.
5. Using the Join command, join the arc at the top and the line at its right.
6. Using the Join command, convert the two arcs at the top and at the bottom to be a full circle.
7. Using the Join command, join all objects except the four circles to form a single polyline.
8. You should get the following shape:

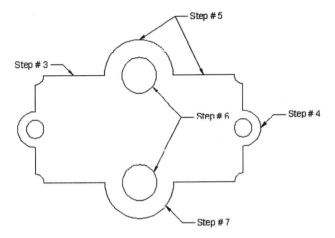

9. Save and close the file.

3.12 USING GRIPS TO EDIT OBJECTS

- Grips are blue squares and rectangles that appear on objects and enable you to modify them using five modifying commands, all using the selected grip as a base point. It is a clever tool to perform modifying tasks faster without compromising the accuracy of the conventional modifying commands previously discussed. Depending on the type of the object, grips will appear at different places. See the following illustration:

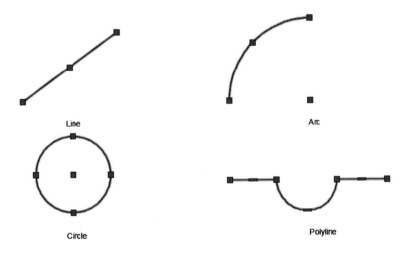

- By default, the color of the grip is blue (cold). If you click it, the grip turns red (hot). Depending on the object and the location of the grip, if you right-click you will see a group of modifying commands that all share a common base point. For example, if you click the grip at the end of line, then right-click, you will see the following menu:

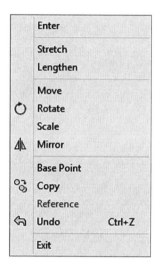

- As you can see, there are six commands. They are:
  - Stretch
  - Lengthen
  - Move
  - Rotate
  - Scale
  - Mirror
- These commands (except Lengthen) all share one thing in common: base point (that is, if you consider the first point of the mirror line equivalent to a base point). Copy here is a mode, not a command so it will work with all other commands.
- Holding [Shift] while selecting the grip enables you to select more than one base point.
- Holding [Ctrl] while specifying a second point (Move and Stretch), rotation angle (Rotate), specifying scale factor (Scale), specifying second point of mirror line (Mirror) allows AutoCAD to remember the last input and repeat it graphically.

- The base point option enables you to select another base point other than the grip selected.
- Let us look at an example. In the following shape, select the two polylines:

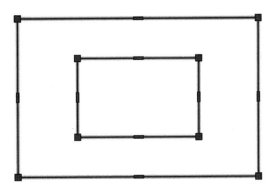

- Click the upper-right corner to make it hot, then right-click to access the menu and select Rotate command:

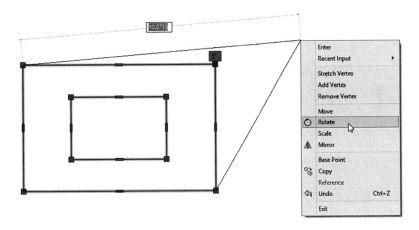

- Now you can rotate the two shapes around the grip (this is considered your base point). Right-click again and select the Base point option to select another base point, which will be the center of the two rectangles (using OSNAP and OTRACK):

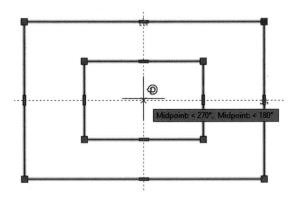

- Right-click again and select the Copy option to copy while rotating. Using Polar tracking specify an angle of 90, then press [Esc]. You will get the following result:

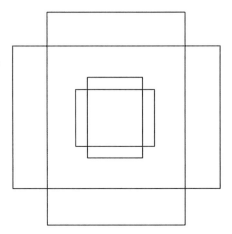

- Specific objects like polylines will show more than the blue squares. They will show rectangles at the midpoint of each line and arc segments. The rectangles at the midpoint of the polyline have multiple functions. Depending upon whether, they are a line segment or arc segment,

AutoCAD will show a different menu. Go to the grip, hover for a second (do not click), and you will see something similar to the following:

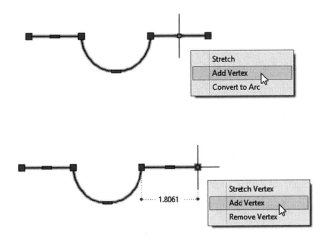

- As you can see, you can stretch the selected segment; add a vertex; convert a line to an arc; or convert an arc to a line.
- When you are done with grips, click [Esc] once or twice, depending on the situation you are in, and this will end the grips mode.

3.13 GRIPS AND DYNAMIC INPUT

- If you stayed on one of the grips (without clicking), Dynamic Input along with grips will aid you to get information about the selected objects based on their type. See the following examples:
- Using line and one of the endpoints, you will see the length and angle with the east, along with two commands Stretch and Lengthen:

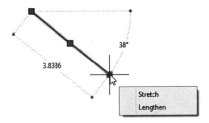

- The two connected lines also show two lengths and two angles:

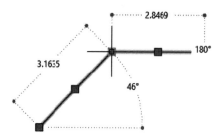

- The midpoint of an arc shows the radius and included angle, along with the two commands, Stretch and Radius:

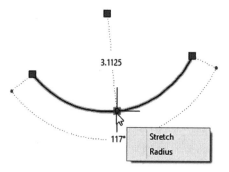

- However, the endpoints of an arc will show radius and angle with the east, along with the Stretch and Lengthen commands:

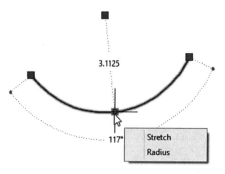

- As for a circle, you will see the radius only:

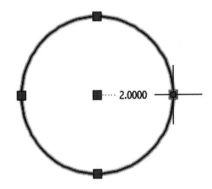

- Shared endpoints between two segments of a polyline show the length of the shared lines, along with a shortcut menu including options such as Stretch Vertex, Add Vertex, and Remove Vertex.

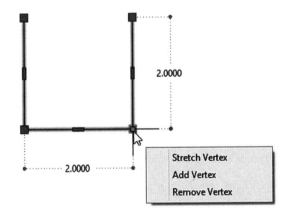

3.14 GRIPS AND PERPENDICULAR AND TANGENT OSNAPs

- Using grips, you can specify perpendicular and tangent OSNAPs, keeping in mind that these two settings are turned on when running OSNAP. See the following two examples.

- Tangent example:

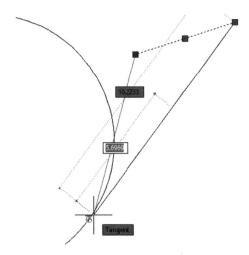

- Perpendicular example:

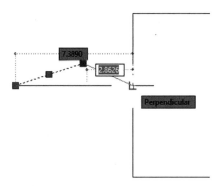

PRACTICE 3-10

Using Grips to Edit Objects

1. Start AutoCAD 2015.

2. Open the file, Practice 3-10.dwg (you should complete this practice using grips techniques only).

3. Select the large circle, make the center hot, and scale it with copying by 1.2 as the scale factor.

4. Select the circle at the right of the drawing and move it from its center to the center of the existing two circles.

5. Mirror the three lines at the right to other side of the part (you should use another base point along with Copy mode).

6. You should now have the following shape:

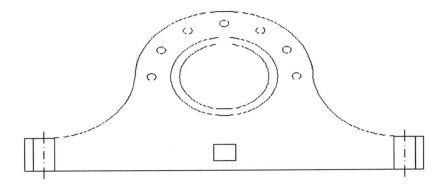

7. Select the polyline below the two big circles. Convert the top horizontal line to an arc by moving up distance = 0.25.

8. Using the rectangle grip at the mid of the lower horizontal line of the polyline, stretch it downward by distance = 0.1.

9. Using the grips and the dynamic input what is the horizontal distance and the vertical distance of the polyline? _____, _____

10. Press [Esc] to clear grips, then re-select the polyline again, select one of the grips (any one), and then right-click to select Move command. Right-click again and select Copy mode, but ensure that Polar tracking is on to help get exact angles. Now move to the right, type 2 as a distance, and then press [Enter]. Next, hold the [Ctrl] key so AutoCAD will remember this distance. While you are still holding [Ctrl] key, make one copy to the right and one copy to the left.

11. Select the lines at the right using Crossing. See the following illustration:

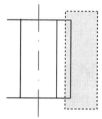

12. Hold the [Shift] key, select the lower-right corner grip, and then select the upper-right corner grip. Next, release the [Shift] key, click any of the two grips, and go to the right by distance = 0.25.

13. Do the same steps to the left side of the drawing.

14. Make sure that Tangent and Perpendicular are both on in the OSNAP dialog box.

15. Click the right green line. Make the lower grip hot; stretch it to be perpendicular on the horizontal line. Do the same thing for the other line.

16. Using the upper grip, make both lines tangent to the inner circle.

17. You should have the following picture:

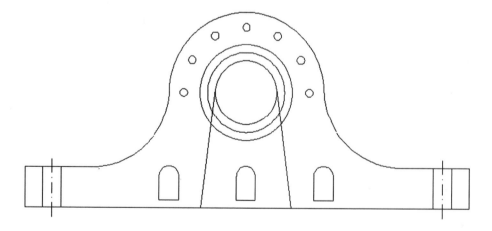

18. Save and close the file.

NOTES:

CHAPTER REVIEW

1. One of the following is <u>not true</u> about the Reference concept.

    a. You can use it with Rotate and Scale.

    b. In Rotate, you have to specify the current and the new angle by specifying four points.

    c. In Scale, you have to specify two distances, the current and the new.

    d. You cannot use it in Stretch command.

2. Stretch command will ask for something that other modifying commands will not.

    a. True

    b. False

3. If you want to select the last selected set, type _____ letter at the command window.

4. In Mirror command, you should draw a line in order to act as a mirror line.

    a. True

    b. False

5. While using grips holding _____ while selecting the grip, will enable you to select more than one base point.

6. _____ will help you control the outcome of text in Mirror command.

7. One of the following commands is not among the commands for grips:

    a. Stretch

    b. Lengthen

    c. Join

    d. Rotate

8. While using the Lengthen command, the option to double the current length without knowing it is:

　a. Delta

　b. Total

　c. Percent

　d. Dynamic

CHAPTER REVIEW ANSWERS

1. b

3. P

5. [Shift]

7. c

CHAPTER 4

MODIFYING COMMANDS PART II

In This Chapter

- How to offset objects
- How to fillet and chamfer objects
- How to trim and extend objects
- How to array objects using three different methods
- How to break objects

4.1 INTRODUCTION

- The commands discussed in this chapter are modifying commands with special abilities. They can build over the shapes you drew. Each one of them has a unique function; two of them can create objects (such as Offset and Array) and others can change an existing object shape (such as Fillet, Chamfer, Trim, and Extend commands).
- The following is a brief description of each command:
  - Offset command: to create copies of an object parallel to the original.
  - Fillet command: to create neat intersection between two objects either by extending/trimming lines or using arcs.
  - Chamfer command: to create a neat intersection but only for lines. The neat intersection is created by extending/trimming the two lines or by creating a new line showing the chamfered edge.
  - Trim command: this command will trim objects using other objects as cutting edges.

- Extend command: this command will extend objects using other objects as boundary edges.
- Array command: this command will create objects using three different methods: rectangular, circular, and using a path.
- Break command: this command will break an object into two objects by specifying two points and deleting the portion between them.

4.2 OFFSETTING OBJECTS

- The offset command allows you to create copies of an object parallel to the original. The new object possesses the same properties of the original object. You can offset this by using offset distance or using a point the new object will pass through. You can start this command by going to the **Home** tab, locating the **Modify** panel, then selecting the **Offset** button:

- You then see the following AutoCAD prompts:

  ```
  Current settings: Erase source=No Layer=Source OFFSET-
  GAPTYPE=0
  Specify   offset   distance   or   [Through/Erase/Layer]
  <Through>:
  ```

4.2.1 Offsetting Using Offset Distance Option

- If you know the distance between the object and the new parallel copy, input the value then select the original object. Next, click on the side you want the new object to go to. Going this route, you see the following prompts:

  ```
  Specify   offset   distance   or   [Through/Erase/Layer]
  <Through>:
  ```

```
Select object to offset or [Exit/Undo] <Exit>:
Specify point on side to offset or [Exit/Multiple/ Undo]
<Exit>:
```

- This will allow you to create a single offset. To create more offsets using the same offset command, select another object, and perform the same steps again. Pressing [Enter] or right-clicking will end the command.

4.2.2 Offsetting Using Through Option

- If you do not know the offset distance but you do know a point in the drawing, the new parallel object will pass through, this option will help you accomplish your mission. You see the following AutoCAD prompts:

```
Specify   offset   distance   or   [Through/Erase/Layer]
<Through>:
Select object to offset or [Exit/Undo] <Exit>:
Specify through point or [Exit/Multiple/Undo] <Exit>:
```

- This allows you to create a single offset. To create more offsets using the same offset command, select another object and perform the same steps again. Pressing [Enter] or right-clicking will end the command.
- Here is an example:

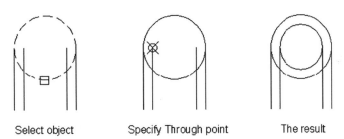

Select object   Specify Through point   The result

4.2.3 Using Multiple Option

- You can use the *Multiple* option to repeat the offset distance or Through option in the same command repeatedly by clicking on the side of offset, or by specifying a new through point. The prompts for multiple options are:

```
Specify through point or [Exit/Multiple/Undo] <Exit>: M
Specify point on side to offset or [Exit/Undo] <next
object>:
```

- While offsetting, keep the following things in mind:
  - If you make a mistake, use the Undo option.
  - AutoCAD remembers the last offset distance used and will save it in the file.
  - When offsetting an arc or circle, the new arc and circle will share the same center point, hence the result will be a smaller or bigger arc or circle.
  - If you offset a closed polyline, the output will be smaller or bigger.
  - Offset has an automatic preview feature, which will show you the result before you click to accept it.

PRACTICE 4-1

Offsetting Objects

1. Start AutoCAD 2015.
2. Open the file, Practice 4-1.dwg.
3. Offset the outer polyline to the outside by distance = 0.3.
4. Using the endpoints of the lines inside the two circles, offset the two circles to the inside using the Through option.
5. You should have the following picture:

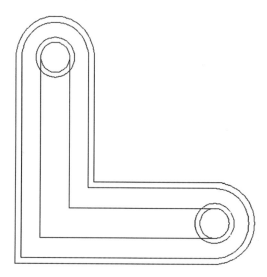

6. Save and close the file.

PRACTICE 4-2

Offsetting Objects

1. Start AutoCAD 2015.
2. Open the file, Practice 4-2.dwg.
3. Offset the polyline which represents the outside edge of the outer wall by distance = 1'-0".
4. Offset the yellow line representing one of the stair steps, 10 times using distance = 1'-6", and using Multiple option.
5. Explode the inner polyline.
6. Offset the vertical line at the right of the exploded polyline to the left, using the Through option to pass through the inner right end point of the inclined line (as shown in the following picture).
7. Offset the newly created line to the right by distance = 6".
8. You will get the following picture:

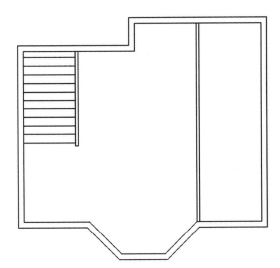

9. Save and close the file.

4.3 FILLETING OBJECTS

- The mission of Fillet command is to create neat intersections. User should set the value of Radius as the first step. If it was 0 (zero), then you can use Fillet only between two lines and Fillet will extend/trim the lines to the proposed intersection point. But if the value of Radius is greater than 0 (zero), then Fillet can use lines and circles to fillet these objects with an arc.

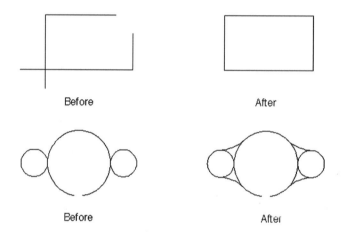

- While working with Radius greater than zero, you can select between: **Trim**, which will allow trimming the original objects and **No trim** will allow the original objects to stay as is. In both cases, you can see the arc when hovering over the second object to ensure this is the correct value.
- Here is an example:

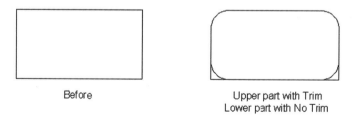

- To start this command go to the **Home** tab, locate the **Modify** panel, then select the **Fillet** button:

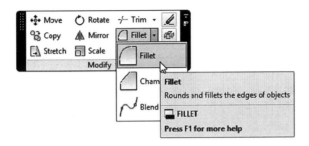

- You then see the following AutoCAD prompts:

  ```
  Current settings: Mode = TRIM, Radius = 0.0000
  Select first object or [Undo/Polyline/Radius/Trim/
  Multiple]:
  ```

- You should always check the first line of this command because it will report the current value of **Radius**. And you canl decide whether to keep it or change it. Type **r** or right-click and select the Radius option to set the new value of Radius. You will see the following prompt:

  ```
  Specify fillet radius <0.0000>:
  ```

- Type **t** or right-click and select **Trim** to change the mode whether **Trim** or **No trim**. You will see the following prompt:

  ```
  Enter Trim mode option [Trim/No trim] <Trim>:
  ```

- Fillet command allows only a single fillet per command. To make a multiple fillets in the same command, simply change the value to **Multiple** mode. If you make a mistake, use the **Undo** option. To end the command, press [Enter].
- You can do two significant things while using the Fillet command:
  - You can fillet two parallel lines regardless of the current Radius value.
  - You can fillet any two lines with radius = 0, regardless of the current value of Radius by holding down the [Shift] key.

NOTE *If you use the Multiple option, you can use different radius values in the same command.*

PRACTICE 4-3

Filleting Objects

1. Start AutoCAD 2015.
2. Open the file, Practice 4-3.dwg.
3. Using the Fillet command try to get the following final result:

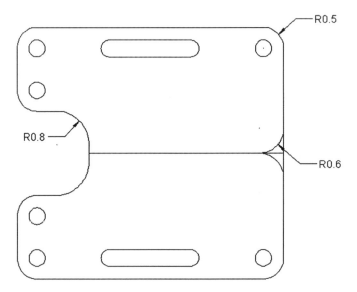

4. Save and close the file.

4.4 CHAMFERING OBJECTS

- Chamfer command allows you to create neat intersections as well, but with this command you should set the value of Distance (or Distance and Angle) as the first step. If it was 0 (zero), then you can use Chamfer to extend/trim the lines to the proposed intersection point. But if the value of Distance is greater than 0 (zero), then Chamfer will create a sloped edge between the two lines. While working with a distance greater than zero, you can select between **Trim**, which will allow trimming the original objects and **No trim**, which allows the original objects to stay as is. In both cases, you will be able to see the chamfer line when hovering over the second object and you can make sure that this is the correct value.

- To create the sloped edge, use one of the available two methods:
  - Distance (two distances)
  - Distance and Angle

4.4.1 Chamfering Using Distance Option

- To chamfer using the Distance option, you will see the following two prompts:

```
Specify first chamfer distance <0.0000>:
Specify second chamfer distance <0.0000>:
```

- There will be two different cases:

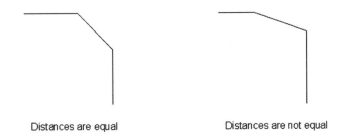

Distances are equal   Distances are not equal

4.4.2 Chamfering Using Distance and Angle

- To chamfer using Distance and Angle, you will see the following two prompts:

```
Specify chamfer length on the first line <0.0000>:
Specify chamfer angle from the first line <0>:
```

- Set the length (which will be cut from the first object selected) and an angle.

- See the following illustration.

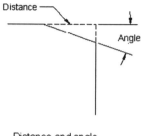

Distance and angle

- To start this command go to the **Home** tab, locate the **Modify** panel, then select the **Chamfer** button:

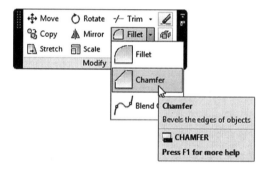

- You then see the following prompts:

  ```
  (TRIM mode) Current chamfer Dist1 = 0.0000, Dist2 = 0.0000
  Select first line or [Undo/Polyline/Distance/Angle /
  Trim/mEthod/Multiple]:
  ```

- You should always check the first line of this command because it will report the current value of method used (whether Distance or Distance and Angle), and the current values. Accordingly, you can choose whether to keep or change it.
- Other options such as Multiple, Trim, and Undo is identical to what we learned in the Fillet command. The Method option is to select the default method to be used in the chamfering process.

- Another similarity to Fillet command is you have to hold down the [Shift] key, which allows you to chamfer by extending/trimming the two lines regardless of the current distance values.

NOTE *Trim and Untrim in Chamfer will affect Fillet command and vice-versa.*

PRACTICE 4-4

Chamfering Objects

1. Start AutoCAD 2015.
2. Open the file, Practice 4-4.dwg.
3. Use the Chamfer command to make the shape look like the following:

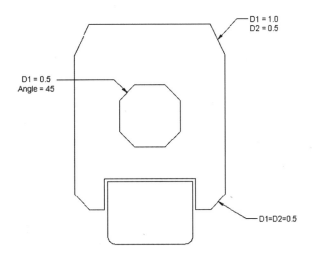

4. Save and close the file.

4.5 TRIMMING OBJECTS

- This command allows you to remove part of an object based on cutting edge(s). In order to successfully complete this command, you have

to select the cutting edges first, press [Enter], and then select the part of the objects you want to remove.
- The following example illustrates the process of trimming:

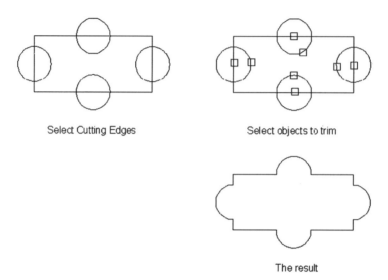

Select Cutting Edges   Select objects to trim

The result

- To issue this command, go to the **Home** tab, locate the **Modify** panel, then select the **Trim** button:

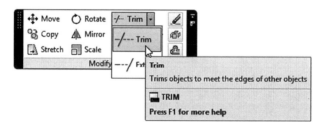

- You then see the following prompts:

```
Current settings: Projection=UCS, Edge=Extend
Select cutting edges ...
Select objects or <select all>:
```

- The first line is a message from AutoCAD telling you the current settings. The second line is asking you to select the cutting edges, which is the first

stage of the Trim command. When finished, press [Enter] or right-click. You can also use the **select all** option, by pressing [Enter] right away. It is usually faster because it selects all the objects to act as cutting edges.
- You will see the following prompt either way:

```
Select object to trim or shift-select to extend or
[Fence/Crossing/Project/Edge/eRase/Undo]:
```

- You can now start clicking on the part of objects you want to remove. You have three ways to do that:
  - Selecting these objects one by one.
  - Click on empty space and go to the left to start Crossing mode to select the objects to be removed collectively.
  - Type F to start the Fence option (discussed in Chapter 3).
- If you made any mistake, simply type u to undo the last trimming process.
- While you are trimming you may get an orphan object as a result, you can get rid of it by issuing the eRase option (use letter r).

PRACTICE 4-5

Trimming Objects

1. Start AutoCAD 2015.
2. Open the file, Practice 4-5.dwg.
3. Using the Trim command try to make the shape look like the following:

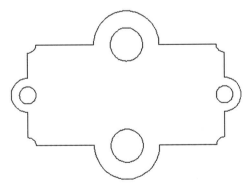

4. Save and close the file.

4.6 EXTENDING OBJECTS

- This command allows you to extend an end of an object to boundary edge(s). In order to successfully complete this command, you have to select the boundary edges first, press [Enter], and then select the objects' ends you want to extend.
- Check the following illustration:

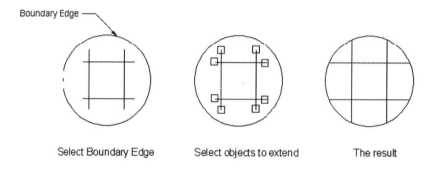

Select Boundary Edge   Select objects to extend   The result

- To issue this command go to the **Home** tab, locate the **Modify** panel, then select the **Extend** button:

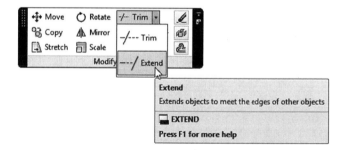

- You then see the following prompts:

```
Current settings: Projection=UCS, Edge=Extend
Select boundary edges ...
Select objects or <select all>:
```

- The first line is a message from AutoCAD telling you the current settings. The second line is asking you to select the boundary edges, which is the first stage of the Extend command. When you are finished, press [Enter] or right-click. You can also use the fastest method, the **select all** option, which can select all the objects to act as boundary edges. This can be done by pressing [Enter]. Either way, you will see the following prompt:

```
Select object to extend or shift-select to trim or
[Fence/Crossing/Project/Edge/Undo]:
```

- You can start now by clicking on the part of objects you want to extend. You have three ways to do that:
  - Selecting these object ends one by one.
  - Click on empty space and go to the left to start Crossing mode to select the objects to be removed collectively.
  - Type F, to start Fence option (discussed in Chapter 3).
- If you made any mistakes, simply type **u** to undo the last extending process.
- The last feature in both the Trim and Extend commands is the ability to use each command while you are using the other. See the following example:

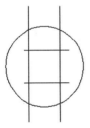 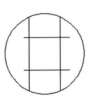

Select the circle as cutting edge   Trim lines as shown   Hold [Shift] to convert Trim command to Extend command and select the other lines

PRACTICE 4-6

Extending Objects

1. Start AutoCAD 2015.
2. Open the file, Practice 4-6.dwg.

3. Using the Extend command (and Trim if needed) correct the architectural plan to look like the following:

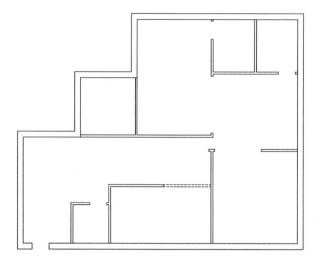

4. Save and close the file.

4.7 ARRAYING OBJECTS – RECTANGULAR ARRAY

- This command allows you to create replicates in matrix fashion using rows and columns. The resultant shape will be one object and can be edited.

4.7.1 The First Step

- To issue this command, go to the **Home** tab, select the **Modify** panel, and then select the **Rectangular Array** button:

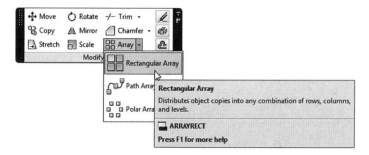

- The following prompt is shown:

  ```
  Select objects: 1 found
  ```

- This prompt asks you to select the desired objects. Once done, press [Enter] and AutoCAD immediately adds a 3 row and 4 columns grid showing grips. See the following:

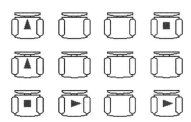

- Subsequently, you see the following context tab called Array Creation:

- The following prompts then appear at the command window:

  ```
  Type = Rectangular Associative = Yes
  Select grip to edit array or [ASsociative/Base point/
  COUnt/Spacing/COLumns/Rows/Levels/eXit]<eXit>:
  ```

- AutoCAD asks you to select the suitable grip to edit the array. You can do this two ways:
  - Using the Array Creation context tab.
  - Using grips to set the number of rows, columns, distances between columns, distances between rows, and direction of arraying (downward, upward, right, or left).

4.7.2 Using Array Creation Context Tab

- Using the Array Creation context tab is more convenient if you know the numbers and distances. You can do this as follows:
  - Using the Column panel input two of three pieces of information: Columns (number of columns), Between (distance between columns), and Total (total distance the columns will occupy). Keep in mind the

following things: You should be consistent, taking the same reference point to measure distances (from left to left, or from center to center). Also, you should consider the direction of arraying; positive distances means upward, and negative distances means downward.
- Use the Rows panel to input two of three pieces of information: Rows (number of rows), Between (distance between rows), and Total (total distance the rows will occupy). Keep in mind the following things: User should be consistent, taking the same reference point to measure distances (from top to top or from center to center). Also, you should consider the direction of arraying; positive distances means going to the right and negative distances means going to the left.
- Ignore Levels for 2D, because this is only for 3D.
- Associative means all objects resultant from the array are considered a single object holding all the information used to build it.
- By default, the first object will be considered the base point for the array, but AutoCAD allows you to select a different one.
- Click the Close Array button to end the command.

4.7.3 Editing Rectangular Array Using Grips

- After you insert a rectangular array, click once on any object in order to edit it. Grips will appear on the objects like the following:

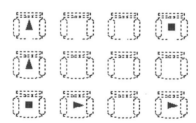

- Each one of these has a function:

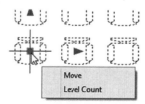

- This grip is for Moving the whole array objects or to control Level count (3D only). Use the [Ctrl] key to browse between these options.

- This grip is for changing the column spacing alone.

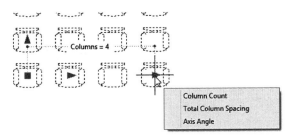

- This grip is for changing Column count, Total Column Spacing, or Axis Angle (to specify another angle other than horizontal and vertical). Use the [Ctrl] key to browse between these options.

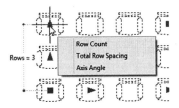

- This grip is for changing Row Count, Row Spacing, or Axis Angle. Use the [Ctrl] key to browse between these options.

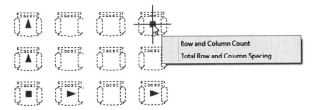

- This grip is for changing Row and Column Count or Total Row and Column Spacing. Use the [Ctrl] key to browse between these options.

4.7.4 Editing Rectangular Array Using Context Tab

- When you click a rectangular array, a context tab named Array appears, similar to the following:

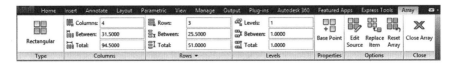

- This tab is almost identical to Array Creation context tab, except for the Options panel which includes three buttons:
  - Edit Source: this command will make changes to one of the objects arrayed, which reflects on all other arrayed objects.
  - Replace Item: this command will replace one or more of the arrayed shapes with another shape.
  - Reset Array: this command reverses the effects of the Replace Items command.

4.7.5 Editing Rectangular Array Using Quick Properties

- When you click a rectangular array, Quick Properties will be shown as in the following:

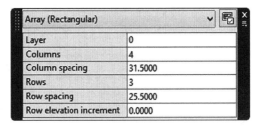

- You can change Column number, Column spacing, Row number, and Row spacing.

PRACTICE 4-7

Arraying Objects Using Rectangular Array

1. Start AutoCAD 2015.
2. Open the file, Practice 4-7.dwg.
3. Using the chair try to create a rectangular array like the following:

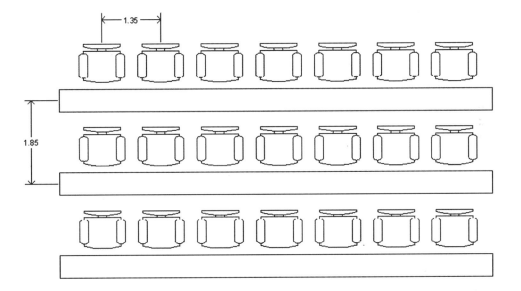

4. Exit the Array command.
5. Click the rectangular array. Using grips make the number of columns = 5 and distance between columns = 2.
6. Make the layer Chair current.
7. Click the rectangular array to select it, click Edit Source button, and then select the upper right chair (you can select any other one if you want). A message will appear, click OK, and then you will see that only the chair you selected is highlighted. Zoom to it and add two diagonal lines.
8. A small panel at the right is displayed titled Edit Array, click it, and then select Save Changes button. You can see all the other chairs holding the new changes.

9. Thaw layer Different.
10. Another chair will appear at the left.
11. Select the rectangular array.
12. Select the Replace Item button.
13. Select the new chair at the left, then press [Enter].
14. When AutoCAD asks about Base point for replacement objects, set the new base point using OSNAP (Midpoint) and OTRACK to locate the center of the rectangle of the new chair.
15. Click all the chairs in the bottom, press [Enter] twice, and then close the array command.
16. Your file should look like the following:

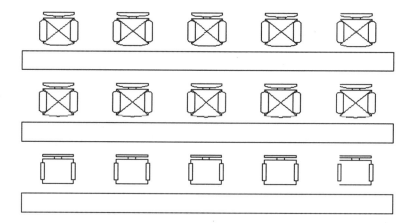

17. Save and close the file.

4.8 ARRAYING OBJECTS – PATH ARRAY

- This command allows you to create an array using an object like polyline, spline, and arc, etc. To issue this command, go to the **Home** tab, locate **Modify** panel, select the **Path Array** button:

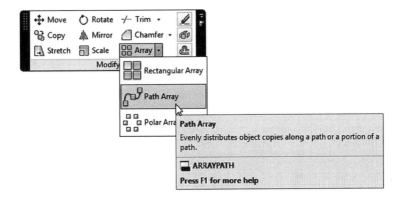

- The following prompts are shown:

```
Select objects:
Type = Path Associative = Yes
Select path curve:
```

- The first prompt asks you to select the desired objects. Once done, press [Enter]. The second prompt shows a message that the type of array is path and the associativity is on. AutoCAD then asks you to select the path curve. Once selected, AutoCAD will show 10 objects arrayed using the path. The following prompt appears:

```
Select grip to edit array or [ASsociative/Method/Base
point/Tangent direction/Items/Rows/Levels/Align items/Z
direction/eXit]<eXit>:
```

- The Array Creation context tab appears and looks like the following:

- When you think about arraying an object using a path, you should consider three things:
  - Base point
  - Aligning objects with the path
  - Measure or Divide

- Let's see how AutoCAD treats the following case.

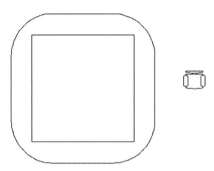

- Select the chair as the objects to be arrayed and the outer polyline to be the path.

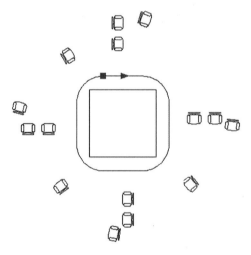

- Now, go to the context tab and select the Base point option to specify a new base point (which will be the center of the chair):

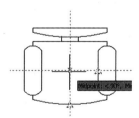

- The following is the result after specifying the new base point:

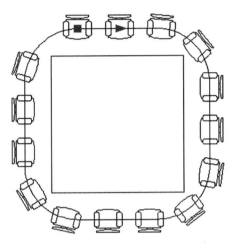

- The Align button is on by default; if you turn it off, this what you will see:

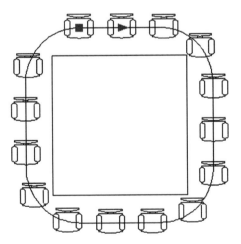

- If you select the Divide option, AutoCAD will divide the path equally for all the objects but if you select Measure, you should specify the distance between each object. The preceding was produced using Measure and the following by using Divide:

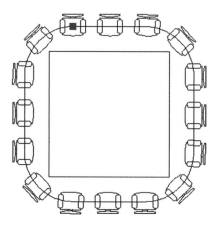

- After you specify the three factors, now you will be able to set the Items (number of items), Between (the distance, will be off in case of Divide), and Total (the total distance). Also, you can add rows. You should see something similar to the following:

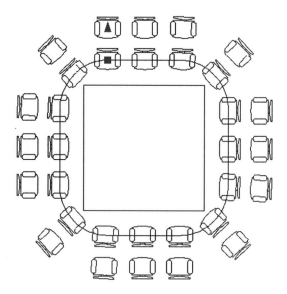

- After you create the path array, you can select the array for editing. You will see three grips (this will be true only for Divide; in Measure, you will see an extra arrow):

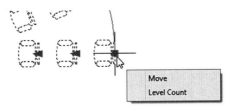

- You can use these to control the movement and the level count.
- The second one is:

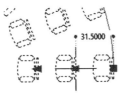

- This grip specifies the distance between rows. The third one is:

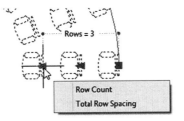

- This grip controls the Row Count and the Total Row Spacing.
- A new context tab called Array appears and looks like the following:

- This is identical to what we discussed earlier.

- Also, when you select the path array, you will see something like the following Quick Properties:

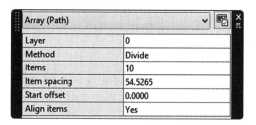

- In the Quick Properties, you can change the Method (Divide or Measure), the number of items, item spacing, Start Offset (not to start from the first point of the path), and whether to align items or not while arraying them.

PRACTICE 4-8

Arraying Objects Using Path

1. Start AutoCAD 2015.
2. Open the file, Practice 4-8.dwg.
3. Using the Path Array try to create an array, bearing in mind the following:
    a. Number of items = 22
    b. Base point = Center of the Chair (use OSNAP + OTRACK)
    c. Divide
    d. Align
    e. Number of rows = 2
    f. Distance between rows = 1.5
4. Erase the outer polyline.

5. You should get the following result:

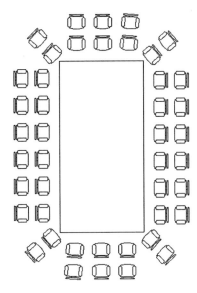

6. Using grips you can make the number of rows = 3 and 4, then back to 2.

7. Save and close the file.

4.9 ARRAYING OBJECTS – POLAR ARRAY

- This command allows you to duplicate objects in a circular fashion. To issue this command go to the **Home** tab, locate the **Modify** panel, then select the **Polar Array** button:

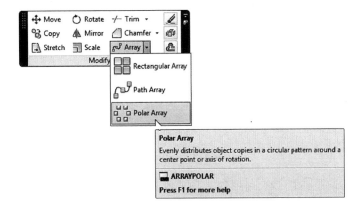

- The following prompts are shown:

```
Select objects:
Type = Polar Associative = Yes
Specify center point of array or [Base point/Axis of
rotation]:
```

- The first prompt asks you to select the desired objects. Once done, press [Enter]. The second prompt shows a message that the type of array is polar and the associativity is on. The third line asks you to specify the center point of array. After you specify the center, AutoCAD creates a polar array with 6 objects filling 360°. You will also see a new context tab titled array Creation that looks like the following:

- Using the context tab, you can specify the number of items, angle between items, and Angle to fill. You are also invited to specify the number of rows, with the distance between rows. Using the Properties panel, input if this will be an associative array or not, specify a new base point for the object to be arrayed, specify whether to rotate items as you are copying them, and, finally, specify the direction of arraying (CW or CCW).
- AutoCAD allows you to select the polar arraying for further editing. Once you select it, you will see something like the following:

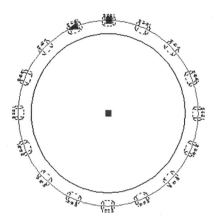

- The first grip will show something similar to the following:

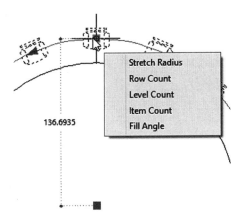

- You will see the current radius value along with a menu to help you: stretching the radius and changing the row count, level count (for 3D only), item count, and fill angle. The following is an example of changing the row count.
- The second arrow grip shows the current angle between the first and second items.

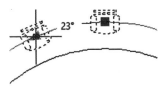

- A new context tab called Array appears and looks like the following:

- This is identical to what we discussed earlier.

- Also, when you click the polar array, you will see the following Quick Properties:

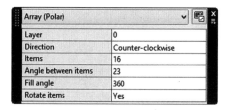

- In Quick Properties, you can change the direction of the array (CW or CCW), change the number of items, angle between items, total fill angle, and whether or not to rotate items while they are copied.

PRACTICE 4-9

Arraying Objects Using Polar Array

1. Start AutoCAD 2015.
2. Open the file, Practice 4-9.dwg.
3. Using the Polar Array create the following shape:

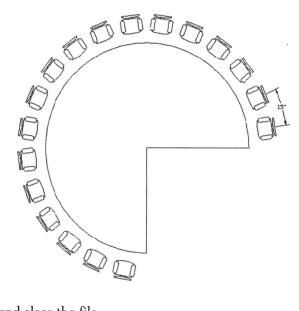

4. Save and close the file.

4.10 BREAK COMMAND

- This command allows you to break any object into two objects by removing the portion between two specified points. To issue this command, go to the **Home** tab, locate the **Modify** panel, and select the **Break** button:

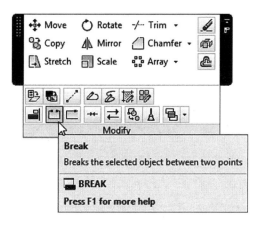

- You then see the following prompts:

```
Select object:
Specify second break point or [First point]:
```

- If selecting the object also specifies the first point, then select the second point. This ends the command. But, if you think selecting is just selecting, and you did not specify the first point, then type **F**. You will see the following two prompts:

```
Specify first break point:
Specify second break point:
```

- Using these two prompts, specify two points on the object and the Break command will end.

- AutoCAD offers the same command using a different technique, breaking on the same point. To issue this command, go to the **Home** tab, locate the **Modify** panel, then select the **Break at Point** button:

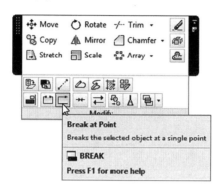

- You will see the following prompts:

```
Select object:
Specify second break point or [First point]: _f
Specify first break point:
Specify second break point: @
```

- After you select the desired object, it will flip automatically to specify the first point. At the second point prompt, AutoCAD responds by inputting @, which means "using the same first point."

NOTE *If the object to break is a circle, make sure to specify the two points counterclockwise. See this example:*

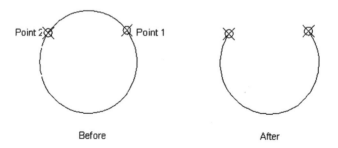

MODIFYING COMMANDS PART II • **135**

PRACTICE 4-10

Breaking Objects

1. Start AutoCAD 2015.
2. Open the file, Practice 4-10.dwg.
3. In order to locate the meeting table exactly at the center of the room horizontally and vertically, we need to break the upper horizontal line and the left vertical line.
4. Using the Break command, break the upper horizontal line from the points shown:

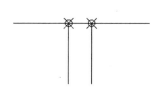

5. Using the Break command, break the left vertical line using one point as shown:

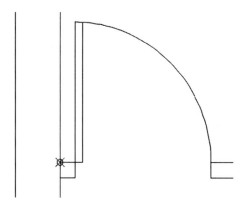

6. Using the Move command, move the meeting table at the center of the room using OSNAP and OTRACK.

7. You will get the following:

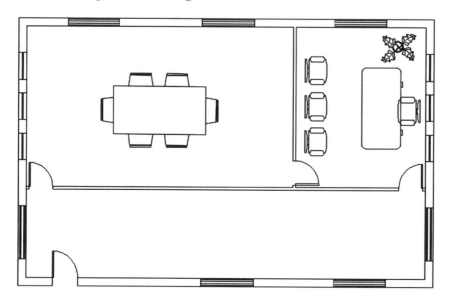

8. Save and close the file.

NOTES:

CHAPTER REVIEW

1. You can fillet using more than one fillet radius using the same command.

    a. True

    b. False

2. There are three types of arrays.

    a. True

    b. False

3. If you hold _____ while filleting, you will get radius = 0.0 regardless of the current radius.

4. Which of the following is not related to Chamfer command?

    a. Distance 1 and Distance 2

    b. Distance and Angle

    c. Distance and Radius

    d. Trim and No trim

5. You can offset using offset distance and _____.

6. Using the Trim command, you will select _____ as a first step, then you will press [Enter].

7. While using the Extend command, if you hold _____ you will convert the command to Trim.

    a. [Ctrl]

    b. [Ctrl] + [Shift]

    c. [Shift]

    d. [Alt] + [Ctrl]

8. While you are using the Break command, you can break at the same point, using which character?

    a. $

    b. #

    c. @

    d. &

CHAPTER REVIEW ANSWERS

1. a
3. [Shift]
5. Through
7. c

CHAPTER 5

LAYERS AND INQUIRY COMMANDS

In This Chapter

- What are layers in AutoCAD?
- How to create and set layer properties
- What are layer controls?
- How to use Quick Properties and Properties
- Inquiry commands

5.1 LAYERS CONCEPT IN AUTOCAD

- Layers are the most important way to organize and control your AutoCAD drawings. Managing layers means managing the drawing. What are layers in AutoCAD? Layers are simulation of a transparent piece of paper, in which you will draw part of the drawing using a certain color, linetype, and lineweight. Each object on a layer will hold the properties of this layer, meaning the object will have the same color, linetype, and lineweight that the layer resides in. This setting is called BYLAYER, which means you control the drawing through controlling layers rather than controlling objects.
- Each layer should have a name, which will be considered as the first step of the creation process. Proper naming should adhere to the following rules:
  - Name length should not exceed 255 characters.

- You can use in the name all letters (small or capital).
- You can use in the name all numbers.
- You can use (-) hyphen, (_) underscore, and ($) dollar sign.

■ A unique layer will exist in all AutoCAD drawings called 0 (zero). This specific layer cannot be deleted or renamed. Other layers can be deleted and renamed.

■ The layer at the top of the pile is the only layer we can draw on so it is called the current layer. As a rule of thumb, make the desired layer current first and then start drawing. To start building your layers, go to the **Home** tab, locate the **Layers** panel, and then click the **Layer Properties** button:

■ You will see the following palette, which is called **Layer Properties Manager**:

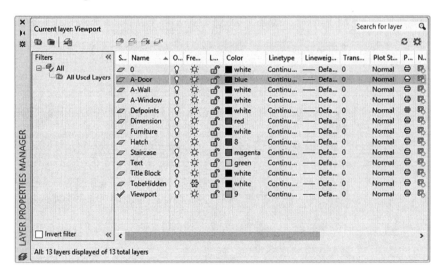

■ Palettes in AutoCAD have more features than the normal dialog box.

- The dialog box has two states, either displayed on the screen or closed. But, Palette can be displayed but hidden, which will spare you the continuous effort of opening and closing.

- Almost all of the dialog box cannot be resized, contrary to palettes, which can be resized horizontally, vertically, and diagonally.

- Dialog boxes cannot be docked but palettes can be docked on the four sides of the screen.

5.2 CREATING AND SETTING LAYER PROPERTIES

In this section, we will:
- How to create a new layer
- How to set a color for a layer(s)
- How to set the linetype for layer(s)
- How to set a lineweight for layer(s)
- How to set the current layer

5.2.1 How to Create a New Layer

- This command allows you to add a new layer to the current drawing. Using the **Layer Properties Manager**, click the **New Layer** button:

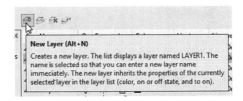

- There will create a new layer with the temporary name *Layer1*. The **Name** field will be highlighted. Type the desired name of the layer. You should always stick to a good naming convention; for example, a layer containing doors should be named Door.

NOTE

All the settings discussed in the following sections, require that you select layer(s) in the Layer Properties Manager. Selecting layers in AutoCAD is just like any other software running on Windows OS. You can press the [Ctrl] key and/or [Shift] to select multiple layers.

5.2.2 How to Set a Color for a Layer(s)

- You can use one of the 256 colors available in AutoCAD. The first seven colors can be set using the name or its number (from color 1 to color 7).
  - Red (1)
  - Yellow (2)
  - Green (3)
  - Cyan (4)
  - Blue (5)
  - Magenta (6)
  - Black/White (7)
- Other colors should be set only using their number.
- To set the color for a layer, complete the following steps:
  - Using the **Layer Properties Manager**, select the desired layer(s).
  - Using the **Color** field, click the icon of the color and you will see the following dialog box:

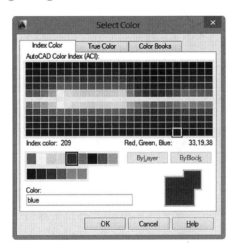

- Select the desired color (or you can type the name/number in the **Color** field) of the color, then click **OK** button to end this action.
■ Another way to set up (or modify) a layer's color is by using the pop-up list in the **Layers** panel, as shown:

5.2.3 How to Set the Linetype for Layer(s)

■ There are two linetype files that come with AutoCAD 2015. They are acad.lin and acadiso.lin. Because these two linetype files are not adequate for all types of engineering, designing, and drafting you should consider buying more linetype files available in the market.
■ Contrary to colors, linetypes are not loaded to the current file. Accordingly, the user should load the desired linetypes when needed. To set the linetype for a layer, do the following steps:
  - Using the **Layer Properties Manager**, select the desired layer(s).
  - Using the **Linetype** field, click the name of the linetype and you will see the following dialog box:

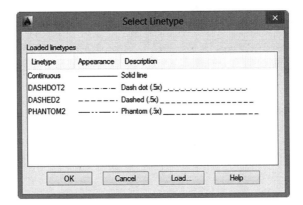

- If the desired linetype is listed, then select it. If not, you need to load it. Click the **Load** button and you will see the following dialog box:

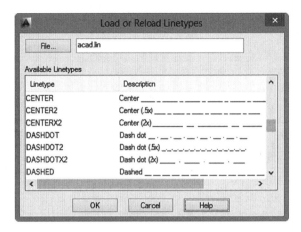

- Browse for your desired linetype and then select it to be loaded, click **OK**. Once the linetype is loaded, you can select it and click **OK**.

5.2.4 How to Set a Lineweight for Layer(s)

- This option will set the lineweight for layer(s). All objects in AutoCAD (except polyline with width) has a lineweight of Default, which is 0 (zero). You can set the lineweight for objects through their layers. To do this, complete the following steps:
  - Using the **Layer Properties Manager** select the desired layer(s).
  - Under the field **Lineweight**, click lineweight icon and the following dialog box appears:

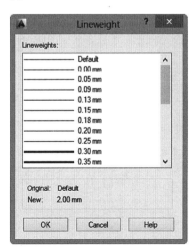

- Select the desired Lineweight and click **OK**.
- To see the lineweight on the screen, use the status bar and click the **Show/Hide Lineweight** button to on.

5.2.5 How to Set the Current Layer

- There are several ways to make a layer a current layer:
- The easiest way is to use the layer pop-up list in the Layers panel, as in the following:

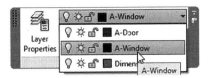

- Another way is to use the **Layer Properties Manager** palette then double-click the Status of the desired layer's name.
- The longer method is to use the **Layer Properties Manager** palette, select the desired layer, and then click the **Set Current** button:

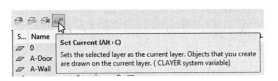

PRACTICE 5-1

Creating and Setting Layer Properties

1. Start AutoCAD 2015.

2. Open the file, Practice 5-1.dwg.

3. Create a new layer and call it Centerlines. The color is yellow and the linetype is Center. Make it current and draw two lines, horizontal and vertical, using the centerline of the circle and OSNAP as well as OTRACK.

4. Create another layer and call it Hidden. The color is 9 and linetype is Dashed. Nake it current and draw a circle using the same center point of the other circles with R = 1.5.

5. You should have the following shape:

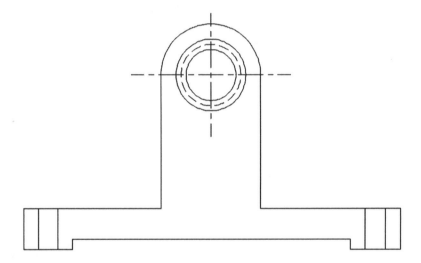

6. Save and close the file.

5.3 LAYER CONTROLS

- The following commands allow you to have full control over the layers because primary tools control the drawing. We will learn, how to control the visibility of layers, locking layers, plotting layers, deletion of layers, renaming of layers, etc.

5.3.1 Controlling Layer Visibility, Locking, and Plotting

- AutoCAD provides controls to show/hide layers (Freeze and Off), lock and unlock layers, as well as plot and no plot layers. See the following examples:

- The first example is layer "Title Block":

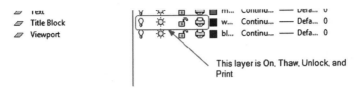

- The second example is layer "Staircase":

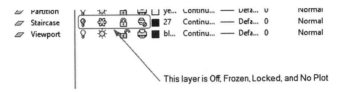

- When new layer is created, it will be On, Thaw, Unlock, and Plot (by default). You can hide the contents of layers by turning them off or freezing them. Freeze has a deeper effect than off. Objects in the frozen layer will not be considered in the drawing so the drawing size will be temporarily smaller (you can use Freeze to decrease the drawing size, if the drawing loads slowly).
- If you tried to freeze the current layer, AutoCAD will show the following message:

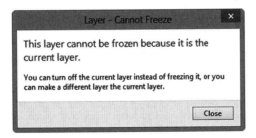

- But if you tried to turn off the current layer, you will see the following message:

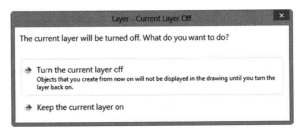

- If you lock a layer, objects reside in it and the layer will not be selected to any modifying commands. Objects in the locked layer will be faded but when you get closet to them, you will see small lock icon appears to tell you that this layer is locked. See the following illustration:

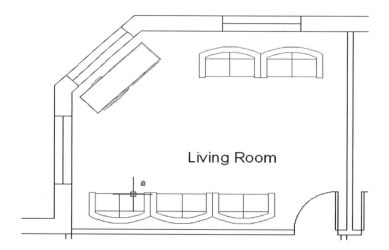

- If you choose No Plot for a layer, then objects will be displayed but not plotted in this layer.
- On/Off, Thaw/Freeze, and Lock/Unlock can be controlled using the pop-up list in the Layer panel and Layer Properties Manager whereas Plot/No Plot can be controlled only in the Layer Properties Manager.

5.3.2 Deleting and Renaming Layers

- AutoCAD will not delete a layer containing objects. AutoCAD will delete only empty layers. In order to delete layers, complete the following steps:
  - In the **Layer Properties Manager** palette, select the desired layer(s).
  - Press [Del] on the keyboard or click the **Delete Layer** button.

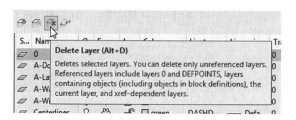

- In the **Layer Properties Manager,** you can rename layers as well. In order to do that, complete these simple steps:
  - In the **Layer Properties Manager** palette, select the desired layer.
  - Click the layer name once and it will be highlighted for editing. Type the new name, then press [Enter].
- If you tried to delete a layer containing objects, you receive the following message:

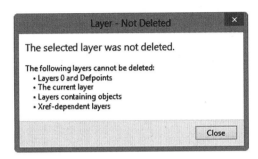

- To rename a layer, do the following steps:
  - Select the desired layer.
  - Click the name once and you will see the following:

  - The name will become editable. Type in the new name and press [Enter].

5.3.3 How to Make an Object's Layer Current Layer

- This is the fastest way to make a layer the current layer. After issuing the command, select an object that resides in this layer. Follow these steps:
  - Go to the **Home** tab, locate the **Layers** panel, then click the **Make Current** button:

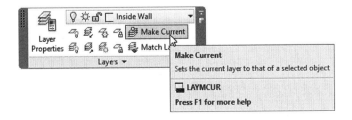

- You then see the following prompt:

```
Select object whose layer will become current:
```

- Select the desired object. See what the current layer is and you will find it became the object's layer (even without you knowing the object's layer).

5.3.4 How to Undo Only Layers Actions

- The function to undo the layers actions only is called **Layer Previous**. This function will help you restore previous states of the layers (like Freeze, Thaw, On, and Off, etc.) without affecting other drawings or modifying commands. Do the following steps:
  - Change the layer states as needed.
  - Go to the **Home** tab, locate the **Layers** panel, and then click the **Previous** button.

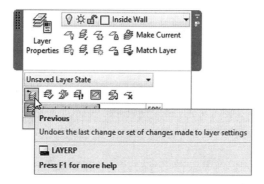

- In the command window, you see the following message:

```
Restored previous layer status
```

5.3.5 Moving Objects from One Layer to Another

- All similar objects must reside in the same layer but mistakes may occur. If you draw on the wrong layer and you want to move the objects to the right layer, you can use **Match** command. Do the following:
  - Go to the **Home** tab, locate the **Layers** panel, then click the **Match** button:

LAYERS AND INQUIRY COMMANDS • 153

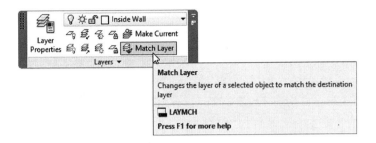

- The following prompt appears:

```
Select objects to be changed:
Select object on destination layer or [Name]:
```

- This command contains two prompts. Using the first, you will select the object mistakenly drawn in the wrong layer. At the second prompt, either you select an object that resides in the right layer or simply type its name.
- When you are done, you will see something similar in the command window:

```
8 objects changed to layer "Dimensions"
```

5.4 USING THE LAYER PROPERTIES MANAGER

- While you are in the Layer Properties Manager palette, you can do several actions. For example, if you select a layer and then right-click, you will see a menu like the following:

- In this menu, you can do all or any of the following:
  - Set the current layer
  - Create a new layer
  - Delete a layer
  - Select All layers
  - Clear the selection
  - Select All but Current
  - Invert Selection (make the selected unselected and vice versa)
- You also have the option to show or hide the Filter Tree:
  - Show Filter Tree (by default it is turned on)
  - Show Filters in Layer List (by default it is turned off)
- If you turned off **Show Filter Tree**, you will be allowed to see more information about your layers. See the following:

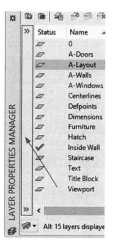

- Another way to do the same thing is by clicking the small arrows at the top-right part of the Filter Tree pane:

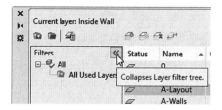

PRACTICE 5-2

Layer Controls

1. Start AutoCAD 2015.
2. Open the file, Practice 5-2.dwg.
3. Make Layer 0 the current layer.
4. Freeze layers: Centerlines and Hatch.
5. Lock layers: Furniture. What happened to the color? _____
6. Get closer to any object in Furniture layer and check that the icon appears when you get closer.
7. Try to erase one of the objects in layer Furniture. What message do you receive? _____
8. Using the Layer Properties Manager, try to freeze layer 0 (the current layer). What message do you receive from AutoCAD? _____
9. Try to rename layer 0. What message do you receive? _____
10. Rename the layer Partition to become Inside Wall.
11. Using the Match command, select the two doors at the bottom (their color is black or white) to match one of the blue doors.
12. Using the Layer Properties Manager, select all the layers, then change their color to black or white). Close the Layer Properties Manager.
13. Click Layer Previous. What happened? _____
14. Click the Make Object's layer current button then click one of the yellow lines. What is the current layer now? _____
15. Save and close the file.

5.5 CHANGING OBJECT'S LAYER, QUICK PROPERTIES, AND PROPERTIES

- We will learn to control object's properties.

5.5.1 Reading Instantaneous Information About an Object

- AutoCAD provides you of an instantaneous information about any object when your mouse hover over this object, check the following illustration:

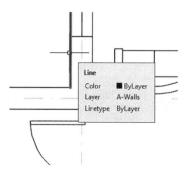

- You can see that AutoCAD is showing the type of object (Line), its color, the layer it is in, and its linetype.

5.5.2 How to Move an Object from a Layer to Another Layer

- Any object in AutoCAD should reside in a layer. To move an object from one layer to another, do the following steps:
  - Click the desired object(s).
  - Go to the **Home** tab, locate the **Layers** panel, and check the layer name that the pop-up list is displaying. You may see this part blank sometimes; this happens when your desired objects reside in different layers. Click the layer's pop-up list and select the new layer name.
  - Press [Esc] once to deselect all the selected objects.

5.5.3 Quick Properties

- This is the first of two commands, which enables you to change the properties of selected objects. Object's properties differ depending on the object type; line properties are different in comparison to arcs, circles, or polylines.

- Quick Properties will pop-up on the screen whenever you select objects without issuing a command (using grips selection). If not, click the following button on the Status bar to switch it on/off:

- When you select object(s), you will see the following (circle as example):

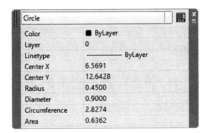

- You will see information like:
  - Color
  - Layer
  - Linetype
  - Coordinates of Center point (X & Y)
  - Radius and Diameter
  - Circumference and Area
- If you select more than one object from the same type, you will see this:

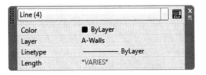

- But if you select more than one object from different types, you will see this:

- You will see All(number) which means you are seeing eight objects (as our example) from different types, but you can also see a breakdown of the selected objects by clicking the pop-up list, as shown here:

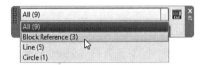

- When you select a single object type, you can change the general and specific object's properties.
- User should note that while you are at Quick Properties, whenever you select a different layer, color, or linetype, etc., you will see the effect of your change concurrently. This applies as well to the next command.

5.5.4 Properties

- As the name indicates, Quick Properties offers a fast way to change the properties of object(s). The Properties command is very comprehensive; all of the rules we discussed in Quick Properties is applicable here as well.
- To issue the **Properties** command, select the desired object(s), right-click, and select **Properties**.
- Another way is to double-click any object to get the Properties palette. This way has two drawbacks; first, it will be for a single object only; the second, is that some objects like polyline, block, hatch, text, will interpret this action as edit. Either way, you will see something like the following:

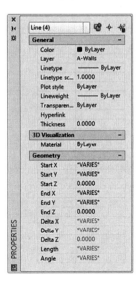

- As you can see, the information displayed here is much more than what is displayed in the Quick Properties, giving you the power to make more changes to the selected object(s).

NOTE *Some of the information is shaded, which means, it is unchangeable.*

PRACTICE 5-3

Changing an Object's Layer, Quick Properties, and Properties

1. Start AutoCAD 2015.
2. Open the file, Practice 5-3.dwg.
3. Hover over one of the circles of the centerlines What is the name of the layer?. _____
4. Select all circles and text inside them, and move them to layer Centerlines.
5. Change the properties of the circles to be Continuous.
6. Delete layer Centerlines-TAGS.
7. Zoom to the lower door. You will find two red lines at the right and at the left, move them from layer Dimensions to layer A-Walls.
8. Save and close the file.

5.6 INQUIRY COMMANDS – INTRODUCTION

- The main purpose of this set of commands is to allow you to measure lengths between two points, inquire the radius of a circle or arc, measure the angle, measure the area, or measure the volume for 3D objects. You will use this set of commands, to make sure that your drawing is correct according to the design intent. To issue these functions, go to the **Home** tab and locate the **Utilities** panel. For all of these commands, the cursor will change to the following:

5.7 MEASURING DISTANCE

- This command measures the distance between two selected points. To issue this command, go to the **Home** tab, locate the **Utilities** panel, and click the **Distance** button:

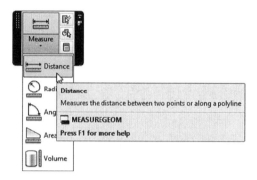

- You will see the following prompts:

```
Specify first point:
Specify second point:
```

- Select the desired points and AutoCAD will display something like the following:

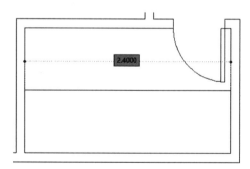

- You will see the following result in the command prompt:

```
Distance = 2.4,   Angle in XY Plane = 0,   Angle from XY
Plane = 0
Delta X = 2.4,   Delta Y = 0.0000,   Delta Z = 0.0000
```

5.8 INQUIRING RADIUS

- This command will check the radius (and parameter as well) of a drawn circle or arc. To issue this command, go to the **Home** tab, locate the **Utilities** panel, and select the **Radius** button:

- AutoCAD will display the following prompt:

```
Select arc or circle:
```

- Select the desired arc or circle. You will see the following:

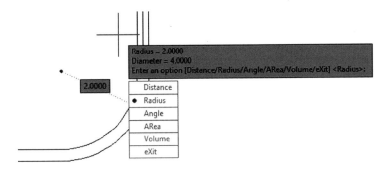

- You will also see the following in the command window:

```
Radius = 2.000
Diameter = 4.000
```

5.9 MEASURING ANGLE

- This command allows you to measure angles (between two lines, the included angle of an arc, or two points and the center of the circle). To issue this command, go to the **Home** tab, locate the **Utilities** panel, then select the **Angle** button:

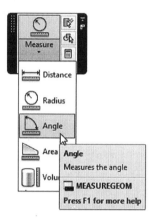

- You will see the following prompt:

    Select arc, circle, line, or <Specify vertex>:

- Select the desired objects (whether two lines, arc, points on a circle) and AutoCAD displays the following:

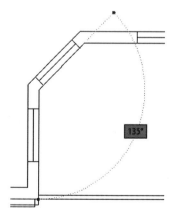

- Also, you will see something like the following in the command window:

```
Angle = 135°
```

5.10 MEASURING AREA

- This command allows you to measure areas, whether simple area (area that has no islands inside) or complex area (areas that have islands inside). AutoCAD can measure areas between points (assuming lines and arcs connecting them), or objects (like circle, closed polyline, etc.). To issue this command go to the **Home** tab, locate the **Utilities** panel, then select the **Area** button:

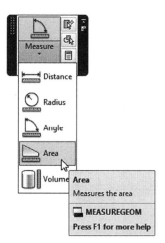

- You will see the following prompt:

```
Specify first corner point or [Object/Add area/ Subtract
area/eXit] <Object>:
```

5.10.1 How to Calculate Simple Area

- The definition of simple area is any closed area without any objects (islands) inside it. AutoCAD will assume you want to measure a simple area if you start by specifying points or selecting objects. If you start by specifying points, AutoCAD will assume there are either lines or arcs connecting them. For lines, you will see the following prompts:

```
Specify next point or [Arc/Length/Undo]:
Specify next point or [Arc/Length/Undo]:
Specify next point or [Arc/Length/Undo/Total] <Total>:
Specify next point or [Arc/Length/Undo/Total] <Total>:
```

- If you see an arc in your area, simply change the mode to Arc and you will see identical prompts to the Polyline command. Keep specifying the point of lines or arcs, up until you press [Enter]. You will see the Total value of the measured area in the command window and the following:

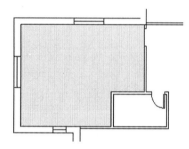

- Also, you will see the following displayed in the command window:

```
Area = 18.8366, Perimeter = 17.0416
```

- If you have a simple area and the parameter is a single object, you can select the object, rather than specifying points. Either right-click and select **Object** option, or type **o** at the command window. You will see the following prompt:

```
Select objects:
```

- Select the object you want to measure and its area, then press [Enter].

5.10.2 How to Calculate Complex Area

- The definition of complex area is any closed area with objects (islands) inside it. To tell AutoCAD you want to calculate a complex area, you have to start with either **Add area** or **Subtract area**.

- If you start with Add area or Subtract area, AutoCAD will start with area = 0, which will add areas, then subtract areas as needed. For Add area mode, you will see the following prompt:

    ```
    Specify first corner point or [Object/Subtract area/eXit]:
    ```

- Specify the area using the same methods previously discussed (points or object) then switch to Subtract area mode. You will see the following prompts:

    ```
    Specify first corner point or [Object/Add area/eXit]:
    ```

- AutoCAD will give you a sub-total after each adding or subtracting areas. When you are done, press [Enter] twice to end the command and get the final net area.
- You will see something like the following picture:

PRACTICE 5-4

Inquiry Commands

1. Start AutoCAD 2015.
2. Open the file, Practice 5-4.dwg.
3. Freeze all layers except A-Walls (make layer A-Walls current, then select all layers except current and freeze them).
4. Measure the length of the slanted wall from inside. What is the information given by AutoCAD?

    a. Length = _____ (1.6971)

    b. Angle in XY plane = _____ (45, or 135)

    c. Delta X = _____ (1.2000)

    d. Delta Y = _____ (1.2000)

5. Thaw layer Partition.
6. Measure the horizontal and vertical lengths of the room at the upper-right part (you can use Nearest and Perpendicular OSNAPs), and type them down:

    a. Horizontal Distance = _____ (5.05)

    b. Vertical Distance = _____ (3.9)

7. Measure the inside area of the room at the lower-right part of the plan. What are the measurements?

    a. Area = _____ (21.8016)

    b. Parameter = _____ (18.2416)

8. Save and close the file.

PRACTICE 5-5

Inquiry Commands

1. Start AutoCAD 2015.
2. Open the file, Practice 5-5.dwg.
3. Calculate the net area of the shape without all the inside objects.
   a. Area = _____ (33.3426)
4. Save and close the file.

NOTES:

CHAPTER REVIEW

1. One of the following is not true about the layer name.

    a. Should not exceed 256 characters.

    b. Space is allowed.

    c. $ is allowed.

    d. $ is not allowed.

2. You cannot _____ the current layer.

3. You can _____ the current layer.

4. You can undo layer actions only.

    a. True

    b. False

5. Area command can calculate only areas without any islands inside it.

    a. True

    b. False

6. The following are facts about layer 0 (zero) except one:

    a. User cannot rename it.

    b. User cannot set a new color for it.

    c. User cannot delete it.

    d. It is in all AutoCAD files.

7. Delta X is one of the information properties in the _____ command.

8. Most objects will respond to _____ to display Properties palette.

CHAPTER REVIEW ANSWERS

1. d

3. Turn off

5. b

7. Measure distance

CHAPTER 6

BLOCKS AND HATCH

In This Chapter
- What are blocks and how to define them?
- How to use (insert) blocks
- How to explode and convert blocks
- How to hatch in AutoCAD
- How to control hatch in AutoCAD
- How to edit hatch

6.1 WHAT ARE BLOCKS?

- In your daily work there will be shape(s) that you need repeatedly. You have two choices, either you to draw it each time or draw it once and save it as a block (block will be a single object). You can then use (insert) it as many times as you wish in the current file and other files as well.
- Using blocks has many benefits, including:
  - The file size will be less because each block will be counted as a single object.
  - Standardization for the same company.
  - Speed of completing a drawing.
  - Use of Design Center and Tool Palettes

6.2 HOW TO CREATE A BLOCK

- To create a block, do the following steps:
  - Draft the shape that you want to create a block from in layer 0 (zero). Layer 0 will enable the block to inherit the properties (color, linetype, lineweight) of the layer in which it will reside.
  - Be sure to control the "Block unit," which will enable AutoCAD to automatically scale the block to appear in the right size in any other drawing.
  - Draft the shape that you want to create a block from in its real life dimensions.
- Let's assume we draw the following shape.

- Now you are ready to issue the command. Go to the **Insert** tab, locate the **Block Definition** panel, and select the **Create Block** button:

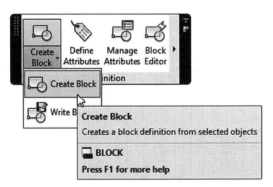

- You will see the following dialog box:

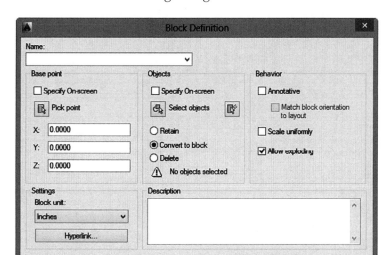

- Do the following:
  - Type the block's name (it should not exceed 255 characters, use only numbers, letters, -, _, $, and spaces).
  - Specify the **Base point**, either by typing X, Y, and Z coordinates or select the **Pick point** button to input the base point graphically.
  - Click the **Select objects** button to select the desired objects.
  - From the drawn objects, AutoCAD will create the needed block, but what should you do with the objects afterwards? You should select one of the three choices available: Retain (leave) objects as they are, Convert them to a block, or Delete them.

- Select whether the block will be Annotative (a feature discussed in Chapter 9), whether to scale the block uniformly in both X and Y, and whether to explode the block later on.

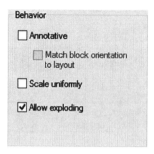

- Select the **Block unit**. This tells AutoCAD that each AutoCAD unit used in this block will equal. This helps AutoCAD in the Automatic scaling feature.

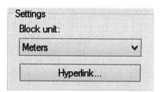

- Type block description.
- Turn the checkbox "Open in a block editor" off because this is an advanced feature used for creating dynamic blocks.
- Once you finished inputting all the preceding data, click **OK**.
■ Later, you will use (insert) the block but this only inserts a copy of the block definition. The original block definition stays intact.

6.3 HOW TO USE (INSERT) BLOCKS

■ After block creation, you are ready to use (insert) the block in the current drawing. You should make sure that you are in the right layer and the drawing is ready to accept the block (make sure the door openings are made before the insertion of the door, for example).

- Now you are ready to issue the command. Go to the **Insert** tab, locate the **Block** panel, then click the **Insert** button:

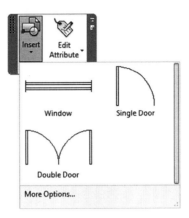

- You will see a list of the current blocks available in your drawing. Using this method, you cannot change the scale or rotation angle. This is a very fast way to insert blocks in your drawing. On the other hand, if you would like to customize the insertion process, then select **More Options** to see the following dialog box:

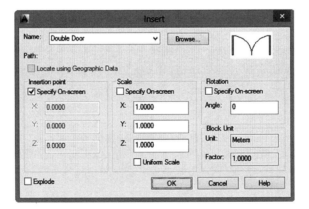

- If the block is created in the current file, click the pop-up list to select it. Also, you can click the **Browse** button to select a file and insert it in the current file as a block.

- You should specify the **Insertion point**, the **Scale**, and the **Rotation** angle, either by using the Specify On-Screen checkbox or by typing the needed value.
- While using the **Scale,** you can insert mirror images of the block by using negative values. See the following illustration:

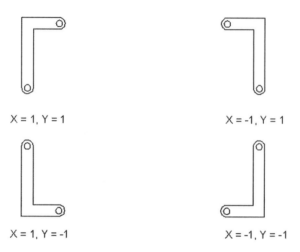

- While you are using the Rotation angle, remember the CCW is always positive.

6.3.1 Block Insertion Point OSNAP

- After inserting a block, click it to see the following:

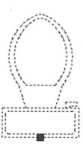

- The whole block is one unit and only one grip is highlighted, which is the insertion point. There is a specific OSNAP to snap to this point called

Insertion (or insert depending where you are looking). See the following illustration:

PRACTICE 6-1

Creating and Inserting Blocks

1. Start AutoCAD 2015.

2. Open the file, Practice 6-1.dwg.

3. There are three shapes at the left drawn in layer 0. From these shapes, create three blocks and name them Window, Single Door, and Double Door, setting the block unit to the default value.

4. Using the Insert command to insert the three blocks in the proper places using the correct layers, just like in the following picture:

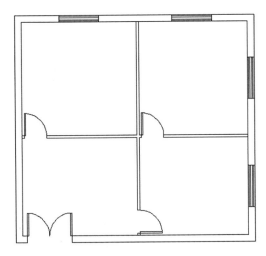

5. Save and close the file.

6.4 EXPLODING BLOCKS AND CONVERTING THEM TO FILES

6.4.1 Exploding Blocks

- You can explode blocks the same way you explode lines and arcs. The explode command will bring them back to their original objects. This practice is not recommended, as we always advocate keeping blocks as one object, not exploding them.
- To issue the command, go to the **Home** tab, locate the **Modify** panel, and select the **Explode** button:

- You will see the following prompt:

```
Select objects:
```

- Select the desired block(s) and press [Enter] to end the command.

6.4.2 Converting Blocks to Files

- This is an old practice used to convert blocks to files in order to use them in other files. This practice was eclipsed by the emergence of Design Center and Tool Palettes.
- To issue the command, go to the **Insert** tab and locate the **Block Definition**, then click **Write Block** button:

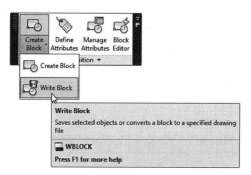

- You will see the following dialog box:

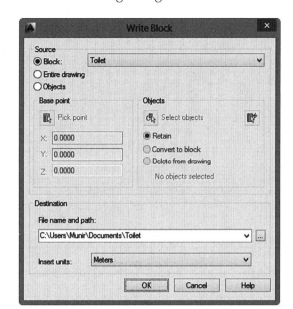

- Select the **Block** option under **Source**. Then select the name of the block. Under **Destination** input the file name and path, and specify the insert unit. You can use the same dialog box to create a file from the entire drawing or from some of the objects in the current drawing.

PRACTICE 6-2

Exploding and Converting

1. Start AutoCAD 2015.

2. Open the file, Practice 6-2.dwg.

3. Start the Write Block command and select the Single Door block to convert it to a file, and save it in your practice folder.

4. Explode one of the single door insertions. Using the Quick Properties, select one of the objects that resulted from the exploding process. In which layer does it reside? _____ Why? _____
   _____

5. Save and close the file.

6.5 HATCHING IN AUTOCAD

- AutoCAD can hatch closed areas and non-closed areas (with maximum distance defined by you).
- There are two hatch pattern files comes with AutoCAD they are *acad.pat* & *acadiso.pat* (as you can see the hatch pattern file's extension is *.PAT).
- There are four pattern types:
  - Solid (single pattern covers the area with single solid color).
  - Gradient (two gradient colors mixed together in several fashions).
  - Pattern (several pre-defined patterns).
  - User defined (most simple pattern: parallel lines).

6.6 HATCH COMMAND: FIRST STEP

- This command will put hatches in the drawing and control all of its properties. The preview is instantaneous. To issue the Hatch command, go to the **Home** tab, locate the **Draw** panel, then click the **Hatch** button:

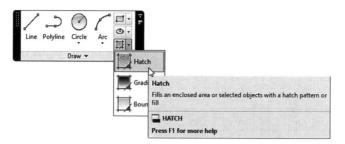

- You will see a new context tab added to the ribbons called **Hatch Creation**. You will see several panels, which will be discussed. Your first step should be locating the **Properties** panel, at the top-left part, and then select the **Hatch Type** as shown:

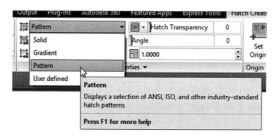

- Once you select the Hatch Type, locate the **Pattern** panel. AutoCAD will take you to the first pattern in the selected type. For instance, if you select **Gradient** in the Hatch Type, the first pattern in the Pattern panel will be GR_LINEAR, which is the first pattern in the gradient patterns:

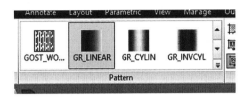

- Next, simply go (without any clicking) to the desired area to be hatched and you will see the area filled. You have two choices:
  - If you liked the result, click to pick the area, then go to the **Close** panel and click **Close Hatch Creation**.
  - If the result did not satisfy you, this means you need to control the properties of the hatch in an attempt to rectify the result.

6.7 CONTROLLING HATCH PROPERTIES

- If you clicked inside the area and you do not like the result, you need to alter the properties of the hatch. All of these functions exist in the **Properties** panel, they are:
  - **Hatch Color:** Specify the color of the hatch or leave it as "Use Current."

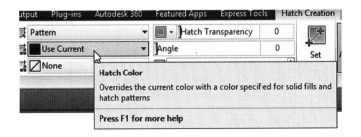

- **Background Color:** Specify the color of the background or leave it as "None."

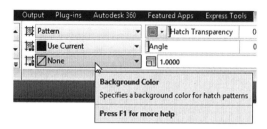

- **Transparency:** By default all colors will be with their normal colors, but you can increase the value of transparency (maximum 90) to decrease the intensity of the color (hatch color and background color).

- **Angle:** Specify the angle of the hatch pattern (no effect with Solid hatching).

- **Scale** (if you chose user-defined type, this will be called Spacing): Input the scale or spacing for the selected hatch pattern.

- **Hatch Layer Override:** By default, hatch will reside in the current layer, using this function you can specify the layer you want the hatch to reside in, regardless of the current layer.

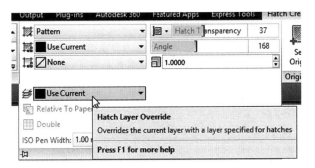

- **Double:** This option is only valid if the hatch type is User-defined. It controls whether the lines are in one direction or cross hatched.

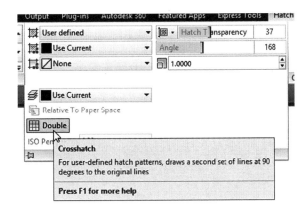

PRACTICE 6-3

Inputting Hatch and Controlling Hatch Properties

1. Start AutoCAD 2015.
2. Open the file, Practice 6-3.dwg.
3. Make layer Hatch the current layer.

4. Start the Hatch command.
5. Make sure of the following:

    a. At the Properties panel, Hatch Type is Pattern.

    b. At the Pattern panel, ANSI31.

6. Hover over any part of the drawing and you will see the preview of the hatch. You will notice that Scale is a little small so increase it to become equal to 2.
7. Select Hatch background = Yellow.
8. Then select the following areas to hatch:

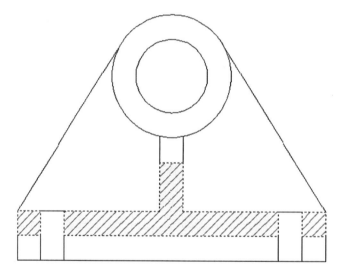

9. Click Close Hatch Creation button at the right.
10. Start the Hatch command again and you will notice that all options used by the previous hatch are still valid. Change the angle to 90o and apply it to the lower portion of the shape as shown. Click the Close Hatch Creation button at the right.

11. You will have the following result:

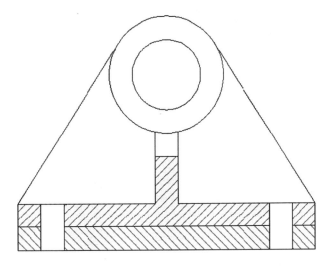

12. Save and close the file.

6.8 SPECIFYING HATCH ORIGIN

- When you want to hatch an area, want the pattern to start from a certain point, and to not abide to the default settings of AutoCAD, then you need to manually set the Hatch Origin. By default, AutoCAD uses 0,0 as the starting point for any hatch, which means you will never know for sure if your hatch will be displayed correctly or not. In order to control the Hatch Origin, make sure you are still in the **Hatch Creation** context tab and locate **Origin** panel. You will see the following:

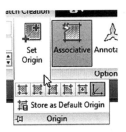

- The apparent button is **Set Origin**, which will ask you to:

```
Specify origin point:
```

- Specify the desired point. You can also use the other pre-defined points (lower-left corner or upper-left corner, etc.). You have the ability to save the point you picked this time, for future use, instead of using 0,0.
- See the following example:

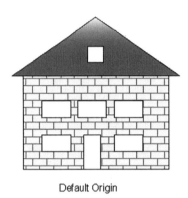
Default Origin

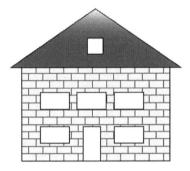
Origin is the lower left corner

6.9 CONTROLLING HATCH OPTIONS

- These options allow you to control the outcome of the hatching process. Using these options you will be able to hatch an open area (Gap Tolerance) or create separate hatches and more:

6.9.1 How to Use Associative Hatching

- Hatch in AutoCAD is associative, which means hatch understands the boundary which it fills. When this boundary changes, hatch will respond appropriately.
- See the following illustration:

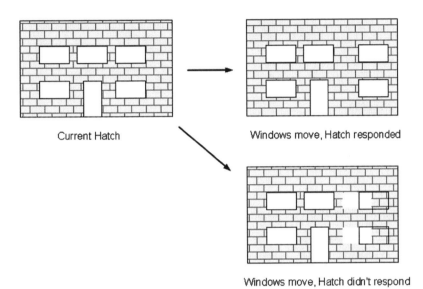

6.9.2 How to Make Your Hatch Annotative

- Annotative is an advanced feature related to printing that will be discussed later in this book.

6.9.3 Using Match Properties to Create Identical Hatches

- This option allows you to create an identical hatch from an existing hatch (it will reside in the same layer; also, it will have the same angle, scale, transparency, etc.).

- This option has two button associated with it: Use current origin or Use source hatch origin. Both options are self-descriptive.

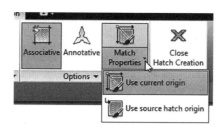

- Select the **Match Properties** button and you will see the following prompt:

```
Select hatch object:
```

- Select the hatch object you desire to mimic and you will see the following prompt:

```
Pick internal point or [Select objects/seTtings]:
```

- Click inside the desired area. Keep selecting the area; when done, press [Enter] to end the command.

6.9.4 Hatching an Open Area

- By default, AutoCAD will hatch only closed areas. But, you can ask AutoCAD to hatch an area with an opening. To tell AutoCAD to allow hatching an open area, simply set the **Gap Tolerance**, which will be considered the maximum allowable opening. Any area with opening bigger than this value will not be hatched.

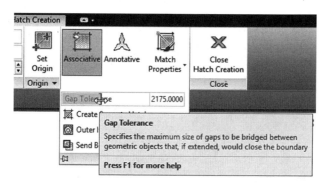

- When you want to hatch this area, the preview is not displayed so you need to click inside the opened area. You will see the following warning message:

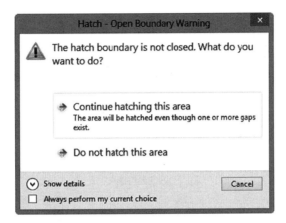

- The message gives you these two self-explanatory options.

6.9.5 Creating Separate Hatches in the Same Command

- Using the same command, if you hatched several separated areas, they will be considered a single hatch (single object).
- You can override this default setting by telling AutoCAD that you want separate hatches for separate areas. Simply click this button on.

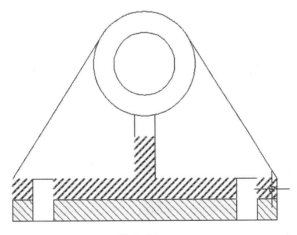

Single Object

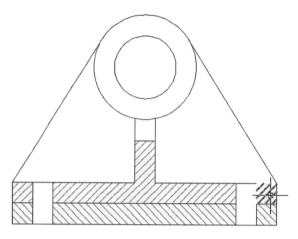

Seperate Hatches

6.9.6 Island Detection

- When you are hatching an area containing several areas (islands) inside it, these areas may contain more islands. We want to know how AutoCAD will treat these islands.
- There are four different ways to do this:

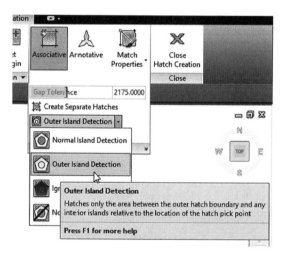

- **Normal Island Detection:** AutoCAD will hatch the first area (the outer one), then leave the second one, hatching the third, and so on.
- **Outer Island Detection:** AutoCAD will hatch the outer area only.

- **Ignore Island Detection:** AutoCAD will ignore all of the inner islands and hatch the outer area, as if there are no areas inside.
- **No Islands detection:** This option will turn off the island detection feature, which is the same result as the Ignore Island Detection option.

6.9.7 Set Hatch Draw Order

- Hatch is just like any other object in AutoCAD. You can set the draw order for it, relative to the other objects. You have five choices to pick from:

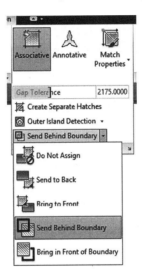

- These are:
    - Do Not Assign, use the default
    - Send to Back
    - Bring to Front
    - Send Behind Boundary
    - Bring in Front of Boundary
- See the following example:

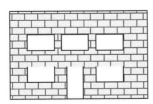

Send Behind Boundary

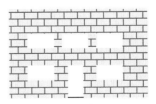

Bring in Front of Boundary

PRACTICE 6-4

Hatch Origin and Options

1. Start AutoCAD 2015.
2. Open the file, Practice 6-4.dwg.
3. Start the Hatch command.
4. Using the Options panel, click the Match Properties button, and select the hatch of the Toilet. Apply it to the Kitchen then press [Enter] to end the command.
5. Start the Hatch command again, set the background color = 40, scale = 4, transparency = 0, and hover over Study room. You will need to change the Origin point to the lower-left corner of the room. Click inside Study and end the command.
6. Zoom to the lower-right corner of the Living Room. You will see the area is open. Start the Hatch command again and set the Gap Tolerance = 0.3 (the opening in this drawing is 0.2 so 0.3 is enough), then set to create separate hatches. When you click inside Living Room, you will see a warning message. Select the **Continue hatching this area** option, then click inside the Sitting Room, and end the command.
7. Thaw layer **A-Doors**, notice the two doors of the Living Room and Study; they are not shown properly. To solve this problem, select the hatch, right-click, select **Draw Order** option, and select the **Send to Back** option.
8. Start Hatch command for the fourth time. For the Hatch Type, select Solid and hatch the outside walls.
9. Zoom to any window of the Kitchen and move it a short distance. What happened to the hatch? _____ Why? _____
10. Save and close the file.

6.10 HATCH BOUNDARY

- If we were discussing older versions of AutoCAD, this panel (Boundary panel) would be the first panel to discuss. However, because of the instantaneous display of the hatch once you are inside the hatch area in recent versions, we discuss this later on because it is less important.

- Depending on what you are doing, creating a new hatch or editing an existing one, some of the buttons will be turned off, and some of them will be on.

- The **Pick Points** button is always on, which will allow you to pick the areas to hatch. The **Select** and **Remove** buttons will add/remove more objects to be included in the hatch boundaries.
- If you are editing a hatch (by clicking it), you will see the **Recreate** button is on. This button allows you to recreate the boundary, if (for any reason) the boundary was deleted. Simply click the hatch without its boundary, then click the **Recreate** button. You will see the following prompts:

```
Enter type of boundary object [Region/Polyline] <Polyline>:
Reassociate hatch with new boundary? [Yes/No] <N>:
```

- The first prompt asks you to select the desired boundary type, then asks you to re-associate the boundary with the hatch.
- If you select any hatch, AutoCAD enables you to highlight the (**Display**) **Boundary Objects**, so the user can edit the boundary. See the following illustration:

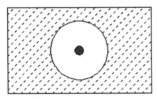
Display Boundary objects = Off

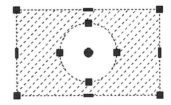
Display Boundary objects = On

- When you create a hatch, AutoCAD normally creates a polyline (or region), which fits the boundary exactly. Once the command ends, AutoCAD will delete it. Using the **Retain Boundary** pop-up list, you can ask AutoCAD to keep it as a polyline or as a region, or not to keep it at all.

- In order for the Hatch command to work, it needs to analyze all the objects in the current viewport (in model space this means the area you are seeing right now). This may take very long time (depending on the number of objects) so you can ask AutoCAD to analyze only the relative objects rather than all objects. Locate the **Select New Boundary Set** button:

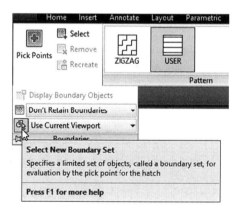

- You then see the following prompt:

```
Select objects:
```

- Select the desired objects, then press [Enter], and the pop-up list will read this time **Use Boundary Set**, instead of the default prompt.

6.11 EDITING HATCH

- Editing hatches in AutoCAD have never been easier. There are two ways to edit a hatch: by clicking (single click) or using Properties (double-click).
- If you click a hatch a single click, three things will take place:
  - You will see a grip (small dot) at the center of the area.
  - The context **Hatch Editor** tab will appear, which includes the same panels Hatch Creation tab does.
  - The Quick Properties palette will appear.
- If you move your mouse to the grip (without clicking), you will see the following menu:

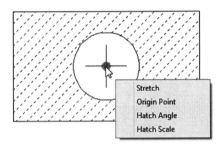

- You can do any or all of the following:
  - Stretch should stretch the hatch but instead moves it. In order to stretch the hatch, it is preferable to display boundary objects as previously discussed.
  - Modify the Origin Point on the spot.
  - Modify the Hatch Angle on the spot.
  - Modify the Hatch Scale on the spot.
- Also, the context tab Hatch Editor can make all the necessary modifications because it contains the same panels as the Hatch Creation tab.
- The **Quick Properties** palette appear as well so you can make the edits needed:

- On the other hand, double-clicking the desired hatch, or selecting the hatch, right-clicking, and selecting **Properties** option, will show **Properties** palette:

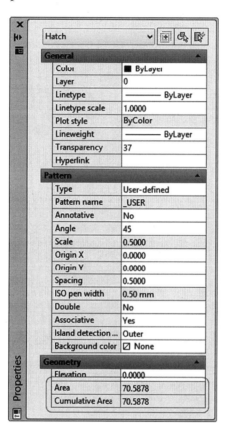

- The data valid for editing is everything related to the hatch properties, options, and boundary. Moreover, the Properties palette provides a single piece of information that other methods do not, which is the Area of the hatch. If you select a single hatch (whether single area or multiple areas hatched as a single area), the **Area** field and the **Cumulative Area** will be the same. But if you select multiple areas created using different commands, you will see only the **Cumulative Area** filled and the **Area** field showing **Varies** as a value.

- If you select the hatch and right-click, you will see some Hatch-related editing commands such as:

- These commands include:
  - **Hatch Edit,** which will show the old Hatch dialog box.
  - **Set Origin**, which will help you set a new origin.
  - **Set Boundary**, which will help you set a new boundary set.
  - **Generate Boundary**, which will regenerate a new boundary for this hatch.

PRACTICE 6-5

Hatch Boundary and Hatch Editing

1. Start AutoCAD 2015.
2. Open the file, Practice 6-5.dwg.
3. Select the existing hatch at the top-left of the drawing.
4. From the **Boundaries** panel, click **Recreate**, then select the **Polyline** and **Yes**. You can see that AutoCAD recreated a new boundary for this orphan hatch.
5. Using the existing hatch at the middle left, select the hatch so you can see the special grip, then change the angle to 90°.
6. Using the existing hatch at the bottom left, select the hatch so you can see the special grip, then change the origin point to be the lower-left corner and the scale to = 0.1
7. Using the existing hatch at the upper-right, select it and right-click, then select the **Properties** option; change the scale to be 0.5 and the background color to Magenta. What is the total area of the hatch?

8. Start the Hatch command and select pattern name = steel, and set the scale= 2.0. Now go to the Options panel and change Island detection to Normal. Try Hatching the shape at the middle right. Change the Island detection to Outer and try it, then Ignore and try it. Finally, get it back to Normal and finish the command.

9. Save and close the file.

NOTES:

CHAPTER REVIEW

1. The command to convert a block to a file is:
    a. Makeblock
    b. Createblock
    c. Wblock
    d. None of the above
2. There are _____ hatch types in AutoCAD.
3. When you create a block in a drawing, you cannot use it in other drawings.
    a. True
    b. False
4. Hatch grips are like normal objects:
    a. True
    b. False
5. Using the Insert command, you can see the blocks available in the current drawing.
    a. True
    b. False

CHAPTER REVIEW ANSWERS

1. c
3. a
5. a

CHAPTER 7

WRITING TEXT

In This Chapter

- How to write using single line text and multiline text
- How to edit text

7.1 WRITING TEXT USING SINGLE LINE TEXT

- This command allows you to create lines of text; each line is independent from the other lines. To issue the command, go to the **Annotate** tab, locate the **Text** panel, then click the **Single Line** button:

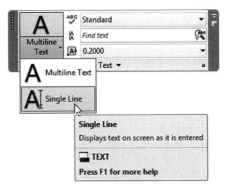

- You will see the following prompt:

```
Current text style: "Standard" Text height: 0.2000 Anno-
tative: No Justify: Left
Specify start point of text or [Justify/Style]:
Specify rotation angle of text <0>:
```

- The first prompt gives you some information about the current settings. The report shares the current style (in our example Standard) and the current height (in our example =0.20), whether this text will be annotative or not, and the justification of the text.
- At the first prompt, you can change the current Justification (discussed in the Multiline text) and Style settings by typing **J** or **S**. You can also specify the start point of the baseline of the text then specify the rotation angle (default value as 0 (zero)). Once you press [Enter], you see a cursor on the screen blinking, meaning it is ready for you to type any text. To finish a line, press [Enter] and to end the command press [Enter] twice.
- Another way of setting the current text style is to go to the **Annotate** tab then locate the **Text** panel. At the top-right, you will see the current text style and you can change it to any desired text style:

PRACTICE 7-1

Creating Text Style and Single Line Text

1. Start AutoCAD 2015.

2. Open the file, Practice 7-1.dwg.

3. Make the Room Names text style current.

4. Make the layer Text current.

5. Type the room names as shown.

6. Make Standard text style current (this text style the height = 0, hence you should set it every time you want to use this style).

7. Make the layer Centerlines current.

8. Zoom to the upper-left centerline and note the letter A is missing.

- Start the Single Line text, right-click, and select the Justify option. From the list, pick MC (Middle Center), select the center of the circle as the Start point. For the height, set it at 0.25, rotation angle = 0, type A, then press [Enter] twice.

9. You should get the following results:

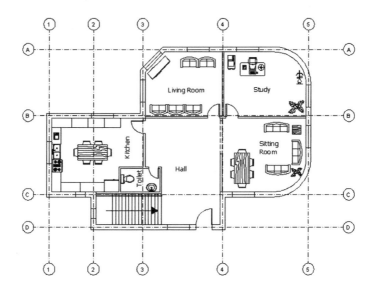

10. Save and close the file.

7.2 WRITING TEXT USING MULTILINE TEXT

- This command enables you to type text in a MS Word similar environment. To issue this command, go to the **Annotate** tab, locate the **Text** panel, then click the **Multiline Text** button:

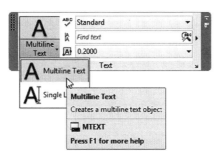

- You will see something like the following in the command window:

  ```
  Current text style: "TNR_05" Text height: 0.500
  Annotative No
  Specify first corner:
  Specify opposite corner or [Height/Justify/Line spacing/
  Rotation/Style/Width]:
  ```

- The first prompt gives you some information about the current settings. The report shares what the current style (in our example TNR_05), the current height (in our example =0.50), and whether this text will be annotative or not. You need to specify an area to write in so the cursor will change to:

- At the command window, you then see the following prompt:

  ```
  Specify first corner:
  ```

- Specify the first corner and you see something like the following:

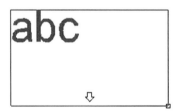

- At the command window, you see the following prompt:

  ```
  Specify opposite corner or [Height/Justify/Line spacing/
  Rotation/Style/Width/ Columns]:
  ```

- Specify the other corner, or select one of the options, and a text editor with a ruler appears:

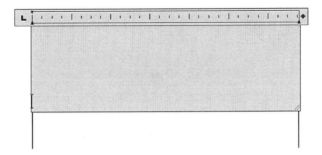

- AutoCAD then shows a context tab called **Text Editor**, which looks like the following:

- Using these panels will allow you to do many things; they are discussed as follows.

7.2.1 Style Panel

- This is the Style panel:

- Use the Style panel to select the current text style you want to use and set the height (this value overwrites the text style height, so be careful). You can also change the **Background Mask**, which you see in the following dialog box:

7.2.2 Formatting Panel

- This is the Formatting Panel:

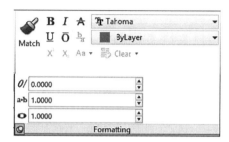

- Use the **Formatting** panel to do any of the following:
  - Match the selected text properties and apply it to other text.
  - Make the selected text **Bold**, **Italic**, **Underlined**, **Overlined**, or **Strikethrough**.
  - Change the **Font** and the **Color** of the selected text.
  - Change the selected text to be **Superscript** or **Subscript**.
  - Change a capital letter to a small letter and vise-versa.
  - Clear the text formatting.
  - Change the **Oblique Angle** of the selected text.
  - Change the **Tracking** (Increases or decreases the spaces between letters. If the value is greater than 1, there is more space between letters and vise-versa).
  - Change the **Width Factor**.

7.2.3 Paragraph Panel

- This is the Paragraph panel:

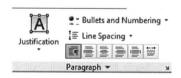

- Use the Paragraph panel to do any of the following:
  - Change the **Justification** of the text relative to the text area selected. Choose one of the following options:

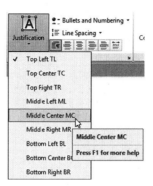

- The following illustration shows the nine points available relative to the text area:

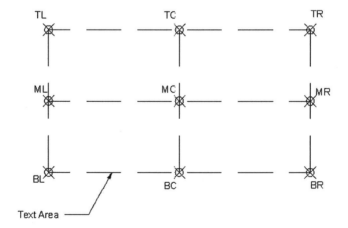

- Change the text selected to use **Bullets and Numbering**; there are three choices listed (Numbered, Lettered, and Bulleted):

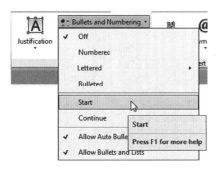

- Change the **Line Spacing** of the paragraph:

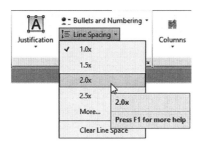

- Change the horizontal justification of the text, by using one of the six buttons as shown:

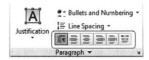

7.2.4 Insert Panel

- This is the Insert Panel:

- In the Insert Panel, you can do any of the following to the selected text:
  - Convert text to two **Columns** or more. If you click the **Columns** button, the following menu appears:

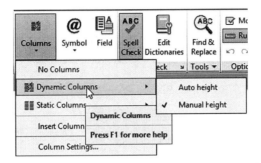

- The preceding shows **Dynamic Columns,** which allows you to select whether you want AutoCAD to specify the height (Auto height) or you want to set the height (Manual height). If you want static columns, select the **Static Columns** option to specify the number of columns:

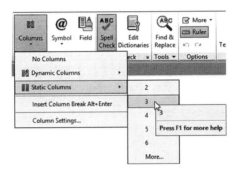

- Select the **Insert Column Break Alt+Enter** option to insert a column break at a certain line. This means the rest of the column will go to the next column.
- Select the **Column Settings** option to show **Column Settings** dialog box, which allows you to do all of the previous settings plus you can set the **Column** width and the **Gutter** distance:

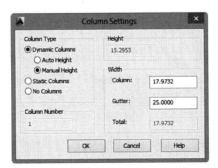

- Here is an example of columns:

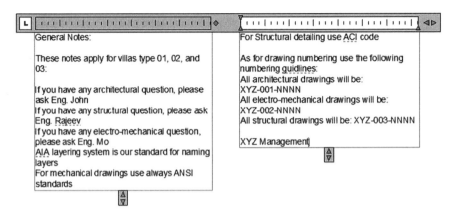

- Select the **Symbols** to add scientific characters to your text. You will see the 20 available symbols:

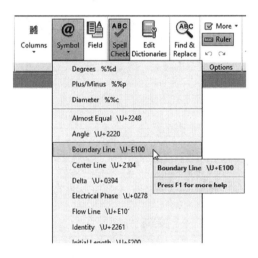

7.2.5 Spell Check Panel

- This is the Spell Check Panel:

- As long as you are in the Text editor, you can leave the **Spell Check** button on to catch any misspelled word. You will see a dotted red line appears underneath it, as shown in the following example:

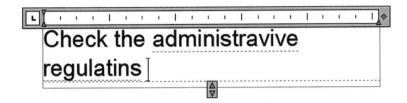

- To get suggestions for the correct word, move to the word, right-click, and you will see something like the following:

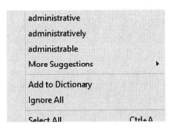

- Select the **Edit Dictionaries** button, which enables you to select another dictionary other than the default.

7.2.6 Tools Panel

- This is the Tools Panel:

- In the **Tools** panel, you can do any of the following:
  - Click the **Find & Replace** button to search for a word and then replace it with another word. You will see the following dialog box:

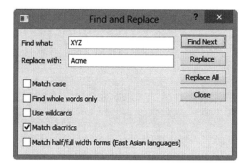

  - Use the **Import Text** button to import text from text files.
  - Use the **AutoCAPS** button to write in the Text editor using capital letters.

7.2.7 Options Panel

- This is the Options Panel:

- Using the Options panel, you can do any of the following:
  - Change the Character set, change the Editor Settings, or learn more about Multiline text through Help:

  - To show the ruler (by default it is displayed) or hide it.
  - Undo and redo text actions.

7.2.8 Close Panel

- This is the Close panel and it contains a single button to close the Text editor and all its panels:

7.2.9 While You Are in the Text Editor

- While you are in the text editor, you can do the following:
  - Using the ruler, you can set the First Line indent and Paragraph indent:

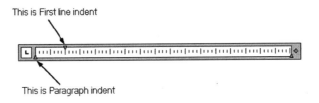

  - If you right-click, you will see the following menu, which includes all the functions previously discussed:

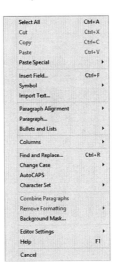

- You can change the width and height of the area using the following controls:

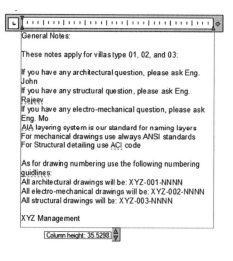

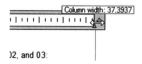

PRACTICE 7-2

Writing Using Multiline Text

1. Start AutoCAD 2015.

2. Open the file, Practice 7-2.dwg.

3. You are now in the layout called "Cover."

4. Make sure that the current layer is Text.

5. Make the text style Title the current text style.

6. Start Multiline text and specify the two corners of the rectangle as your text area.

7. Before you write anything, change the Justification to MC.

8. Type the following as shown:

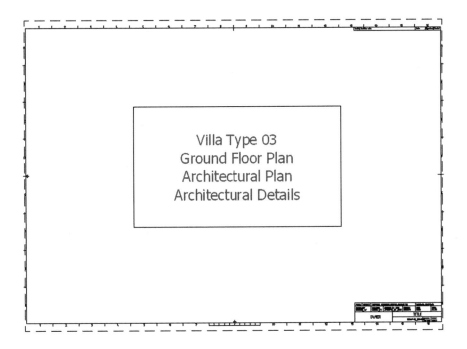

9. Erase the rectangle.

10. Go to the layout called ISO A1 – Overall.

11. Make the Notes text style the current text style.

12. Using the rectangle at the right, specify the two corners of your text area.

13. Using the **Tools** panel, select the **Import Text** command, and select the Notes.text file.

14. Select the General Notes, make it Bold and Underlined, and the size = 5.0.

15. Stretch the width a bit so 03 will be in the same line.

16. Select the text starting from, "If you have …. " to "ACI code", select Bullets and Numbering, then select Numbered.

17. Select the text starting from, "As for …. " until the last "NNNN" and make them Bulleted with line spacing = 1.5 x.

18. Select the last line "XYZ Management" then make it centered and Italic.

19. Close the text editor.
20. Erase the rectangle.
21. Go to ISO A1 - Architectural Details.
22. Using the rectangle, start Multiline text, select two opposite corners, and import the same file: Notes.txt.
23. Do the same thing you did previously and make the text numbered and bulleted.
24. Start the Column command and set two static columns; then using the Column Settings dialog box, set Column Width =125 and Gutter = 22.5.
25. Using the two arrows at the lower-left corner, make sure that all the six points are in the first column.
26. Erase the rectangle.
27. This what you should get:

General Notes:

These notes apply for villas type 01, 02, and 03:

1. If you have any architectural question, please ask Eng. John
2. If you have any structural question, please ask Eng. Rajeev
3. If you have any electro-mechanical question, please ask Eng. Mo
4. AIA layering system is our standard for naming layers
5. For mechanical drawings use always ANSI standards
6. For Structural detailing use ACI code

- As for drawing numbering use the following numbering guidlines:
- All architectural drawings will be: XYZ-001-NNNN
- All electro-mechanical drawings will be: XYZ-002-NNNN
- All structural drawings will be: XYZ-003-NNNN

XYZ Management

28. Save and close the file.

7.3 TEXT EDITING

- There are several ways to edit the text, including double-click and using Quick Properties, Properties, and Grips.

7.3.1 Double-Click Text

- To edit the text whether it is Single Line Text or Multiline text, simply double-click it. If it is Single Line Text, you will see the text selected and you will be able to add to it or modify it. If it is Multiline text, you will see the Text Editor reopen and the Text Edit context tab appear at the top. Make all of your desired changes.

7.3.2 Quick Properties and Properties

- To show the Quick Properties for either Single Line Text or Multiline Text, simply click the desired text. You will see something like the following (Text in the following is Single Line Text and MText is Multiline Text):

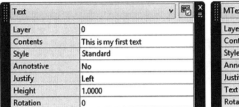 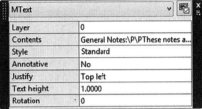

- You can change the following: Layer, Contents, Style, Annotative (Yes or No), Justification, Height, and Rotation angle.
- If you select Single Line Text or Multiline Text, right-click then select the Properties option. You will see something like the following (Text in the the following is Single Line Text and MText is Multiline Text):

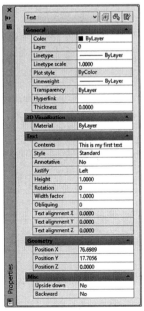 

- As you can see, you can change anything related to the selected text.

7.3.3 Editing Using Grips

- If you click a Single Line text, you will see a grip showing at the start point of the baseline and another one at the justification point (you may see only one, if both points coincide):

.This is my first text

- You will see a single grip appear at the Justification point selected when the Multiline is selected and then you will see two triangles, one at the lower part, and one at the right side. The lower part triangle allows you to cut your Multiline Text to columns, simply stretch it up, and see the text cut into two columns. The triangle at the right allows you to increase/decrease the horizontal distance of the text, thereby increasing/decreasing the height of the text.

▪These notes apply for villas type ▶
01, 02, and 03:
▼

- If you click any multi-column text, you will see the same thing for the first column at the left, except the arrow at the bottom is pointing downward instead of upward. For the other columns, you will see an arrow at the right side. For the last column at the right, you will see an extra to the right of the normal arrow and an arrow at the lower-right corner, which allows for the whole text to increase/decrease in width and height. See the following illustration:

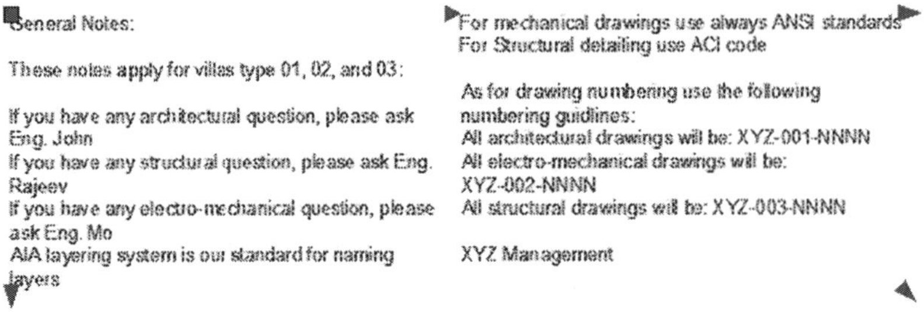

▪General Notes:

These notes apply for villas type 01, 02, and 03:

If you have any architectural question, please ask Eng. John
If you have any structural question, please ask Eng. Rajeev
If you have any electro-mechanical question, please ask Eng. Mo
AIA layering system is our standard for naming layers
▼

▶For mechanical drawings use always ANSI standards▶
For Structural detailing use ACI code

As for drawing numbering use the following numbering guidlines:
All architectural drawings will be: XYZ-001-NNNN
All electro-mechanical drawings will be: XYZ-002-NNNN
All structural drawings will be: XYZ-003-NNNN

XYZ Management

◀

7.4 SPELL CHECK AND FIND AND REPLACE

- When you are in the Text Editor, you can spell Check and Find and Replace. But what about if you are not in the Text editor? Can you Spell Check and Find and Replace? The answer to that question is yes! AutoCAD can spell check the whole drawing and not only the current text but also the current space and/or layout (layout is discussed in Chapter 9) plus any desired selected text. To issue the Check Spelling command, go to the **Annotate** tab, locate the **Text** panel, and click the **Check Spelling** button:

- You will see the following dialog box:

- If AutoCAD finds any misspelled word, it will give you suggestions to select the proper replacement from or you can ignore AutoCAD's findings. Also, AutoCAD can find any word or part of the word and replace it in the entire drawing file.

- To issue the **Find and Replace** command, go to the **Annotate** tab, locate the **Text** panel, type the desired word you want to replace in the **Find text** field, and then click the small button at the right:

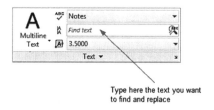

Type here the text you want to find and replace

- You will see the following dialog box:

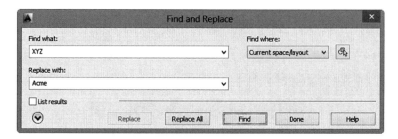

- Under **Replace with**, type the new word(s) you want to replace. Select one of the following choices: **Find**, **Replace**, and **Replace All**. When you are finished, click **Done**.

PRACTICE 7-3

Editing Text

1. Start AutoCAD 2015.
2. Open the file, Practice 7-3.dwg.
3. You are now in Cover layout. Click the text then using the arrow at the bottom, drag it upward to cut the text into two column, each holding two lines.
4. Using the arrow at the right of the second column, drag it till the two columns touch each other. Press [Esc] to end the editing process

5. Go to layout ISO A1 – Overall.
6. Zoom to the text at the right, select it, right-click, choose the Properties option, change the following:
   a. Change the style from Notes to Standard.
   b. Change Justify to be Middle Center.
   c. Change Text Height to be 4.5.
7. Using the arrow at the right, stretch the text to the right by 10 units (if OSNAP is annoying you switch it off).
8. Press [Esc] to end the editing process.
9. Go to the Annotate tab, locate the Text panel in the Find and Replace field, type XYZ, then click the small button at the right. When the dialog box comes up in the Replace field, type ACME, then click Replace All.
10. Go to layout ISO A1 Architectural Details, double-click the multi-column text, and convert it to single column text.
11. Press [Esc] to end the editing process.
12. Go to Model space.
13. Click the word "Hall." The Quick Properties will appear and set the Justification to be Middle Center. Using the Move command, try to put this word at the middle center of the space.
14. Save and close the file.

NOTES:

CHAPTER REVIEW

1. There are two types of text in AutoCAD; single line and multiline:

    a. True

    b. False

2. One of the following is <u>not</u> a panel in Text Editor context tab:

    a. Insert

    b. Options

    c. Text

    d. Paragraph

3. When you start the Single line text command, AutoCAD will inform you of the current text style:

    a. True

    b. False

4. While you are in the multiline command, select _____ option to make the rest of the column go to the next column.

5. In multiline text command, you can set the background for a text.

    a. True

    b. False

6. Degrees, Centerline, and Not Equal are _____ you can insert in the multiline text command.

CHAPTER REVIEW ANSWERS

1. a

3. a

5. a

CHAPTER 8

DIMENSIONS

In This Chapter
- Dimensioning and dimension types
- How to insert different types of dimensions
- How to edit dimension blocks

8.1 WHAT IS DIMENSIONING IN AUTOCAD?

- Dimensioning in AutoCAD is just like Text. You should create your dimension style first then you can use it to insert dimensions. The Dimension Style controls the overall outcome of the dimension block generated by the different types of the dimension commands.
- To insert a dimension, depending on the type of the dimension, you should specify points or select objects and then a dimension block will be added to the drawing. For example, in order to add a linear dimension the user will select two points representing the distance to be measured and a third point as the location of the dimension block. See the illustration:

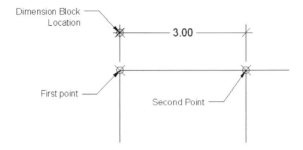

- The generated block consists of three portions. They are:
  - Dimension line
  - Extension lines
  - Dimension Text
- See the following illustration:

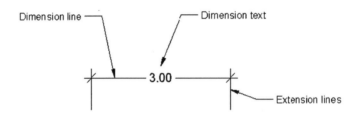

8.2 DIMENSION TYPES

- The following are the dimension types in AutoCAD:

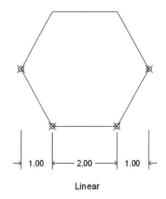

Linear

Dimensions • 227

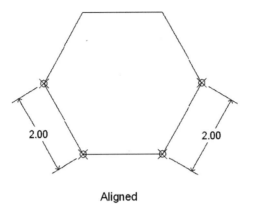

Aligned

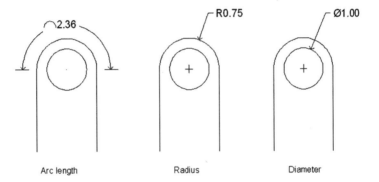

Arc length   Radius   Diameter

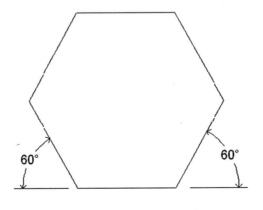

Angular

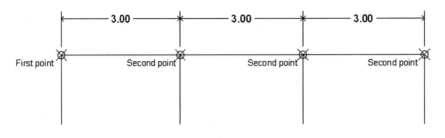

Continuous Dimension

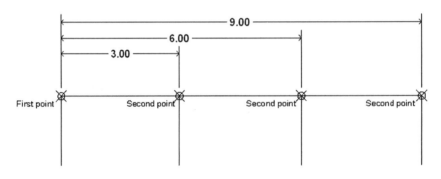

Baseline Dimension

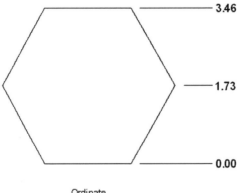

Ordinate

8.3 HOW TO INSERT A LINEAR DIMENSION

- This command allows you to create a horizontal or vertical dimension. To start the Linear command, go to the **Annotate** tab, locate the **Dimensions** panel, then select the **Linear** button:

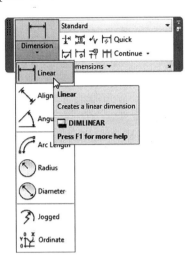

- You will see the following prompts:

```
Specify first extension line origin or <select object>:
Specify second extension line origin:
Specify dimension line location or
[Mtext/Text/Angle/Horizontal/Vertical/Rotated]:
```

- Specify the first point and second point of the dimension distance to be measured, and then specify the location of the dimension block by specifying the location of the dimension line. The following is the result:

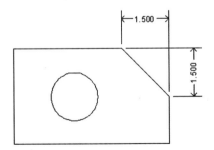

- Prompts contain other options. They are:
  - Mtext
  - Text
  - Angle
  - Horizontal
  - Vertical
  - Rotated
- Mtext edits the measured distance in MTEXT mode, whereas Text edits the measured distance in DTEXT (Single line) mode. Angle allows you to change the angle of the text (default is 0 (zero)). Horizontal and Vertical allows you to force the dimension to be either horizontal or vertical (default will specify either by the movement of the mouse). Finally, Rotated allows you to create a dimension line parallel to another angle given by you.

8.4 HOW TO INSERT AN ALIGNED DIMENSION

- This command allows you to create a dimension parallel to the two points specified. To start this command, go to the **Annotate** tab, locate the **Dimensions** panel, and then select the **Aligned** button:

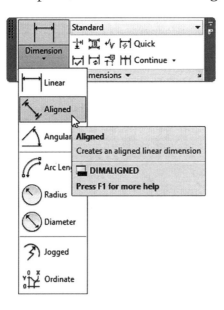

- The following prompt appears:

  ```
  Specify first extension line origin or <select object>:
  Specify second extension line origin:
  Specify dimension line location or
  [Mtext/Text/Angle]:
  ```

- Specify the first and second point of the dimension distance to be measured, then specify the location of the dimension block by specifying the location of the dimension line. See the following illustration:

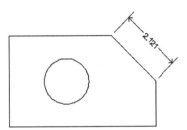

- The rest of the options are identical to the **Linear** command prompts.

8.5 HOW TO INSERT AN ANGULAR DIMENSION

- This command allows you to insert an angular dimension between two lines, included the angle of a circular arc, two points and the center of a circle, or three points. To start this command, go to the **Annotate** tab, locate the **Dimensions** panel, then select the **Angular** button:

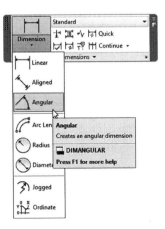

- AutoCAD may use one of the following methods based on the selected objects:
  - If you select a circular arc, it measures the included angle.
  - If you select a circle, your selecting points are the first point, the center of the circle is the second point, and then you select third point.
  - If you select a line, AutoCAD will ask you to select a second line.
  - If you select a point, it will be considered as the center point then you specify two more points.

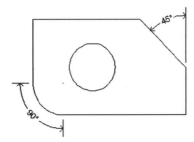

- Based on the previous discussion, when you start the command you see the following prompts (this example is when an arc is selected):

```
Select arc, circle, line, or <specify vertex>:
Specify dimension arc line location or [Mtext/Text/
Angle]:
```

PRACTICE 8-1

Inserting Linear, Aligned, and Angular Dimension

1. Start AutoCAD 2015.
2. Open the file, Practice 8-1.dwg.
3. Make layer Dimension the current layer.
4. Make sure that Dimension Style = Part.
5. Insert the dimension as shown:

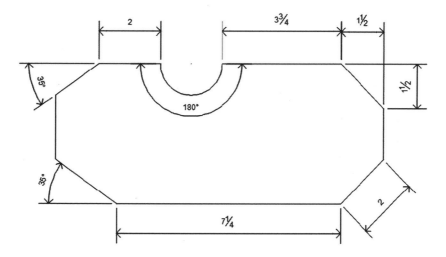

6. Save and close the file.

8.6 HOW TO INSERT AN ARC LENGTH DIMENSION

- This command allows you to create a dimension measuring the length of an arc. To start this command, go to the **Annotate** tab, locate the **Dimensions** panel, then select the **Arc Length** button.

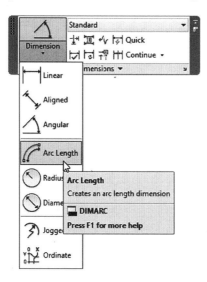

- You will see the following prompts:

```
Select arc or polyline arc segment:
Specify arc length dimension location, or [Mtext/Text/
Angle/Partial/Leader]:
```

- Select the desired arc and then locate the dimension block, either inside or outside the arc. You will see something like the following:

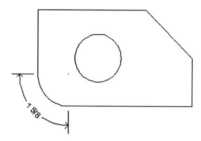

- The options, Mtext, Text, and Angle, were already tackled previously. The partial option allows you to insert an arc length dimension on part of the arc. You select the arc and then select two internal points on the arc. You will get something like the following:

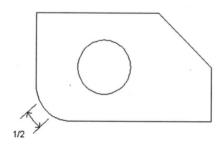

8.7 HOW TO INSERT A RADIUS DIMENSION

- This command allows you to insert a radius dimension on an arc or circle. To start this command, go to the **Annotate** tab, locate the **Dimensions** panel, then select the **Radius** button:

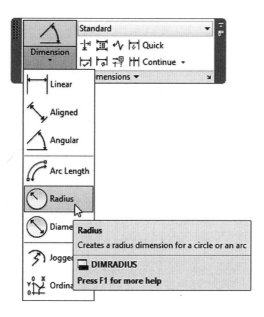

- You will see the following prompts:

```
Select arc or circle:
Specify dimension line location or [Mtext/Text/ Angle]:
```

- Select the desired arc or circle and then locate the dimension block. You will see something like the following:

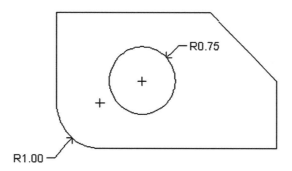

8.8 HOW TO INSERT A DIAMETER DIMENSION

- This command allows you to insert a diameter dimension on an arc or circle. To start this command, go to the **Annotate** tab, locate the **Dimensions** panel, then select the **Diameter** button:

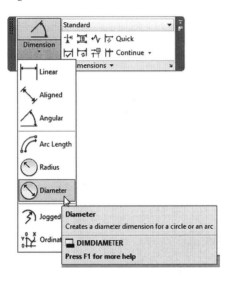

- You will see the following prompts:

```
Select arc or circle:
Specify dimension line location or [Mtext/Text/ Angle]:
```

- Select the desired arc or circle and then locate the dimension block. You will get something like the following:

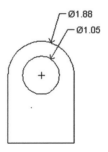

PRACTICE 8-2

Inserting Arc Length, Radius, and Diameter Dimension

1. Start AutoCAD 2015.
2. Open the file, Practice 8-2.dwg.
3. Make the layer Dimension the current layer.
4. Make sure that the current Dimension Style is Part.
5. Insert the dimension as shown:

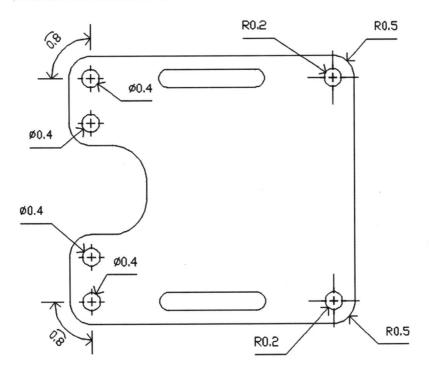

6. Save and close the file.

8.9 HOW TO INSERT A JOGGED DIMENSION

- This command allows you to insert a jogged arc dimension for a big arc, simulating a new center point. To start this command, go to the **Annotate** tab, locate the **Dimensions** panel, and select the **Jogged** button:

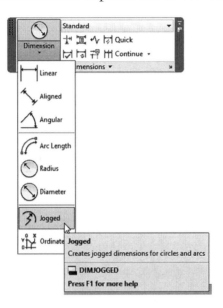

- The following prompts will appear:

```
Select arc or circle:
Specify center location override:
Dimension text = 1.5
Specify dimension line location or [Mtext/Text/ Angle]:
Specify jog location:
```

- First, select an arc or circle, specify the point which will be the new center point (AutoCAD calls it override), locate the dimension line, then specify jog's location. You will see something like the following:

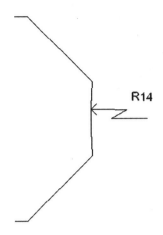

8.10 HOW TO INSERT ORDINATE DIMENSION

- This command allows you to insert dimensions relative to a datum, either in X or in Y. See the following illustration:

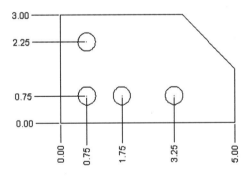

NOTE *Use the UCS command and Origin option to relocate the origin to one side of the shape so the values in both X, Y will be right. If you leave the origin to the current UCS origin, the values inserted may be wrong.*

- To start this command, go to the **Annotate** tab, locate the **Dimensions** panel, then select the **Ordinate** button:

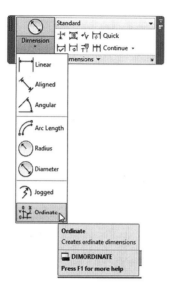

- You will see the following prompts:

```
Specify feature location:
Specify leader endpoint or [Xdatum/Ydatum/Mtext/Text/
Angle]:
```

- First, select the desired point. By default, AutoCAD will give you the freedom to go in the direction of X or Y. If you want to force the mouse to measure points related to the X axis, then select **Xdatum** option; the same applies for Y axis. We have already discussed the rest of the options.

PRACTICE 8-3

Inserting Dimensions

1. Start AutoCAD 2015.
2. Open Practice 8-3.dwg.
3. Make the layer Dimension current.
4. Make sure that the Dimension Style is Part.
5. Insert the dimensions as shown:

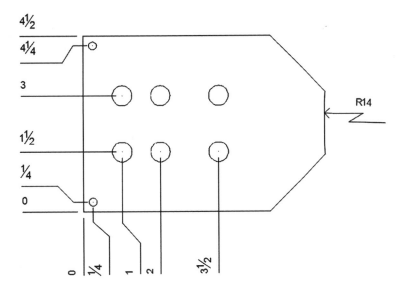

6. Save and close the file.

8.11 INSERTING SERIES OF DIMENSIONS USING CONTINUE COMMAND

- AutoCAD allows you to input series of dimensions using the Continue command. The Continue command follows the last dimension command by asking to input the second point, assuming that the last point of the last dimension will be considered the first point for the coming one. To start this command, go to the **Annotate** tab, locate the **Dimensions** panel, then select the **Continue** button:

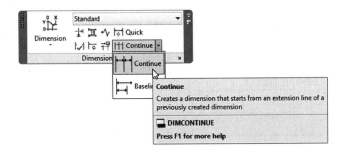

- There are two scenarios when using the Continue command:
  - If there is not a dimension command issued in an AutoCAD session, AutoCAD asks you to select an existing dimension (linear, ordinate, or angular). You then see the following prompt:

```
Select continued dimension:
```

  - If there is a dimension command issued in an AutoCAD session, AutoCAD prompts you to continue this command by asking you to specify the second point. Also, you can select an existing dimension block to continue it or you can undo the last continue command. You then see the following prompts:

```
Specify a second extension line origin or [Undo/ Select]
<Select>:
```

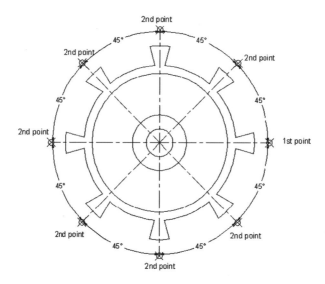

8.12 INSERTING SERIES OF DIMENSIONS USING BASELINE COMMAND

- This command is identical to the Continue command, except all dimensions are measured based on the first point specified by you as the baseline. To start this command, go to the **Annotate** tab, locate the **Dimensions** panel, then select the **Baseline** button:

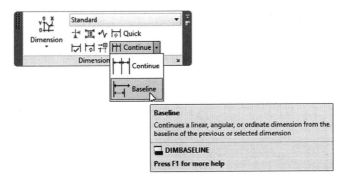

- We do not need to discuss the prompts of this command because it resembles the Continue command prompts. You will see something like the following:

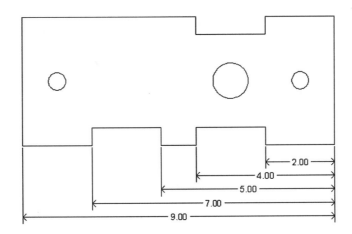

PRACTICE 8-4

Continue Command

1. Start AutoCAD 2015.
2. Open the file, Practice 8-4.dwg.

3. Create an angular dimension, then using the Continue command, complete the shape as follows:

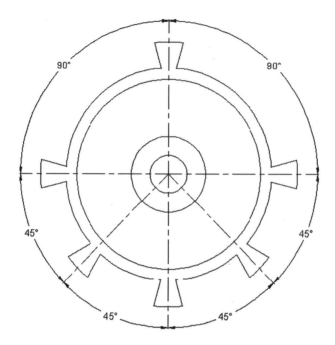

4. Save and close the file.

PRACTICE 8-5

Baseline Command

1. Start AutoCAD 2015.
2. Open the file, Practice 8-5.dwg.
3. Create the linear dimension, then using the Baseline command, complete the shape as follows:

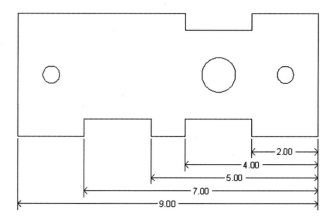

4. Save and close the file.

8.13 USING THE QUICK DIMENSION COMMAND

- This command allows you to insert a group of dimensions in one shot. To start this command, go to the **Annotate** tab, locate the **Dimensions** panel, then select the **Quick Dimension** button:

- You will see the following prompt:

```
Associative dimension priority = Endpoint
Select geometry to dimension:
Specify   dimension   line   position,   or   [Continuous/
Staggered/Baseline/Ordinate/Radius/           Diameter/
datumPoint/Edit/seTtings] <Continuous>:
```

- First, select the desired geometry you want to dimension, using clicking, window mode, crossing mode, or any other mode you know. If this is the first time you are using the command in this AutoCAD session, AutoCAD will use Continuous as the default option. But, if you right-click, you will see the following menu:

- Using this shortcut menu, select the desired dimension type and then specify the dimension line location; consequently, a set of dimensions will be inserted.
- See the following examples:

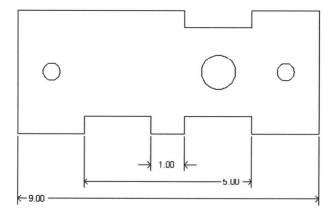

Staggered

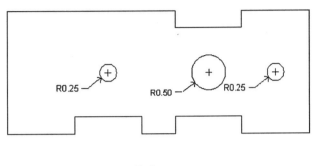

Radius

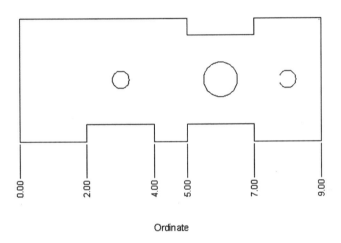

Ordinate

- If you select the Settings option, you will see the following menu:

- This option enables you to set the default object snap for specifying extension line origins.

8.14 EDITING A DIMENSION BLOCK USING GRIPS

- After inserting the dimension block, it is easy to edit it using grips or by right-clicking. Depending on the type of the dimension, if you click any dimension block you will see grips in certain places. The following are some examples:
    - For linear and aligned dimensions, grips appear in five places; at the two points measured, the two ends of the dimension line, and at the dimension text. If you *hover* over the text grip, you will see the following:

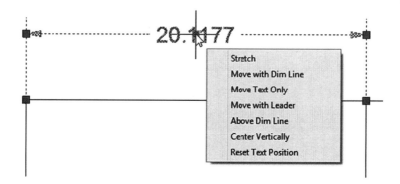

    - AutoCAD allows you to Stretch, Move with Dim Line, Move Text Only, etc., while holding this grip. If you hover over the two grips at the ends, you will see:

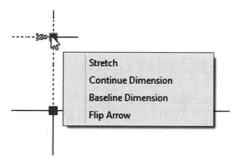

    - The list of actions include Stretch, creating a continuous or baseline dimension based on the selected type of dimension, or you can flip the arrow to the nearest selected grip.

- Angular dimension grips appear at four places: at the endpoints of the two lines involved, the dimension line, and at the dimension text. If you hover over the text grip, you will see the following:

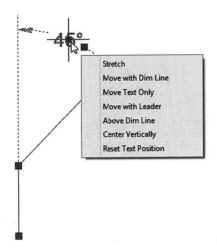

- These commands are the as those discussed for linear and aligned dimensions. If you hover over the two ends grip, you will see the following:

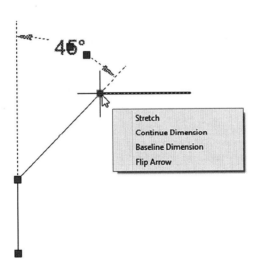

- These are the same commands mentioned for linear and aligned dimensions.

- Ordinate dimension grips appear at four places: the origin point and measured point, the dimension line, and finally at the dimension text. If you hover over the text grip, you will see the following:

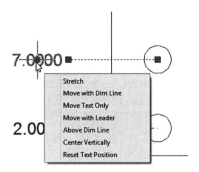

- The same set of commands were previously discussed. Hovering over the end of the line grip shows the following:

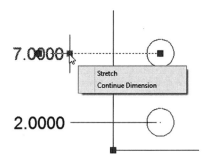

- Ordinate does not work with Baseline and does not have arrows so these two options are not given for this type of dimension.
- Radius and diameter dimension grips appear at three places: at the selected point, center, and the dimension text. If you hover over the text grip, you will see the following:

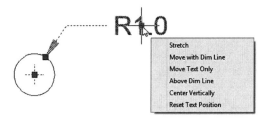

- This the same list of commands previously mentioned. If you hover over the grip at the end of the arrow, you will see the following:

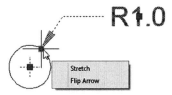

- Because Radius and Diameter do not have the ability to use Continue or Baselines, these two commands are not mentioned. You can use only the Stretch and Flip Arrow.
- Arc length will show four grips; one at the two ends of the arc, one at the text, and, finally, one near the text. If you hover over the text grip, you will see the following:

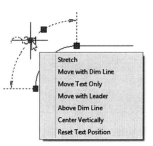

- This is the same list of commands as previously discussed. If you hover over the end point grip, you will see the following:

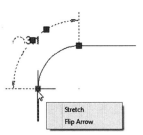

- Because Arc Length does not work with Continue and Baseline, only the Stretch and Flip Arrow appears.

8.15 EDITING A DIMENSION BLOCK USING RIGHT-CLICK MENU

- On the other hand, if you select a dimension block and right-click, you will see the following shortcut menu:

- You can change the dimension style used to insert the selected dimension block. Or better yet, the changes you made can be saved on a new dimension style, as shown:

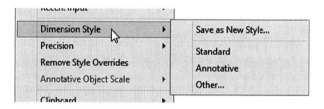

- You can also change the Precision of the dimension text:

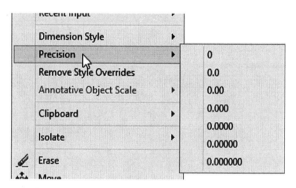

8.16 EDITING A DIMENSION BLOCK USING QUICK PROPERTIES AND PROPERTIES

- If you select a dimension block, the Quick Properties comes up automatically. You will see something like the following:

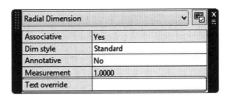

- You have the ability to modify the dimension style and the annotative feature (discussed in the next chapter). Also, you can find the exact measurement of the selected dimension and change the number by using the Text override field.
- Changing Properties, on the other hand, makes global changes to the selected dimension blocks. Simply select the dimension block(s), right-click, and then select the **Properties** option (you can use double-click as well). You then see something like the following:

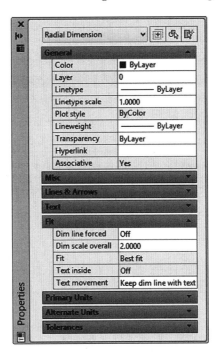

- Using the Properties palette, you can change everything related to the dimension block or the dimension style of this block (the previous example, showed the variables you can control under the Fit category).

PRACTICE 8-6

Quick Dimension and Editing

1. Start AutoCAD 2015.
2. Open the file, Practice 8-6.dwg.
3. Make Dimension the current layer.
4. Make Part the current Dimension Style.
5. Using Quick Dimension create continuous for the top part of the shape.
6. Using Quick Dimension create baseline for the bottom part of the shape.
7. Using Quick Dimension create Radius for all circles inside, making the radius dimension point up and right.
8. Using grips get all Radius dimensions inside the shape.
9. Select the only dimension shown at the left to show its grips. Hover over the grip and when the menu comes up select Continue Dimension, then add two more dimensions.
10. Zoom to the dimension reading 1.5". Click it to show the grips, right-click, select Dimension Style, then select Standard. What happened to this specific dimension block? _____
11. Using grips change the dimension reading from 2.5" to Center Vertically.
12. Change the Precision of the dimension reading from 3.5 to 0.000.
13. Using Quick Properties change the dimension reading from 6.3" to 6.5".
14. You should have something similar to the following:

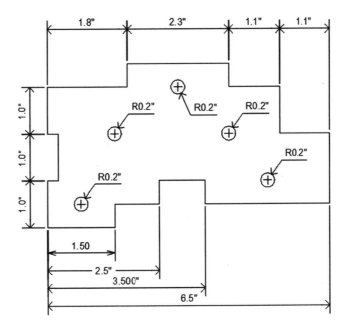

15. Save and close the file.

NOTES:

CHAPTER REVIEW

1. Arc length command should be used only with arcs and polylines, not with circles.

    a. True

    b. False

2. Continue and Baseline cannot work with which of the following?

    a. Linear

    b. Radius

    c. Ordinate

    d. Angular

3. _____ will insert dimensions relative to a datum, either in X or Y.

4. Inputs to use this dimensioning command may be three points or two lines.

    a. Linear

    b. Radius

    c. Ordinate

    d. Angular

5. One of the following is not among the commands of Quick Dimension.

    a. Continuous

    b. Radius

    c. Staggered

    d. Angular

CHAPTER REVIEW ANSWERS

1. a

3. Ordinate

5. d

CHAPTER 9

PLOTTING

In This Chapter

- The difference between Model Space and Paper Space
- How to create a new layout using different methods
- How to create and control viewports

9.1 WHAT ARE MODEL SPACE AND PAPER SPACE?

- AutoCAD provides two spaces; one for creating your drawing, called Model Space, and the other for plotting your drawing, called Paper Space. There is only one Model space in each drawing; in contrast, there are an infinite number of Paper spaces per file, and each one is called layout.
- Each layout is linked to a Page Setup (which is similar to other computer software). You should specify everything related to plotting in Page Setup where you can specify the plotter, paper size, and paper orientation (Portrait or Landscape).
- After you create the layout and link it to page setup, you can insert a title block and add viewports (which represents a portion of the Model space). You can also set up the scale of the viewports.

9.2 INTRODUCTION TO LAYOUTS

- Layout is the place you where you plot your drawing from. Each layout will be linked to a Page Setup, Objects (like title block), text, dimensions, and finally **Viewports**, which are covered separately in the coming discussion. See the following roadmap:

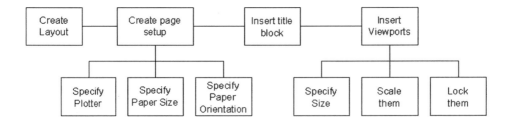

- Each Layout should have a name. By default, when you create a new drawing using *acad.dwt* template, two layouts are included called Layout1 and Layout2. These two preset layouts do not include anything so if you want to use them, complete the following steps:
  - Rename the file to the desired name.
  - Create the page setup and link the layout to it or you can simply link an existing page setup.
  - Insert the title block.
  - Insert Viewports, scale them, and lock them.

9.3 STEPS TO CREATE A NEW LAYOUT FROM SCRATCH

- This method allows you to create a new layout from scratch. It includes the following steps:
  - At the right of the names of the existing layouts there is a (+) sign, click it to add a new layout. See the following illustration:

- Right-click on any existing layout name (the tab at the lower-left corner of the screen and above the command window) and a shortcut menu will appear. Select the **New Layout** option:

- A new layout will be added with temporary name. Rename it using the right-click and selecting **Rename** option (you can double-click the temporary name, the name will be editable, input the new name)

- Link the new layout with a page setup. To do this, right-click the name of the layout and then select **Page Setup Manager**.

- The following dialog box is shown:

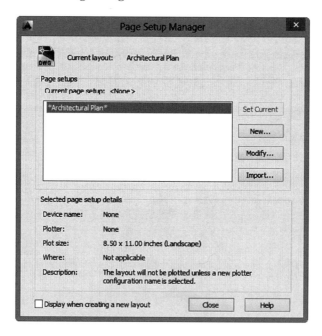

- At the top of the dialog box, you can see the name of **Current layout**. On the other hand, at the bottom you will see **Selected page setup**

details, which is a summary of the current page setup. Finally, you will see a check box that says **Display when creating a new layout**; this checkbox will force this dialog box to appear every time you go to a newly created layout.
- By default, AutoCAD will create a page setup with the same name of the layout. Click the **Modify** button to modify it, and you will see the following dialog box:

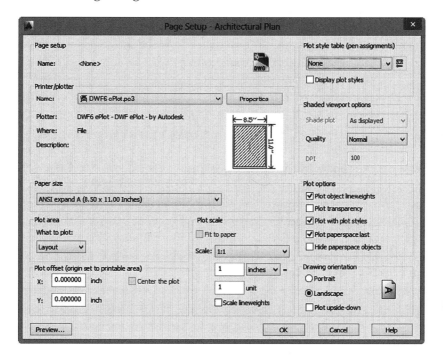

- Select the desired plotter (this plotter should be connected and configured).
- Select the desired **Paper Size**.
- Specify **What to plot**, always leave it as **Layout** (the other options is for Model space printing).
- Input **Plot Offset**, if you are printing from layout, leave the values as zeros.
- Input **Plot Scale**. If you are printing from layout then you will use viewports (this is the next topic); each viewport will hold its own scale. Set the layout plot scale to 1=1. Also, specify if you want to **scale lineweights** or not.

- Select **Plot style table (pen assignment)** (this topic will be discussed at the end of this chapter).
- **Shaded viewport options** are for 3D modeling in AutoCAD.
- Leave **Plot options** to default values.
- Select **Drawing orientation**, whether **Portrait** or **Landscape**.
- Plotter will print from top-to-bottom. Select the check box if you want it otherwise.
- Click **OK**. The Page Setup you create will be available for all layouts in the current drawing file.
- You will return to the first dialog box; select the page setup and click **Set Current** (also, you can double click the name of the Page Setup). Now, the current layout is linked to page setup you selected.
- To modify the settings of an existing page setup, click **Modify**. To bring a saved Page Setup from an existing file click **Import**.

9.4 STEPS TO CREATE A NEW LAYOUT USING A TEMPLATE

- This procedure allows you to import any layout from a template file, including the page setup and any other contents like title block, viewports, text, etc. To do this, complete the following steps:
  - Right-click on any existing layout and you will see the following menu. Select **From template** option:

- The normal Open file dialog box will pop up. Select the desired template, click **Open**, and you will see the following dialog box:

- Click one of the listed layouts and click **OK**. You will see the newly imported layout in your drawing.

9.5 CREATING LAYOUTS USING COPYING

- These two methods allow you to create copies of an existing layout. The first method is similar to one used in Microsoft Excel. Do the following steps:
  - Select the desired layout.
  - Hold the [Ctrl] key on the keyboard then hold and drag the mouse to the new position of the newly copied layout.

- Rename the new layout.

- Another way to copy layouts is to select the desired layout and then right-click. A shortcut menu will appear. Select the **Move or Copy** option:

- You will see the following dialog box:

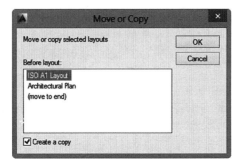

- A list of the current layouts is displayed; select one of them and click the checkbox **Create a copy** on. A copy will be created; you can rename it and make the necessary changes.
- Using the same option, you can move a layout from its current position to the left or right. Alternatively, you can move a layout without this command, by clicking the layout name then holding and dragging to the desired location.

NOTE *To re-arrange layouts position, simply hold the name in the tab click and drag.*

- By now, you have accomplished the following steps:
  - Created a new layout.
  - Created a Page Setup.
  - Linked a Page Setup to the layout.
- When you select the newly created layout, you will see something like the following:

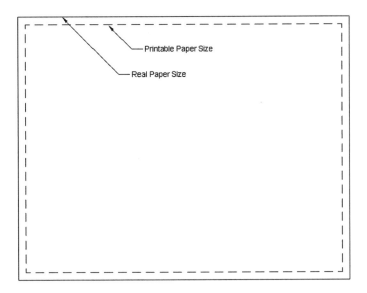

- The outer frame is the real paper size and the inside frame (the dashed line) is the printable paper size, which is the paper size minus the printer's margins. Based on this layout, you will see exactly what will be printed and what is not. Any object outside the dashed line will not be printed; this proves that printing from layouts is WYSIWYG (What You See Is What You Get).
- You can reach some of the commands previously mentioned if you go to the **Layout** tab (this tab will not be visible unless you are at one of the layouts), locate the **Layout** panel. You will see the following buttons:

- In this panel, you can create a new layout from scratch or from the template. The second button will access the page setup dialog box.

PRACTICE 9-1

Creating New Layouts

1. Start AutoCAD 2015.
2. Open the file, Practice 9-1.dwg.
3. Select Layout 1 and delete the existing viewport (select it from its border lines).
4. Rename the layout to be Architectural Plan (A3 Size).
5. Start Page Setup Manager for this layout.
6. Create a new page setup using the following information:
   a. Plotter = DWF6 ePlot.pc3
   b. Paper Size = ISO A3 (420.00 x 297.00 MM)
   c. What to plot = Layout
   d. Plot scale = 1:1
   e. Drawing orientation = Landscape
7. Click OK to end the creation process.
8. Click Close to close the dialog box.
9. Make the layer Title Block the current layer.
10. Insert <u>the file</u> called A3 Size Title Block.dwg to be your title block, using the insertion point of 0,0.
11. Create a copy of the newly created layout and name it Mechanical.
12. Delete Layout 2.
13. Right-click any existing layout and select From the template option.
14. Import ISO A1 Layout from Tutorial-mArch.dwt.
15. Move ISO A1 Layout to be the first layout after Model tab.
16. Save and close the file.

9.6 CREATING VIEWPORTS

- When you visit a layout for the first time (just after creation), you will see a single viewport appears at the center of the paper size. **Viewport** is a window containing the view of your Model Space, initially scaled to the size of the window.
- Viewports inserted in layouts can be tiled, scaled, and printed.
- You can add viewports to layouts using different methods:
  - Adding a single rectangular viewport
  - Adding multiple rectangular viewports
  - Adding a single polygonal viewport
  - Converting an object to be a viewport
  - Clipping an existing viewport

NOTE *Layout tab appears only when you are in a layout.*

9.6.1 Adding Single Rectangular Viewports

- This command allows you to add single rectangular viewports in a layout. You should specify two opposite corners to specify the area of the viewport. To issue this command, go to the **Layout** tab, locate the **Layout Viewports** panel, then select the **Rectangular** button:

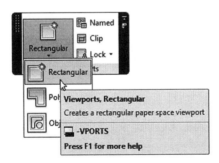

- The following prompts will be shown:

```
Specify corner of viewport or [ON/OFF/Fit/Shadeplot/
Lock/Object/Polygonal/Restore/LAyer/2/3/4] <Fit>:
Specify opposite corner:
```

- Specify the two opposite corners to create a single rectangular viewport.

- This is what you will get.

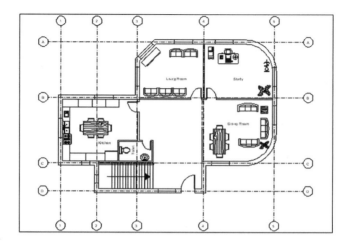

9.6.2 Adding Multiple Rectangular Viewports

- This command will add multiple rectangular viewports in a layout. You should specify two opposite corners to specify the area of the viewports. To issue this command, go to the **Layout** tab, locate the **Layout Viewports** panel, and then select the **Named** button. When you see the following dialog box, select the New Viewports tab:

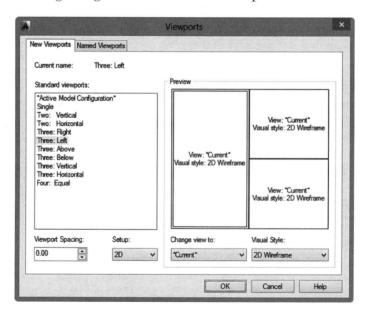

- Select the desired display from the list. By default, the **Viewport Spacing** value = 0 (zero), which means viewports will be tiled. If you want them separated, input a value greater than 0 (zero). Click **OK** and you will see the following prompt:

```
Specify first corner or [Fit] <Fit>:
Specify opposite corner:
```

- See the following illustration:

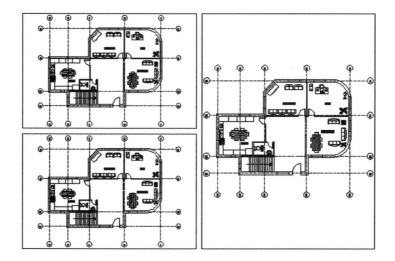

9.6.3 Adding Polygonal Viewport

- This command will add a polygonal viewport that consists of both straight lines and arcs. To start this command, go to the **Layout** tab, locate the **Layout Viewports** panel, then select the **Polygonal** button:

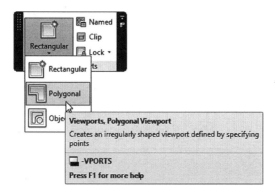

- You will see the following prompts:

  ```
  Specify start point:
  Specify next point or [Arc/Length/Undo]:
  Specify next point or [Arc/Close/Length/Undo]:
  ```

- It looks just like the Polyline command prompts.
- See the following:

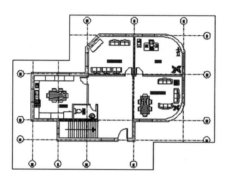

9.6.4 Creating Viewports by Converting Existing Objects

- This command allows you to convert any existing object to viewport. (It should be a single object, like a polyline or circle. Lines and arcs are not allowed.) To start this command, go to the **Layout** tab, locate the **Layout Viewports** panel, then select the **From Object** button:

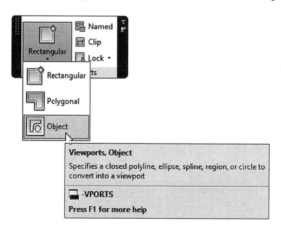

- You will see the following prompt:

  ```
  Select object to clip viewport:
  ```

- See the following:

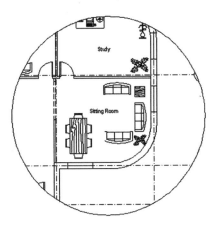

9.6.5 Creating Viewports by Clipping Existing Viewports

- This command allows you to clip an existing viewport, creating a new shape. To start this command, go to the **Layout** tab, locate the **Layout Viewports** panel, then select the **Clip** button:

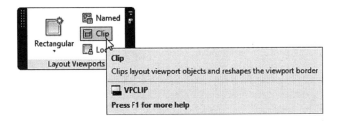

- The following prompt appears:

  ```
  Select viewport to clip:
  Select clipping object or [Polygonal] <Polygonal>:
  Specify start point:
  Specify next point or [Arc/Length/Undo]:
  Specify next point or [Arc/Close/Length/Undo]:
  ```

- First, select an existing viewport. You can draw a polyline previously then select it or you can draw any irregular shape using the **Polygonal** option (which is identical to Polygonal viewport command previously discussed). See the following:

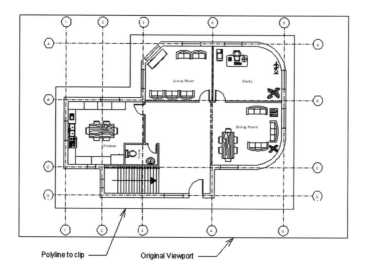

9.6.6 Dealing with Viewports After Creation

- There are two modes to work with viewports:
  - The first mode is: you are outside the viewport, which means you will deal with it as any other object in your drawing. You can select the viewport from its frame and erase, copy, move, scale, stretch, rotate, etc.

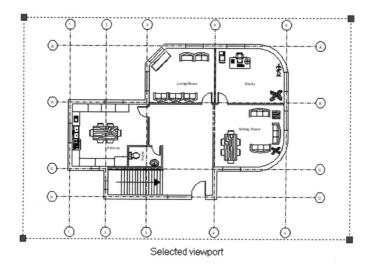

- The second mode is: you are inside the viewport. You achieve this by double-clicking inside the viewport. This mode will allow you to zoom, pan, scale, etc., the objects inside the viewport. To return to the first mode, double-click outside the viewport.

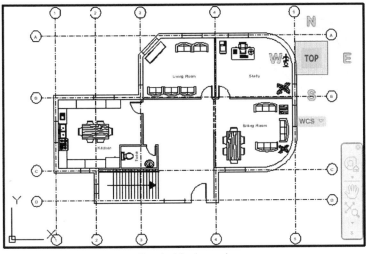

User inside viewport

9.7 SCALING AND MAXIMIZING VIEWPORTS

- By default when you insert a viewport, AutoCAD will zoom the whole drawing into the area you specified as the viewport size. This viewport is *not-to-scale*, and you should set the scale relative to the Model Space units. Do the following steps:
  - Double-click inside the desired viewport (or you can select only the viewport's frame).
  - Look at the right side of the Status bar, you will see Viewport Scale list as shown:

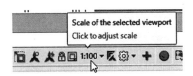

- Click the list which contains all scales, similar to the following:

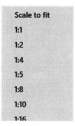

- Select the desired scale for your viewport. If you did not find your desired scale, select the **Custom** option. The following dialog box will be displayed:

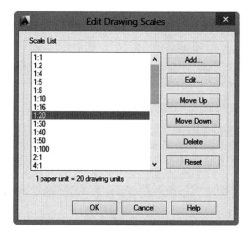

- To add a new scale, select the **Add** button. The following dialog box will be displayed:

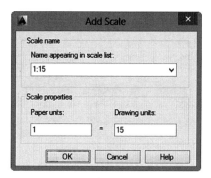

- Input the desired scale and then click **OK** twice.
- After the scaling process is finished, you will see whether the scale is perfect for the area of the viewport or it is either too big or small. The solution is to change the size of the viewport or to change the scale.
- After setting the scale, it is permissible to use the Pan command but not the Zoom command, because this will ruin the scale. To avoid this problem, you can lock the display of the viewport by selecting viewport(s), then clicking the golden opened lock in the status bar (you have to be inside the viewport or select the border of the viewport); the golden lock will change to blue lock, and the viewports will be locked.

- Maximize function will maximize the viewport to fit the size of the screen temporarily. This will give you the needed space to edit objects as you wish without leaving the layout and going to the Model space. Go to the status bar and click the **Maximize Viewport** button as shown:

- The same button will restore the original size of the viewport. Or, you can go to the **Layout** tab or **Layout Viewports** panel and use the following two buttons:

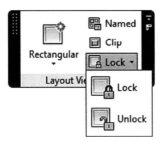

9.8 FREEZING LAYERS IN VIEWPORT

- The freezing function allows you to freeze layer(s) in Model space and all viewports in all layouts. If you want to freeze a layer in a certain viewport, you have to do the following steps:
  - Double-click the desired viewport.
  - Go to the **Home** tab, locate the **Layers** panel, select the layer list, and click the icon **Freeze or thaw in current viewport** for the desired layer. See the following:

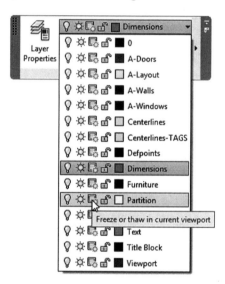

- Also, you have the ability to freeze/thaw layers in all viewports except the current viewport. To do this, start the Layer Properties Manager, select the desired layer(s), and then right-click. Select the **VP Freeze layer** then **In All Viewports Except Current** option, as shown:

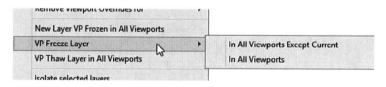

9.9 LAYER OVERRIDE IN VIEWPORT

- In all viewports, a layer will be displayed with the same color, linetype, lineweight, and plot style. You can change these setting in one viewport to display it differently. This is called layer override. Do the following steps:
    - Double-click inside the desired viewport.
    - Issue the Layer Properties Manager command
    - Under VP Color, VP Linetype, VP Lineweight, or VP Plot Style make the desired changes.
    - These changes will take place only in the current viewport.
- See the following:

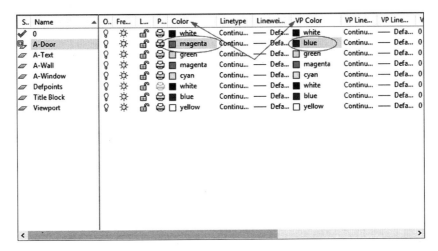

- In the previous example, the layer **A-Door** has a color (applied in the Model Space and all other viewports) that is equal to Magenta and an override color that equals blue in the current viewport.
- Also, check that the layer **A-Door** row is shaded with a different color.

PRACTICE 9-2

Creating and Controlling Viewports

1. Start AutoCAD 2015.
2. Open Practice 9-2.dwg.
3. Make the layer Viewport the current layer.
4. Switch to the D-Size Architectural Plan layout.
5. Insert a single viewport to fill the entire area.
6. Set the scale to ½" = 1', and lock the viewport.
7. Switch to the D-Size Arch Details layout.
8. Insert viewports using Three: Right, with Viewport Spacing = 0.25, and select an area to fill the paper size.
9. Select the borders of the big viewport at the right; scale it to be ¼" = 1'.
10. Using the grips, shrink the area of the viewport to something suitable for the scale chosen.
11. Select the two viewports at the left and set the scale to ¾" = 1'.
12. For the top viewport and using the Pan command, set the view to show the Kitchen and Toilet.
13. For the bottom viewport and using the Pan command, set the view to show the Living Room.
14. Select the three viewports and lock them.
15. Switch to ANSI B Size Architectural Plan layout.
16. Convert the big circle to be a viewport.
17. Double-click inside the new viewport and set the scale to ¾" = 1', then pan to show the Toilet.
18. Click inside the rectangle viewport and freeze in the current viewport the following layers: Centerline, Centerline-TAGS, and Dimensions.
19. Switch to the other layouts to make sure that these three layers were frozen only in this viewport.
20. Change the color of layer A-Walls in this viewport to red.
21. Save and close the file.

9.10 PLOT COMMAND

- This will be our final step. This command mission is to send whatever we sat in the layout to the specified plotter. To issue this command, go to **Output** tab, locate **Plot** panel, then select **Plot** icon:

- You will see the following dialog box:

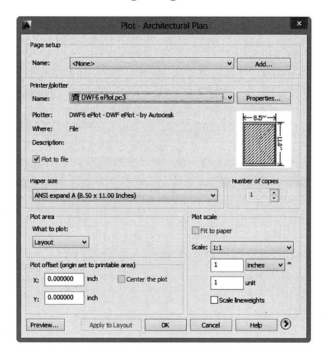

- This dialog box is identical to the Page setup settings. If you modify any of these settings, AutoCAD will separate the Page setup from the current layout. To send the layout to the plotter, click **OK**. To save the settings of this dialog box with the layout, select the **Apply to Layout** button. Click the **Preview** button to see the final printed drawing on the screen before the real printout.

- You can preview your drawing from outside this dialog box by going to the **Output** tab, locating the **Plot** panel, then selecting the **Preview** button:

NOTES:

CHAPTER REVIEW

1. Which of the following is true about creating a new layout?

    a. You can bring a layout from an existing template.

    b. You can create a new layout using the right-click menu at the name of an existing layout.

    c. You can click the (+) sign beside the last layout name at the right.

    d. All of the above.

2. Objects to be converted to a viewport should be _____ objects like Circle and Polyline.

3. Using the Pan command after setting the scale of a viewport will ruin the scaling process.

    a. True

    b. False

4. You can freeze a layer in a certain viewport. You can change the color of a layer in a certain viewport.

    a. The first statement is correct, but the second is wrong.

    b. The two statements are correct.

    c. The two statements are wrong.

    d. The first statement is wrong, yet the second is correct

5. _____ involves specifying plotter, page size, and orientation.

CHAPTER REVIEW ANSWERS

1. d

3. b

5. Page Setup

CHAPTER 10

PROJECTS

In This Chapter
- Prepare Drawing for New Project
- Architectural Projects
- Mechanical Projects

10.1 HOW TO PREPARE YOUR DRAWING FOR A NEW PROJECT

- In order to prepare your drawing for a new project, do the following steps:
    - Start a new drawing based on acad.dwt or acadiso.dwt.
    - Set up Drawing Units. From the Application Menu, select **Drawing Utilities/Units**.

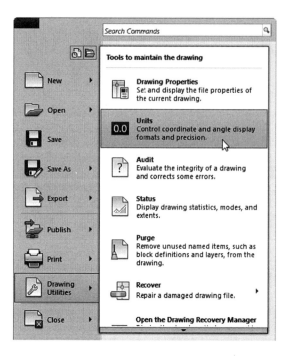

- You will see the following dialog box:

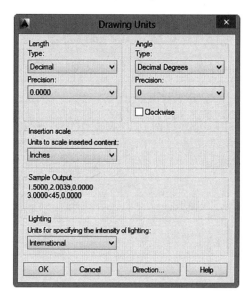

- Pick the desired **Length Type**, select one of the following: Architectural (example: 2'-4 8/16"), Decimal (example: 25.5697), Engineering

(example: 3'-5.6688"), Fractional (example: 16 3/16), and Scientific (example: 8.9643E+03).
- Pick the desired **Angle Type** then select one of the following: Decimal Degrees (example: 36.7), Deg/Min/Sec (example: 47d25'31"), Grads (example: 60.7g), Radians (example: 0.6r), Surveyor's Units (example: N 51d25'31" E).
- Setup the **Precision** for both length and angle units. For example, pick the number of decimals for Decimal units between no decimal places and eight decimal places.
- The Default AutoCAD setting for angles is Counterclockwise but you can switch to be **Clockwise.**
- Under the **Insertion scale**, specify **Units to scale inserted content**, (as discussed in Chapter 6).
- Select the **Direction** button, you will see the following dialog box:

- Change the East to be 0 (zero) angle, which changes the other angles as well (it is recommended to leave this setting as is).
- This step will end the Units command.
- Setup Drawing Limits. Drawing limits is your working area. You will specify it using two opposite corners, the lower-left corner and upper- right corner. To setup the drawing limits right, answer these two questions: what is the longest dimension in the drawing in both X and Y? What AutoCAD unit is being used (meter, centimeter, inch, or foot)?
- If the menu bar is showing, choose **Format/Drawing Limits** or you can type **limits** in the command window. You then see the following prompts:

```
Specify lower left corner or [ON/OFF] <0,0>:
Specify upper right corner <12,9>:
```

- Specify the lower-left corner and right-upper corner by typing or clicking. To forbid yourself from using any area outside these limits, AutoCAD gives you the ability to turn it on and off.
- Create layers.
- Start drafting.

10.2 ARCHITECTURAL PROJECT (IMPERIAL)

- Do the following steps:

1. From the designated folder, open the file Ground Floor_Starter.dwg.
2. Switch off the grid.
3. Set the units to the following:

    a. Architectural

    b. Precision 0'- ½".

    c. Units to scale inserted contents = inch

4. Set up the drawing limits:

    a. Lower-left corner = 0,0

    b. Upper-right corner = 50',50'

5. Double-click the mouse wheel in order to see the new limits.
6. Create the following layers:

Layer Name	Layer Color
A-Door	Blue
A-Wall	White
A-Window	White
Dimension	Red
Furniture	White
Staircase	Magenta
Text	Green
Title Block	White
Viewport	9
Hatch	8

7. Save your file in Chapter 10\Imperial folder and name it Ground Floor.dwg.

8. Make the layer A-Wall current.

9. Draw the following architectural plan and the partitions inside using these guidelines:

   a. Draw the outer shape using polyline.

   b. Offset it to the inside using 6" as the distance.

   c. Explode the two polylines.

   d. Use the outer wall to draw the inner walls using all the commands you learned in this book. The inner wall is 4".

10. This is the architectural plan:

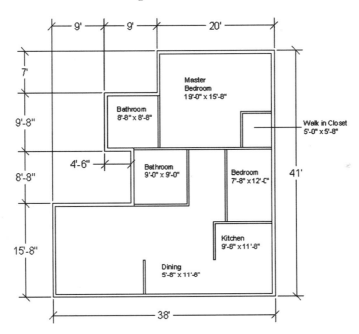

11. Create a 36" door opening as follows (you can always take 4" clearance from the wall). The main entrance, master bedroom, and walk-in closet door are in the middle of the wall.

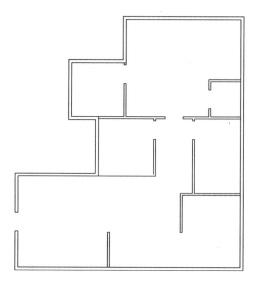

12. Make the layer 0 (zero) current.

13. Create the following door blocks using the name beneath (base point is the lower-left point of jamb):

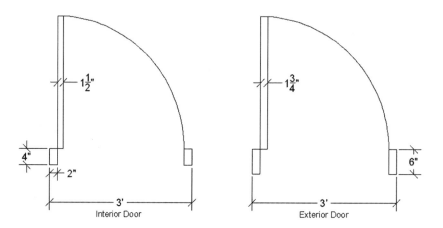

14. Create the following door block using the name beneath (base point is the lower-left point of jamb):

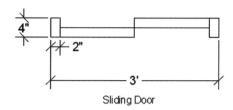

Sliding Door

15. Create the following window blocks using the name beneath (base point is the lower-left point of jamb):

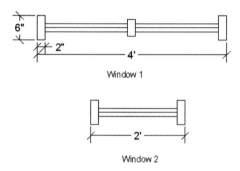

Window 1

Window 2

16. Insert the doors and windows in their respective layers to get the following result:

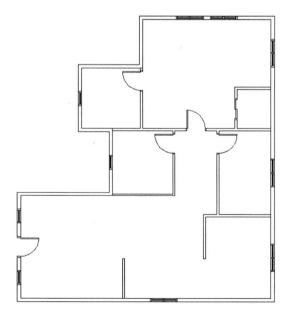

17. Make layer Furniture current.

18. Using the Insert command insert the following blocks.

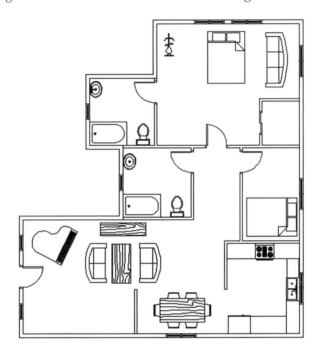

19. Make layer Hatch current.

20. Using Solid hatching, hatch both the outside and the inside wall.

21. Using ANSI37 and scale = 100, hatch the kitchen (hint: draw a line to separate the kitchen from the adjacent room).

22. Using your defined hatch (switch Double checkbox on) and scale = 20, hatch both bathrooms.

23. You should have something like the following:

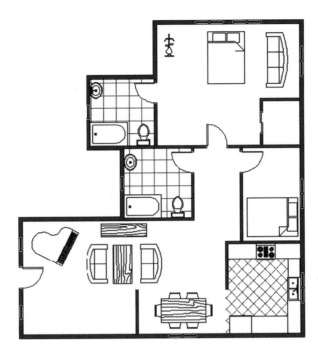

24. Make layer Text current and freeze layer Hatch.

25. Make Room Titles the current text style then write text using Multiline text to add the room titles. As in the following, Justify Middle Center:

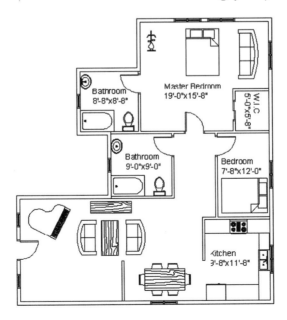

26. Thaw layer Hatch.

27. Select the hatch of one of the two bathrooms. When the context tab appears, locate the **Boundaries** panel, click the **Select** button, and then choose the text. Press [Enter] and then press [Esc]. Do the same for other bathroom and the kitchen.

28. Make the Outside Walls the current dimension style.

29. Make layer Dimension current.

30. Insert the dimensions as shown in the following (use the Continue command whenever possible):

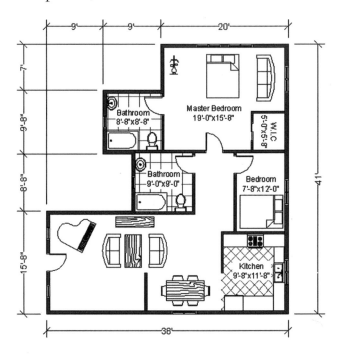

31. Go to Layout1 and rename it to Full Plan.

32. Using the Page Setup Manager modify the existing page setup to be as follows:

　a. Printer = DWF6 ePlot.pc3

　b. Paper = ANSI B (17x11 inch)

　c. Drawing orientation = Landscape

33. Erase the existing viewport.

34. Make layer Title Block current.
35. Insert the file ANSI B Landscape Title Block.dwg in the layout, using 0,0,0 insertion point.
36. Create a copy of the layout and name it Details.
37. Erase Layout 2.
38. Go to layout1 named Full Plan.
39. Make layer Viewport current.
40. Insert a single viewport to fill the space and set the viewport scale to be 3/16" = 1', then lock the viewport.
41. Go to the layout named Details.
42. Create a single viewport to occupy half of the space of the paper.
43. Set the scale to be ½" = 1", and lock the viewport.
44. Pan to the entrance making sure to show the dimension, the two windows, and the door.
45. You are still in Details layout. Create another viewport to occupy half of the remaining area of the paper.
46. Set the scale to ¼" = 1', then lock the viewport.
47. Double-click inside the viewport and pan to the two windows of the master bedroom.
48. Save and close the file.

10.3 ARCHITECTURAL PROJECT (METRIC)

- Do the following steps:
1. From the designated folder, open the file Ground Floor_Starter.dwg.
2. Switch off the grid.
3. Set the units to the following:
    a. Decimal
    b. Precision = 0
    c. Units to scale inserted contents = Millimeters

4. Setup drawing limits to be:

    a. Lower left corner = 0,0

    b. Upper right corner = 15000,15000

5. Double-click the mouse wheel in order to see the new limits.

6. Create the following layers:

Layer Name	Layer Color
A-Door	Blue
A-Wall	White
A-Window	White
Dimension	Red
Furniture	White
Staircase	Magenta
Text	Green
Title Block	White
Viewport	9
Hatch	8

7. Save your file in Chapter 10\Metric folder and name it Ground Floor.dwg.

8. Make layer A-Wall current.

9. Draw the following architectural plan and the partitions inside using the following guidelines:

    a. Draw the outer shape using polyline.

    b. Offset it to the inside using 150 as distance.

    c. Explode the two polylines.

    d. Use the outer wall to draw the inner walls using all the commands you learned in this book. The inner wall is 100.

10. This is the architectural plan:

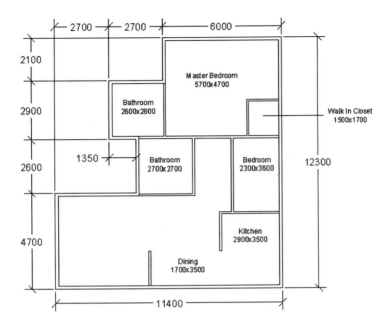

11. Create a 900 door opening as follows (you can always take 100 clearance from the wall). The main entrance, master bedroom, and walk-in closet door are in the middle of the wall.

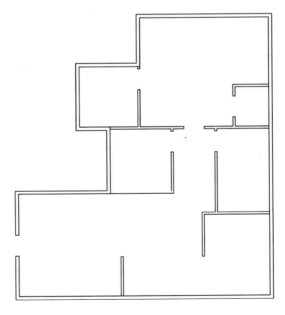

12. Make layer 0 (zero) current.

13. Create the following door blocks using the name beneath (base point is the lower-left point of jamb):

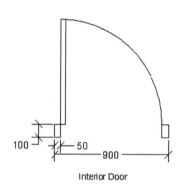
Interior Door

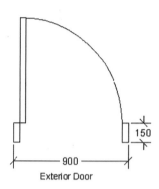
Exterior Door

14. Create the following door block using the name beneath (base point is the lower-left point of jamb):

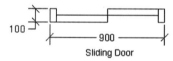
Sliding Door

15. Create the following window blocks using the name beneath (base point is the lower-left point of jamb):

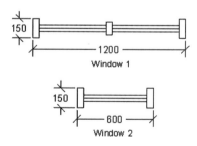

16. Insert the doors and windows in their respective layers to get the following result:

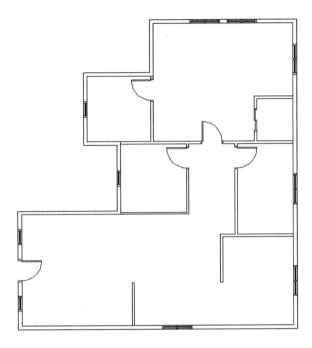

17. Make layer Furniture current.

18. Using the Insert command insert the following blocks.

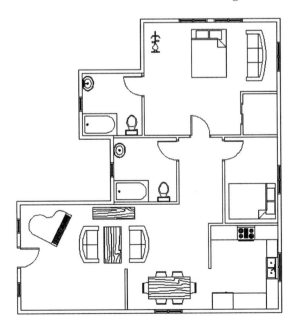

19. Make layer Hatch current.

20. Using the Solid hatching, hatch both the outside and the inside wall.

21. Using ANSI37 and scale = 2000, hatch the kitchen (hint: draw a line to separate the kitchen from the adjacent room).

22. Using the User defined hatch (switch Double checkbox on) and scale = 500, hatch both bathrooms.

23. You should have something similar to the following:

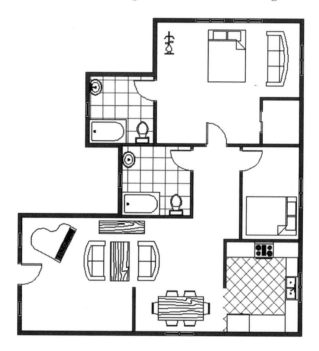

24. Make layer Text current and freeze layer Hatch.

25. Make Room Titles the current text style.

26. Write text using the Multiline text to add the room titles just like the following and Justify Middle Center:

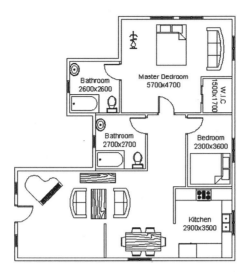

27. Thaw layer Hatch.

28. Select the hatch of one of the two bathrooms. When the context tab appears, locate the **Boundaries** panel, and click the **Select** button, choose the text, press [Enter], and then press [Esc]. Do the same for other bathroom and the kitchen.

29. Make the Outside Walls the current dimension style.

30. Make layer Dimension current.

31. Insert the dimensions as shown (use the Continue command whenever possible):

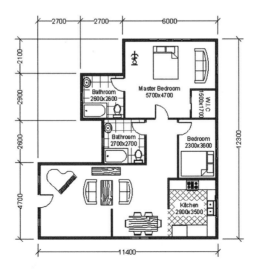

32. Go to Layout1 and rename it to Full Plan.
33. Using the Page Setup Manager modify the existing page setup to the following:
    a. Printer = DWF6 ePlot.pc3
    b. Paper = ISO A3 (420x297 MM)
    c. Drawing orientation = Landscape
34. Erase the existing viewport.
35. Make layer Title Block current.
36. Insert the file ISO A3 Landscape Title Block.dwg in the layout, using 0,0,0 insertion point.
37. Create a copy of the layout and name it Details.
38. Erase Layout 2.
39. Go to the layout named Full Plan.
40. Make layer Viewport current.
41. Insert a single viewport to fill the space and set the viewport scale to be 1:100, then lock the viewport.
42. Go to layout named Details.
43. Create a single viewport to occupy half of the space of the paper.
44. Set the scale to be 1:20 and lock the viewport.
45. Pan to the entrance making sure to show the dimension, the two windows, and the door.
46. Make layer Viewport current.
47. You are still in the Details layout. Create another viewport to occupy half of the remaining area of the paper.
48. Set the scale to 1:40 then lock the viewport.
49. Double-click inside the viewport and pan to the two windows of the master bedroom.
50. Save and close the file.

10.4 MECHANICAL PROJECT – I (METRIC)

- Do the following steps:
1. From the designated folder, open the file Mechanical-1_Starter.dwg.
2. Switch off the grid.
3. Set the units to the following:
   a. Decimal
   b. Precision 0
   c. Units to scale inserted contents = Millimeters
4. Set up drawing limits to be:
   a. Lower left corner = 0,0
   b. Upper right corner = 350,250
5. Double-click the mouse wheel in order to see the new limits.
6. Create the following layers:

Layer Name	Layer Color	Layer Linetype
Centerline	Green	Centerx2
Dimension	Red	Continuous
Hatch	8	Continuous
Hidden	Cyan	Hiddenx2
Part	White	Continuous
Text	Magenta	Continuous
Title Block	White	Continuous
Viewport	9	Continuous

7. Save your file in Chapter 10\Metric folder and name it: Mechanical–1.dwg.
8. Make layer Part current.
9. Draw the below plan, section, and elevation of the mechanical part using the following guidelines:
   a. All lines of the shape in layer Part.
   b. All centerlines in layer Centerline.

c. All hidden lines in layer Hidden.

d. Change the **Linetype scale** using **Properties** palette for both centerlines and hidden lines to be 5, except the two holes at the right and left make the Linetype scale to be 2.

10. Draw the shape without dimensioning for now:

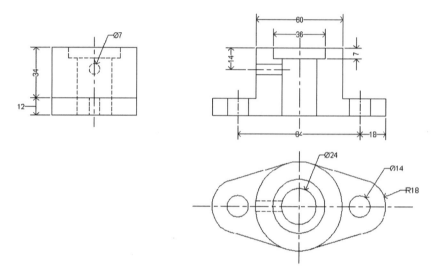

11. Make layer Hatch current.

12. Using ANSI31 with scale = 20, hatch the shape as shown:

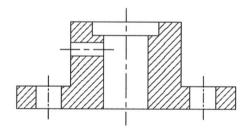

13. Make Part Dim the current dimension style.

14. Make layer Dimension current then insert the dimensions as shown:

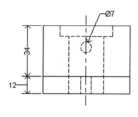
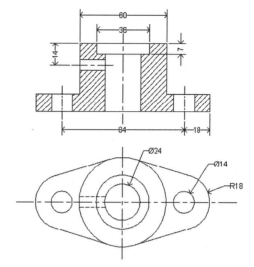

15. Go to Layout1 and rename it to Plan.
16. Using Page Setup Manager modify the existing page setup to be as follows:
    a. Printer = DWF6 ePlot.pc3
    b. Paper = ISO A3 (420x297 MM)
    c. Drawing orientation = Landscape
    d. Make sure Scale = 1:1
17. Erase the existing viewport.
18. Make layer Title Block current.
19. Insert the file ISO A3 Landscape Title Block.dwg in the layout, using 0,0,0 insertion point.
20. Make two copies of Plan layout and rename it Section and Elevation.
21. Delete Layout 2.
22. Make layer Viewport current.
23. Insert a single viewport like the following:
    a. Set the scale to 2:1.
    b. Lock the viewport.

c. Pan to the shape as shown.

d. You will notice that hidden lines and centerlines look like continuous lines. To solve this problem, type **psltscale** in the command window command and set this variable to 0. Then type the **regenall** command to regenerate all viewports. You will see the following result:

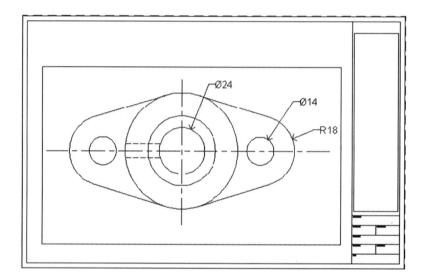

24. Repeat the same procedure to create the section viewport in Section layout and elevation viewport in Elevation layout.

25. Save and close the file.

10.5 MECHANICAL PROJECT – I (IMPERIAL)

- Do the following steps:
1. From the designated folder, open the file Mechanical-1_Starter.dwg.

2. Switch off the grid.

3. Set the units to the following:

   a. Fractional

b. Precision 0 – 1/16

c. Units to scale inserted contents = Inches

4. Set up these drawing limits:

    a. Lower left corner = 0,0

    b. Upper right corner = 18",9"

5. Double-click the mouse wheel in order to see the new limits.

6. Create the following layers:

Layer Name	Layer Color	Layer Linetype
Centerline	Green	Center2
Dimension	Red	Continuous
Hatch	8	Continuous
Hidden	Cyan	Hidden2
Part	White	Continuous
Text	Magenta	Continuous
Title Block	White	Continuous
Viewport	9	Continuous

7. Save your file in Chapter 10\Imperial folder and name it: Mechanical–1.dwg.

8. Make layer Part current.

9. Draw the following plan, section, and elevation of the mechanical part using the following guidelines:

    a. All lines of the shape in layer Part.

    b. All centerlines in layer Centerline.

    c. All hidden lines in layer Hidden.

    d. Change the **Linetype scale** using **Properties** palette to be 0.5 for any line you like.

10. Draw the shape without dimensions for now:

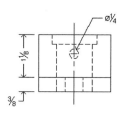
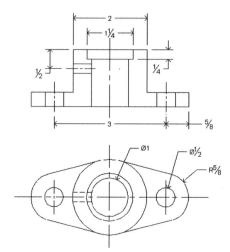

11. Make layer Hatch current.

12. Using ANSI31 with scale = 1, hatch the shape as shown:

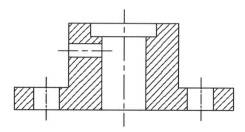

13. Make Part Dim the current dimension style.

14. Make layer Dimension current then insert the dimensions as shown:

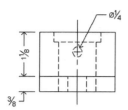
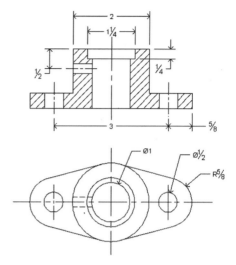

15. Go to Layout1 and rename it to Details.

16. Using Page Set up Manager modify the existing page setup to be as follows:

   a. Printer = DWF6 ePlot.pc3

   b. Paper = ANSI B (17x11 in)

   c. Drawing orientation = Landscape

   d. Make sure Scale = 1:1

17. Erase the existing viewport.

18. Make layer Title Block current.

19. Insert the file ANSI B Landscape Title Block.dwg in the layout, using 0,0,0 insertion point.

20. Make two copies of Plan layout and rename it Section and Elevation.

21. Delete Layout 2.

22. Make layer Viewports current.

23. Insert a three single viewport like the following, doing the following steps:

   a. Set the scale for the three viewports to be 1' = 1'.

   b. Lock the viewports.

c. Pan to the shape as shown.

d. If you noticed that hidden lines and centerlines look like continuous lines, type **psltscale** in the command window and set this variable to 0. Then type the **regenall** command to regenerate all viewports. You will get the following result:

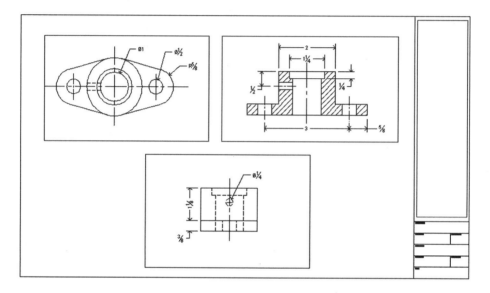

24. Save and close the file.

10.6 MECHANICAL PROJECT – II (METRIC)

- Using the same methodology, we used in Mechanical Project – I (Metric) draw the following project using Mechanical-1_Starter.dwg:

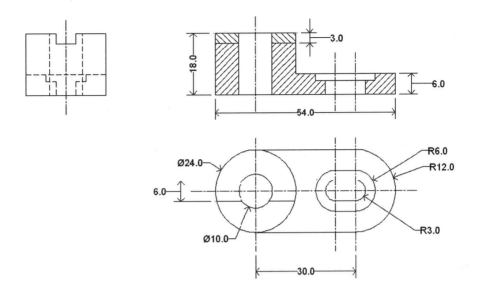

10.7 MECHANICAL PROJECT – II (IMPERIAL)

- Using the same methodology we used in Mechanical Project – I (Imperial), draw the following project using Mechanical-1_Starter.dwg:

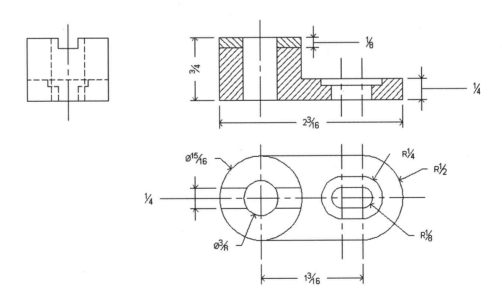

NOTES:

CHAPTER 11

MORE ON 2D OBJECTS

In This Chapter

- Polyline command and other drafting and editing commands
- How to use both Constructions line and Ray commands
- How to use a point with different styles, along with Divide and Measure
- Using the Spline and Ellipse
- Using the Boundary and Region commands along with Boolean operations

11.1 INTRODUCTION

- This chapter is dedicated to the rest of the 2D objects not discussed in Chapter 2. It starts with discussion of the Polyline command then covers all the other 2D objects based on polyline such as the special features of polylines in some of the editing commands. The chapter also discusses other 2D commands such as Spline and Ellipse. These two commands have some unique features that allow you to draw exact 2D curves. Other commands such as Point (Divide and Measure) and Revision Cloud are discussed as well. The close of this chapter focuses on the Boundary and Region commands.

11.2 DRAWING LINES AND ARCS USING THE POLYLINE COMMAND – REVISION

- The Polyline command will do any of the following:
  - Draw both line segments and arc segments.
  - Draw single object in the same command rather than drawing segments of lines and arcs, like in Line and Arc commands.
  - Draw lines and arcs with starting and ending widths.
- To use this command, go to the **Home** tab, locate the **Draw** panel, then select the **Polyline** button:

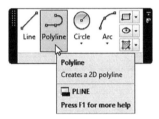

- The following prompt appears:

```
Specify start point:
Current line-width is 1.0000
Specify next point or [Arc/Halfwidth/Length/Undo/ Width]:
```

- AutoCAD asks you to specify the first point; when you do, AutoCAD reports to you the current line-width. If you like it, continue specifying points using the same method previously learned in regard to the Line command. If not, change the width as a first step, type letter **W**, or right-click and select **Width** option. You then see the following prompt:

```
Specify starting width <1.0000>:
Specify ending width <1.0000>:
```

- Specify the starting width, press [Enter], and then specify the ending width. Next time when you use the same file, AutoCAD will report these values for you when you issue the Polyline command. Halfwidth is the same, but instead of specifying the full width, you specify halfwidth.
- The Undo, and Close options are identical to the ones at the Line command.
- Length will specify the length of the line using the angle of the last segment.

- Arc will draw an arc attached to the line segment. You will see the following prompt:

```
Specify endpoint of arc or [Angle/CEnter/CLose/ Direction/ Halfwidth/Line/Radius/Second pt/Undo/Width]:
```

- Arc is attached to the last segment of the line or the first object in a Polyline command, using either method. The first point of the arc is already known, so you need two more pieces. AutoCAD will make an assumption (which you have the right to reject); AutoCAD will also assume that the angle of the last line segment will be considered the direction (tangent) of the arc. If you accept this assumption, you should specify the endpoint. If not, choose from the following to specify the second piece of information:
  - The Angle of the arc
  - The Center point of the arc
  - Another Direction to the arc
  - The Radius of the arc
  - The Second point which can be any point on the parameter of the arc
- Based on the information selected as the second point, AutoCAD will ask you to supply the third piece of information.

11.3 CONVERTING POLYLINES TO LINES AND ARCS, AND VICE-VERSA

- This is a very essential technique that converts any polyline to lines and arcs, and convert lines and arcs to polylines.

11.3.1 Converting Polylines to Lines and Arcs

- The Command **Explode** allows you to explode a polyline to lines and arcs. To issue this command, go to the **Home** tab, locate the **Modify** panel, then select the **Explode** button:

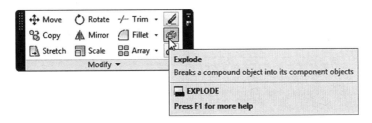

- AutoCAD shows the following prompt:

```
Select objects:
```

- Select the desired polylines and press [Enter] when done. Check the new shape; you will discover it is lines and arcs.

11.3.2 Joining Lines and Arcs to Form a Polyline

- This section discusses an option called **Join** within a command called **Edit Polyline**. To issue this command, go to the **Home** tab, locate the **Modify** panel, then select the **Edit Polyline** button:

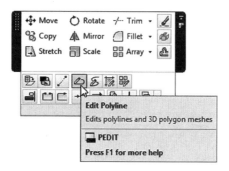

- You will see the following prompts:

```
Select polyline or [Multiple]:
Object selected is not a polyline
Do you want to turn it into one? <Y>
Enter  an  option  [Close/Join/Width/Edit  vertex/Fit/
Spline/Decurve/Ltype gen/Reverse/Undo]: J
```

- Start first by selecting one of the lines or the arcs you want to convert. AutoCAD will respond by telling you that the selected object is not a polyline and giving you the option to convert this specific line or arc to a polyline. If you accept this, options will appear and one of these options will be **Join**. Select the **Join** option then select the rest of the lines and arcs. At the end, press [Enter] twice. The objects were converted to a polyline.

PRACTICE 11-1

Drawing Polylines and Converting

1. Start AutoCAD 2015.
2. Open the file, Practice 11-1.dwg.
3. Draw the following polyline using the start point of 18,5 and width = 0.1.

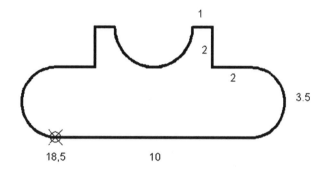

4. Explode the polyline. As evidence, width will disappear.
5. Check the objects after exploding; they are lines and arcs.
6. Save and close the file.

11.4 DRAWING USING THE RECTANGLE COMMAND

- This command allows you to draw a rectangle or square shape. The Rectangle command will use polyline as an object. To issue this command, go to the **Home** tab, locate the **Draw** panel, then select the **Rectangle** button:

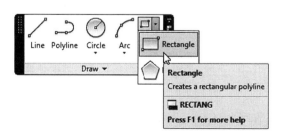

- You see the following prompts:

  ```
  Specify first corner point or [Chamfer/Elevation/ Fillet /
  Thickness/Width]:
  Specify other corner point or [Area/Dimensions/ Rota-
  tion]:
  ```

- By default, you can draw a rectangle by specifying two opposite corners. The other options are the following:

11.4.1 Chamfer Option

- This option allows you to draw a rectangle with chamfered edges. You then see the following prompts:

  ```
  Specify first chamfer distance for rectangles <0.00>:
  Specify second chamfer distance for rectangles <0.2>:
  ```

- Specify the first and second distance, then the Rectangle command will continue with normal prompts. See the following example:

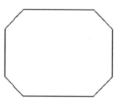

11.4.2 Elevation Option

- This option is dedicated for 3D only.

11.4.3 Fillet Option

- This option is identical to the Chamfer option, except you have to input the Radius instead of the Distance. You will see the following prompt:

  ```
  Specify fillet radius for rectangles <0.0000>:
  ```

- Specify the fillet radius then the Rectangle command will continue with its normal prompts. See the following example:

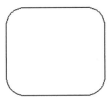

11.4.4 Thickness Option

- This option is dedicated for 3D only.

11.4.5 Width Option

- This option allows you to draw a rectangle with width. You then see the following prompt:

```
Specify line width for rectangles <0.0000>:
```

- Specify the width value and the rectangle command will continue normally. See the following example:

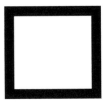

11.4.6 Area Option

- This option allows you to specify the total area of the rectangle, prior to specifying the second corner. You then see the following prompts:

```
Enter area of rectangle in current units <25.0000>:
Calculate rectangle dimensions based on [Length/Width] <Length>:
Enter rectangle length <10.0000>:
```

- Here AutoCAD asks to input the total area and then asks to input either the length (in X-axis) or width (in Y-axis). AutoCAD then draws a rectangle above and to the right of the first corner.

11.4.7 Dimensions Option

- This option allows you to draw a rectangle by specifying length (in X-axis) and width (in Y-axis). You then see the following prompts:

```
Specify length for rectangles <10.0000>:
Specify width for rectangles <10.0000>:
Specify other corner point or [Area/Dimensions/ Rotation]:
```

- Here AutoCAD asks you to input the length and width. The final prompt asks you to input the position of the second point.

11.4.8 Rotation Option

- This option allows you to draw a rectangle with a rotation angle. You then see the following prompt:

```
Specify rotation angle or [Pick points] <0>:
```

- Specify the rotation angle either by typing the value or by specifying points.

11.5 DRAWING USING THE POLYGON COMMAND

- This command allows you to draw an equilateral polygon with 3 sides to 1024 sides. Polygon uses polyline as an object. To issue this command, go to the **Home** tab, locate the **Draw** panel, then select the **Polygon** button:

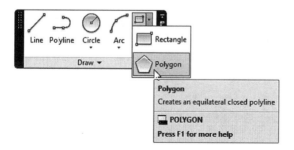

MORE ON 2D OBJECTS • 321

- You will see the following prompt:

  ```
  Enter number of sides <6>:
  ```

- Input the number of sides for your polygon. Then, you see the following prompt:

  ```
  Specify center of polygon or [Edge]:
  ```

- AutoCAD offers two methods to draw a polygon. Either by using an imaginary circle or by specifying the length and angle of one of the sides.

11.5.1 Using an Imaginary Circle

- This method depends on an imaginary circle. The polygon is either inscribed inside it or circumscribed about it. The center of the circle and the polygon coincide, hence the radius of the circle will decide the size of the polygon. The question is: when should you use this or that method? The answer to this question is the available information. If you know the distance between one of the edges and the polygon's center, then you can use the Inscribed option. But, if you know the distance between the midpoint of one of the edges and the center, then the Circumscribed option is the solution. See the following picture.

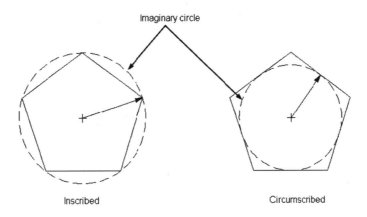

- You will see the following prompts:

  ```
  Enter an option [Inscribed in circle/Circumscribed about circle] <I>:
  Specify radius of circle:
  ```

11.5.2 Using Length and Angle of One of the Edges

- If you do not know the center of the polygon, then you cannot use the preceding method. Alternatively, if you specify the length of one side, the other side's length will be known automatically. While you are specifying the two points as a length of one of the sides, you are also specifying the angle of this side. Accordingly, the angles of the other sides will be defined. You will see the following prompts:

```
Enter number of sides <4>:
Specify center of polygon or [Edge]:
Specify first endpoint of edge:
Specify second endpoint of edge:
```

PRACTICE 11-2

Drawing Rectangles and Polygons

1. Start AutoCAD 2015.

2. Open the file, Practice 11-2.dwg.

3. Use the Rectangle and Polygon commands to complete the practice. Use OSNAP = Node to select the two points. You will be need the commands in the upper Right such as Explode, Extend, and Trim to get the correct results:

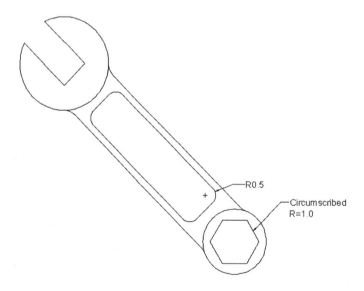

4. Freeze layer Points.

5. Save and close the file.

11.6 DRAWING USING THE DONUT COMMAND

- This command allows you to draw either a circle with width or filled circle. Donut uses polyline as an object. To issue this command, go to the **Home** tab, locate the **Draw** panel, then select the **Donut** button:

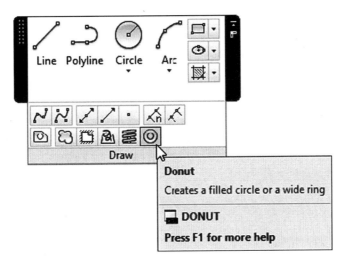

- The following prompt will appear:

```
Specify inside diameter of donut <0.5000>:
Specify outside diameter of donut <1.0000>:
Specify center of donut or <exit>:
```

- AutoCAD is asking you to input the inside and outside diameter, then to specify the center of the donut. You can insert as many donuts as needed.

11.7 DRAWING USING THE REVISION CLOUD COMMAND

- This command allows you to draw the revision cloud using polyline arcs. To issue this command, go to the **Home** tab, locate the **Draw** panel, then select the **Revision Cloud** button:

- You will see the following prompts:

```
Minimum arc length: 15 Maximum arc length: 15 Style:
Normal
Specify start point or [Arc length/Object/Style] <Object>:
```

- You should input the minimum and maximum arc lengths then specify the revision cloud style. If you select the **Arc length**, you will see the following prompt:

```
Specify minimum length of arc <15>:
Specify maximum length of arc <30>:
```

- If you select Style, you will see the following prompt:

```
Select arc style [Normal/Calligraphy] <Normal>:
```

- To know the difference between the two styles, see the following illustration:

- When these two settings are done, simply specify the first point of the revision cloud, then move your mouse (do not click) to the desired direction. AutoCAD will close the shape automatically and end the command once you get closer to the start point.
- Another way to draw a revision cloud is by converting a closed 2D object (circle, polyline, ellipse, etc.). You will see the following prompts:

```
Select object:
Reverse direction [Yes/No] <No>:
```

PRACTICE 11-3

Drawing Using Donut and Revision Cloud

1. Start AutoCAD 2015.
2. Open the file, Practice 11-3.dwg.
3. Using the horizontal line and the vertical lines insert donuts with inside diameter = 0 and outside diameter = 0.25.
4. Freeze Scratch layer.
5. Make Redline the current layer.

6. Using the Revision Cloud command, change the min and max arc length to be 1. Then, draw and convert the circle to get the following result:

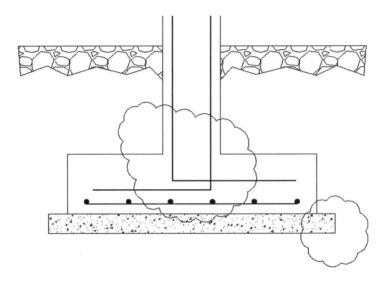

7. Save and close the file.

11.8 USING THE EDIT POLYLINE COMMAND

- This is a special editing command that can deal only with polylines. It can do certain things that the normal modifying commands cannot. You learned about some of the power in the previous discussion on converting lines and arcs to polylines. To issue this command, go to the **Home** tab, locate the **Modify** panel, then select the **Edit Polyline** button:

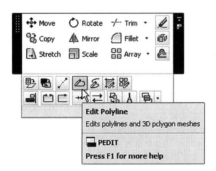

- Another way is to double-click the polyline. You will see the following prompts:

```
Select polyline or [Multiple]:
Enter an option [Close/Join/Width/Edit vertex/Fit/
Spline/Decurve/Ltype gen/Reverse/Undo]:
```

- As you can see, AutoCAD is asking to select a single polyline to perform one of many editing options. If not, AutoCAD is able to deal with **Multiple** polylines. The following section focuses on editing options for single polyline then shifts to options for multiple polylines:

11.8.1 Open and Close Options

- Open option will be displayed if the polyline selected is closed and vice-versa. There are no prompts for these two options, as AutoCAD remembers the last segment drawn, and will erase it to create an opened polyline.

11.8.2 Join Option

- This option allows you to join lines and arcs to the first selected polyline. You then see the following prompt:

```
Select objects:
```

- You are invited to select objects to join them to the selected polyline.

11.8.3 Width Option

- This option gives a width to the selected polyline. You then see the following prompt:

```
Specify new width for all segments:
```

11.8.4 Edit Vertex Option

- This option will select a vertex in the polyline then perform an editing option on this vertex. Many AutoCAD users agree this option is very lengthy, tedious, and difficult. Instead, you can explode the polyline, perform all/any normal modifying commands, and then join the lines and arcs in a single polyline. You will see the following prompt:

```
[Next/Previous/Break/Insert/Move/Regen/Straighten/
Tangent/Width/eXit] <N>:
```

NOTE *You can use Grips and Properties palette to edit vertices.*

11.8.5 Fit, Spline, and Decurve Options

- Fit and Spline both allow you to convert a straight-line segment polyline to curved polyline using two different methods. See the following illustration:

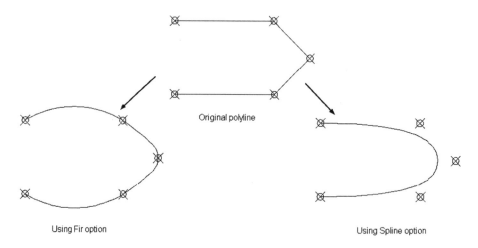

- As you can see from the preceding illustration, each command is handling the process in a different way:
  - The **Fit** option allows you to use the same vertices and connects them using a curve. It is considered to be an approximate method.
  - The **Spline** option allows you to use the vertices as controlling points to draw the needed curve. This method displays a more accurate curve.
- The **Decurve** option allows you to convert back from a curved shape polyline to straight lines polyline.

11.8.6 Ltype gen Option

- This option allows you to convert polylines from straight lines to curve to retain the original linetype. See the following illustration:

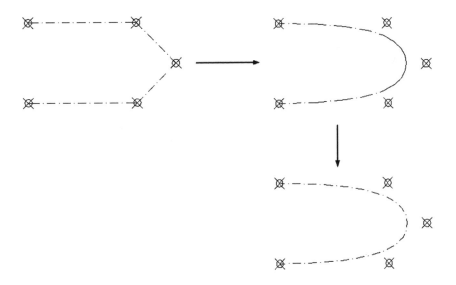

- You will see the following prompt:

    ```
    Enter polyline linetype generation option [ON/OFF] <Off>:
    ```

- In order to retain the linetype, input ON as an answer for this prompt.

11.8.7 Reverse Option

- This option allows you to reverse the order of vertices in a polyline. It will be evident using a special linetype like the following:

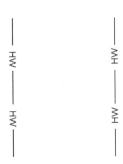

- There is a system variable PLINEREVRESEWIDTHS that controls whether to reverse the polyline width or not. If the value is 0 (zero), the polyline will not be reversed, but if the value is 1, the polyline width will be reversed. See the following illustration:

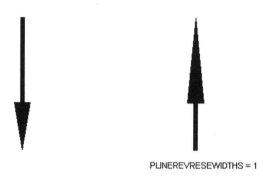

PLINEREVRESEWIDTHS = 1

11.8.8 Multiple Option

- This command allows you to modify multiple polylines when using the single modifying command. Another mission for this command is to join the polylines together. This is different than the join option previously discussed, which is for single polyline joining lines and arcs. This option allows you to join polylines together in a single polyline. You then see the following prompts:

```
Select objects:
Select objects:
Enter an option [Close/Open/Join/Width/Fit/Spline /
Decurve/Ltype gen/Reverse/Undo]:
```

- These prompts are identical to editing a single polyline, except for the Edit Vertex option. If you select the Join option, you see the following prompt:

```
Join Type = Extend
Enter fuzz distance or [Jointype] <0.0000>:
```

- The first line is showing you the current value for Join Type, which is Extend. In order to change it, invoke the **Jointype** option to see the following prompt:

```
Enter join type [Extend/Add/Both] <Extend>:
```

- There are three types of joining. They are:
  - Extend: AutoCAD will extend the two ends to each other to join the multiple polyline.
  - Add: AutoCAD will add a line between the two ends.
  - Both: AutoCAD will join both methods.
- AutoCAD also needs to know the **Fuzz distance**, which is the maximum acceptable distance between the ends of the two polylines to join. AutoCAD will not join polylines greater than this value. See the following illustration:

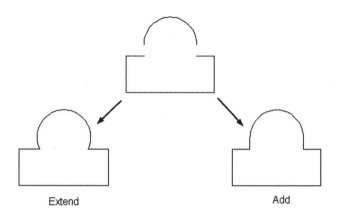

- One final note for the Polyline Edit command, you may tend to cut steps while drafting and editing but AutoCAD can help you with this. System variable **PEDITACCEPT** can be used to assume the answer to the following prompt while selecting the first object in Polyline command to always be yes:

```
Object selected is not a polyline
Do you want to turn it into one? <Y>
```

- PEDITACCEPT has two values:
  - 0 (zero): AutoCAD asks the question and waits for you to confirm.
  - 1: AutoCAD assumes the answer to always be Yes.

PRACTICE 11-4

Using the Polyline Edit Command

1. Start AutoCAD 2015.
2. Open the file, Practice 11-4.dwg.
3. Check the objects drawn. You will find that all objects are polylines, except the outer contour, it is lines. Using the Polyline Edit command convert lines to polylines.
4. Using the Polyline Edit command convert the straight line segments to spline segments.
5. Using the Polyline Edit command retain the dashed line of the converted polylines.
6. Zoom to the two small buildings inside the contours and you will find them as polylines. But, the arc is not reaching the line segments. Measure the void between the arc and the lines, and input a proper Fuzz distance. Then, join the polylines, using Extend at the left shape and Add at the right shape.
7. You should get the following picture:

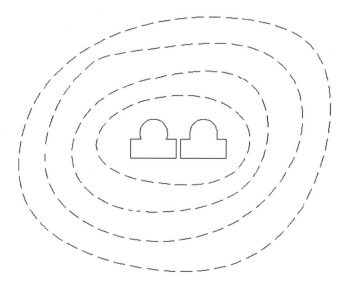

8. Save and close the file.

11.9 USING CONSTRUCTION LINES AND RAYS

- These two commands allow you to produce objects as helping tools to draw accurate drawings; they are not wanted for their own. Construction lines are objects extending beyond the screen in the two directions and can be drawn using different methods. Yet, Rays in AutoCAD are objects that have a known starting point, extending beyond the screen in the one direction. The following is a discussion of both commands.

11.9.1 Construction Lines

- This command will help you to draw a construction line, which extends beyond the screen in two directions. To issue this command, go to the **Home** tab, locate the **Draw** panel, then select the **Construction Line** button:

- You will see the following prompt:

```
Specify a point or [Hor/Ver/Ang/Bisect/Offset]:
```

- There are six methods to specify the angle of the construction line. They are:
    - The first method is the default method, which is to specify two points. Once you specified the first point, you will see the following prompt:

```
Specify through point:
```

   - The **Hor** and **Ver** options allow you to draw horizontal or vertical construction lines. To complete the command, specify the through point. You will see the following prompt:

```
Specify through point:
```

- The **Ang** option allows you draw a construction line using an angle. You will see the following prompts:

```
Enter angle of xline (0) or [Reference]:
Specify through point:
```

- The **Bisect** option allows you to specify three points. It will pass through the first point and bisect the angle formed between the second and third points. You will see the following prompts:

```
Specify angle vertex point:
Specify angle start point:
Specify angle end point:
```

- The **Offset** option allows you to produce a construction line parallel to an existing line. You will see the following prompts:

```
Specify offset distance or [Through] <Through>:
Select a line object:
Specify side to offset:
```

11.9.2 Rays

- This command allows you to draw a ray, which has a starting point and the end extend beyond the screen. To issue this command, go to the **Home** tab, locate the **Draw** panel, and select the **Ray** button:

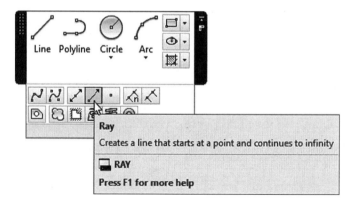

- You then see the following prompts:

  ```
  Specify start point:
  Specify through point:
  ```

- AutoCAD is asking to specify two points; the first point is the starting point and the second will define the angle of the ray. You can define as many rays as you want, using the same starting point.

PRACTICE 11-5

Using Construction Lines and Rays

1. Start AutoCAD 2015.
2. Open the file, Practice 11-5.dwg.
3. Make Construction layer current.
4. Insert vertical and horizontal construction lines using the center of the circle.
5. Insert two construction lines using offset option, using the two vertical lines, with distance = 0.5 to the inside.
6. Using the Ray command set the starting point to be the center of the circle and the second point to be at angle = 60 using Polar Tracking.
7. Repeat the same command, using angles 120, 240, and 300.
8. Draw a new circle with its center coincide with the existing circle, using R=2.
9. You should have the following illustration:

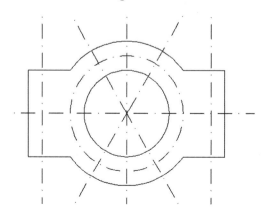

10. Make layer Circle current and make sure Intersection in OSNAP is turned on.

11. Draw circles using the intersection of the four rays and the circle, along with the vertical construction line and the circle, using R=0.2

12. Draw two circles at the intersections of the horizontal and vertical construction lines.

13. Freeze Construction layer.

14. You should have the following illustration:

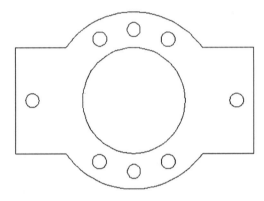

15. Save and close the file.

11.10 USING THE POINT STYLE AND POINT COMMANDS

- The Point style command allows you to set the shape of the point and the Point command will insert a point in the drawing. You can change the point style, as many times as you wish, and points already inserted will shift to the new shape.

11.10.1 Point Style Command

- This command will set the point style. To issue this command, go to the **Home** tab, locate the **Utilities** panel, then select then **Point Style** button:

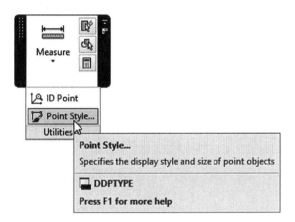

- You will see the following dialog box:

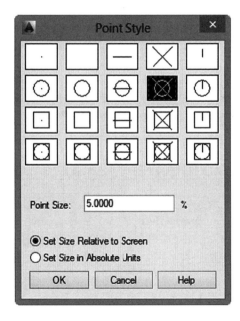

- Pick one of the twenty shapes available then set the point size, either relative to the screen or in absolute units.

11.10.2 Point Command

- This command will insert points in the drawing for as many as you wish.
- To issue this command, go to the **Home** tab, locate the **Draw** panel, then select the **Multiple Points** button:

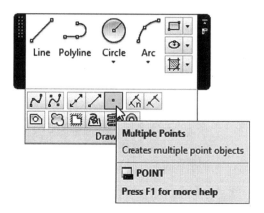

- You will see the following prompts:

```
Specify a point:
```

- AutoCAD is asking you to start specifying points. Once done, press the [Esc] key. In order to pick inserted the points precisely, use the **NODE** OSNAP.

11.11 USING THE DIVIDE AND MEASURE COMMANDS

- The Divide command will cut an object into equal spaced intervals input by you, using points, whereas the Measure command will cut an object into chucks with user specified distance, using points.

11.11.1 Divide Command

- This command allows you to divide an object with equal spaced intervals specified by you. To issue this command, go to the **Home** tab, locate the **Draw** panel, then select the **Divide** button:

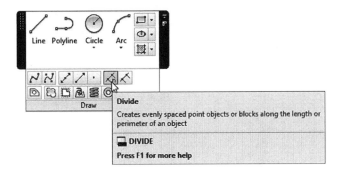

- You will see the following prompts:

```
Select object to divide:
Enter the number of segments or [Block]:
```

- AutoCAD asks to select the desired object, then input the desired number of segments.

11.11.2 Measure Command

- This command allows you to cut an object into segments with user specified distance, using points. To issue this command go to the **Home** tab, locate the **Draw** panel, then select the **Measure** button:

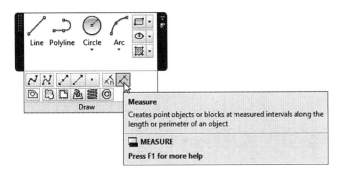

- You will see the following prompts:

  ```
  Select object to measure:
  Specify length of segment or [Block]:
  ```

- AutoCAD is asking that you first select the desired object, then specify the desired length. When you select the object, AutoCAD will start measuring from the end nearest to the selection so be careful.

11.11.3 Divide and Measure Commands with Block Option

- In both commands, you can use Block instead of a point. The following prompts will appear:

  ```
  Enter the number of segments or [Block]:
  Enter name of block to insert:
  Align block with object? [Yes/No] <Y>:
  Enter the number of segments:
  ```

- You should respond with the Block option to the first prompt. Then input the name of the block, with either aligned or not. Finally, enter the number of segments or the distance depending on the used command. To understand the Aligning concept, see the following illustration:

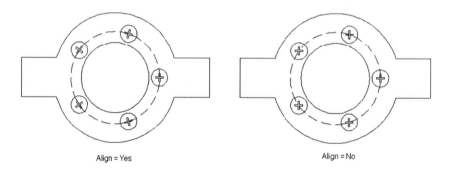

Align = Yes   Align = No

PRACTICE 11-6

Using Point Style, Point, Divide, and Measure

1. Start AutoCAD 2015.
2. Open the file, Practice 11-6.dwg.

3. Change the Point Style to ⊠.

4. Zoom to the upper horizontal lines. The length of the inner horizontal line is 5700. Using the Measure command, add points at 1425 starting from the left end.

5. Erase the point at the far right.

6. Make sure that the Node is on.

7. Using the Insert command, insert block = Window 1, using the three points.

8. Erase the three points.

9. Make layer Furniture current.

10. At the middle of the room, draw a circle with R = 1200. Offset the circle by 250 to the outside.

11. Using the Divide command, add the block = Chair using the outside circle, using eight chairs

12. Erase the outside circle. You should have something like the following:

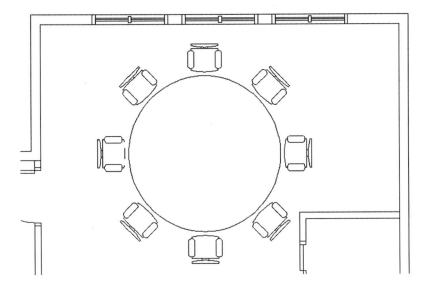

13. Save and close the file.

11.12 USING THE SPLINE COMMAND

- This command allows you to draw smooth curves based on more than two points. This command will draw a spline curve based on exact mathematical equations. There is one command, but two keys invoke two different methods, They are: Fit Points or Control Vertices. To issue these two commands, go to the **Home** tab, locate the **Draw** panel, then select one of the following two buttons:

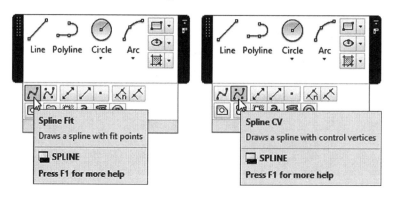

11.12.1 Fit Points Method

- This method allows you to draw a spline with fit points coinciding with it. The following prompts appear:

```
Specify first point or [Method/Knots/Object]:
Enter next point or [start Tangency/toLerance]:
Enter next point or [end Tangency/toLerance/Undo]:
```

- AutoCAD is asking you to specify the desired points to draw the spline with an option to close the shape automatically. In old versions of AutoCAD, you used to have to specify start tangency and end tangency, but in this version there is no need as AutoCAD will make this based on the points specified. But, AutoCAD also prompts will allow you to specify a start and end tangency. AutoCAD will draw a curve connecting the points you select. See the following illustration:

- You have the ability to specify tolerance for points other than the start and end points. See the following illustration:

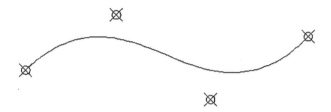

- AutoCAD as well can convert any polyline that was treated using Polyline Edit command and fit in a spline to be a real spline. See the following illustration:

Polyline   Converted to Spline using Edit Polyline command   Converted to Spline using Object option

11.12.2 Control Vertices Method

- This method allows you to draw a spline using control vertices, which will define a control frame. Control frames provide a convenient method to shape the spline. The following prompts appear:

```
Specify first point or [Method/Degree/Object]:
Enter next point:
Enter next point or [Undo]:
Enter next point or [Close/Undo]:
```

- AutoCAD is asking you to specify the desired points to draw the spline with an option to close the shape automatically. Meanwhile, you can specify the degree of the spline, which sets the polynomial degree of the resulting spline. You can input degree 1 (linear), degree

2 (quadratic), and degree 3 (cubic), and so on up to degree 10. You will get something like the following:

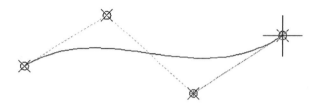

11.12.3 Editing Spline

- When you click a spline produced by Fit Points method, you will see the fit points, along with the triangle, which allow you to show either the Fit Points or Control Vertices:

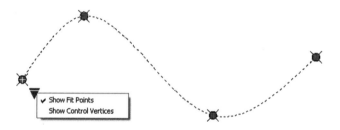

- If you stay at one of the fit points, you will see the following menu:

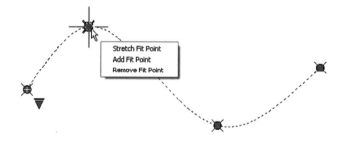

- This menu will Stretch the current fit point, add a new fit point, or remove the current fit point. You will see an extra option if you stay at the start or end point of the spline which is Tangent Direction. This will change the tangent direction of the spline.

- If you clicked a spline drawn using Control Vertices, you will see CV along with the triangle, which displays either the fit points or CV:

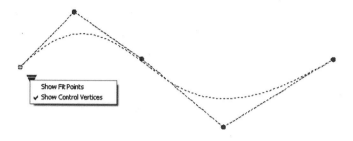

- If you hover one of the CV, you will see the following menu:

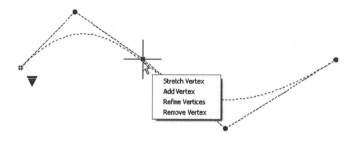

- This will stretch the current vertex, add a vertex, or remove a vertex. The Refine Vertex option will replace the current vertex with two vertices:

PRACTICE 11-7

Using the Spline Command

1. Start AutoCAD 2015.
2. Open the file, Practice 11-7.dwg.

3. Make sure Layer Contour is current.
4. In OSNAP, make sure that Node is on.
5. Using the points at the left, draw an open spline using the Fit Points option.
6. Using the points at the middle, draw a closed spline using the Control Vertices option.
7. Thaw layer Hidden Points.
8. Using grips and the Add Fit Points option, add the two new points at the middle.
9. Stretch the first Fit Point (the one at the left) to the new point to its left.
10. Now spline does not pass through the old first point, using grips add it up.
11. Remove the point indicated in the following illustration:

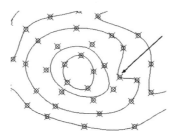

12. Freeze both Points and the Hidden Points layers.
13. Save and close the file.

11.13 USING THE ELLIPSE COMMAND

- This command allows you to draw an elliptical shape or elliptical arc. To issue this command, go to the **Home** tab, locate the **Draw** panel, select the **Ellipse** button, and select one of the methods:

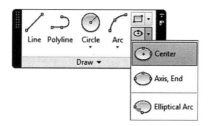

- There are three options to choose from:
  - Center
  - Axis, End
  - Elliptical Arc
- The first two options draw an ellipse. The third option draws an elliptical arc. The following discusses each one of them.

11.13.1 Drawing an Ellipse Using the Center Option

- Using this method, you should specify three points. They are:
  - Center point of the ellipse
  - Endpoint of one of the two axes
  - Endpoint of the other axis
- The following illustrates this concept:

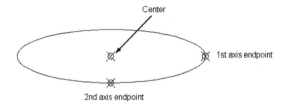

- You will see the following prompts:

```
Specify axis endpoint of ellipse or [Arc/Center]: C
Specify center of ellipse:
Specify endpoint of axis:
Specify distance to other axis or [Rotation]:
```

11.13.2 Drawing an Ellipse Using Axis Points

- You should specify three points. They are:
  - Point on one end of one of the axis
  - Point on the other end of the same axis
  - Point on the other axis
- The following illustrates this concept:

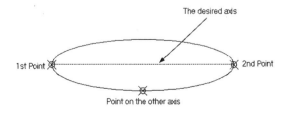

- You will see the following prompts:

    ```
    Specify axis endpoint of ellipse or [Arc/Center]:
    Specify other endpoint of axis:
    Specify distance to other axis or [Rotation]:
    ```

- Using either method, the last step will include an option called Rotation. What is rotation? After you define two points, you will draw a circle. Imagine this circle is in a plane and the plane is rotating; you will get an ellipse. See the following illustration:

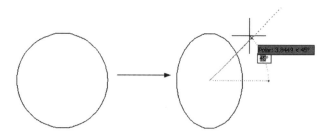

11.13.3 Drawing an Elliptical Arc

- The first three steps to draw an elliptical arc is identical to drawing the ellipse itself, which was previously discussed. Afterward, AutoCAD will ask you to specify starting angle and ending angle, counterclockwise. Another way is after you specify the first angle, you can input the included angle and not the ending angle.
- You will see the following prompts:

    ```
    Specify start angle or [Parameter]:
    Specify end angle or [Parameter/Included angle]:
    ```

PRACTICE 11-8

Using the Ellipse Command

1. Start AutoCAD 2015.
2. Open the file, Practice 11-8.dwg.
3. Make layer Table current.
4. Make sure Node is on.
5. Using the Axis, End method draw an elliptical table using the points displayed.

6. Freeze Points layer.

7. Make layer Window current.

8. Using the Elliptical Arc, complete the window. For the outer elliptical arc, use the endpoints of the two vertical lines; the other axis distance is 3.5. Use the offset command using distance = 0.5.

9. You should have something like the following:

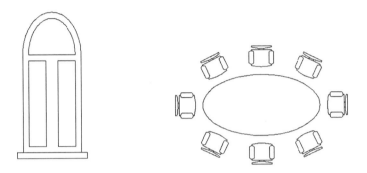

10. Save and close the file.

11.14 BOUNDARY COMMAND

- If you have several intersecting 2D objects (lines, arcs, circles, polylines, and ellipses, etc.) and you want to calculate the net area of these objects, this command helps you more than any other. You can choose between polylines and regions as a resultant object. To issue this command, go to the **Home** tab, locate the **Draw** panel, then select the **Boundary** button:

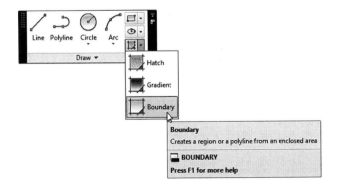

- You will see the following dialog box:

- This command depends on a simple click inside the desired area, in which you want to create a polyline inside it. In this case, start with **Pick Points** button. When you are done, click OK to end the command and create the polyline (or region) desired. You can move it (them) outside to calculate areas or any other desired commands. Meanwhile, you can make some amendments to the command to get different results. You have the following options:
  - Island detection: this option control whether AutoCAD should identify an object within the area.
  - Polyline or Region: you can pick the desired object type.
  - Boundary set: whether all objects are involved in the creation (Current viewport option) or only select objects (click the New button).

PRACTICE 11-9

Using the Boundary Command

1. Start AutoCAD 2015.
2. Open the file, Practice 11-9.dwg.
3. Zoom to Shape 01.
4. Using the Boundary command, and without changing anything, click inside the area. To see the resultant shape, freeze layer Shape 01.

5. Zoom to Shape 02.

6. Using the Boundary command, click off the Island detection checkbox then click inside the area. To see the resultant shape, freeze layer Shape 02.

7. Zoom to Shape 03.

8. Using the Boundary command, click the New button and select all Shape 03 except the four circles at the edges; then, click inside the area. To see the resultant shape, freeze layer Shape.03.

Shape 01   Shape 02   Shape 03

9. Save and close the file.

11.15 USING THE REGION COMMAND

- Assume I bring in some wires, and I ask you to create a rectangle and circle from the wires then place the circle in the center of the rectangle. Although the circle is in the center of the rectangle, there is no relationship between them because if you move one of the two shapes the other will stand still.
- On the other hand, if I bring in a piece of paper and a pair of scissors, and ask you to cut a rectangle with a hole like the shape of a circle.
- This is exactly the difference between the polyline and region in AutoCAD.
- To issue this command, go to the **Home** tab, locate the **Draw** panel, then select the **Region** button:

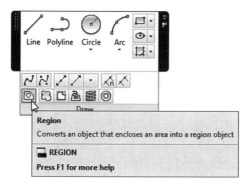

- You can create a region using the previous command Boundary.
- To convert wireframe 2D objects such as lines, arcs, circles, polylines, etc., you have to make sure that they are formulating closed shapes only. You will see the following prompt:

```
Select objects:
```

- Select the desired objects and press [Enter]. When done, objects are converted right away.

11.15.1 Performing Boolean Operation on Regions

- Because regions are real 2D objects, AutoCAD can perform on them Boolean operations. These operations are: Union, Subtract, and Intersect. To issue these three commands, you have to switch to the **3D Basics** workspace, go to the **Home** tab, locate the **Edit** panel, then click one of the following buttons:

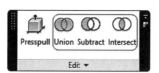

- In the Union and Intersect commands, you can select objects in any order, yet in the Subtract command, select the first region(s) you want to **subtract from**, press [Enter], then select the region(s) to be **subtracted**.

PRACTICE 11-10

Using the Region Command

1. Start AutoCAD 2015.
2. Open the file, Practice 11-10.dwg.
3. Using the Region command convert all objects to regions.
4. Click any object to show the Quick Properties and make sure that it is now a region.
5. Using the Boolean operation, create the following shape:

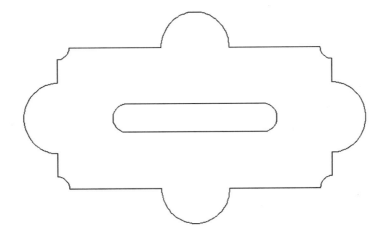

6. Using the Properties palette, what is the total area? _____
7. Save and close the file.

NOTES:

CHAPTER REVIEW

1. In reality donuts, polygons, revision clouds, and rectangles are all:
   a. Splines
   b. Regions
   c. Polylines
   d. None of the above.

2. In Polyline Edit, you have to select the _____ option first to join polylines to polylines.

3. Construction line and Ray will produce objects as helping tools to draw accurate drawings.
   a. True
   b. False

4. You can use Boolean operations with polyline objects.
   a. True
   b. False

5. One of the following statements is not true:
   a. The Divide command will cut any object to equally spaced intervals using blocks.
   b. The Measure command will cut any object to equally spaced intervals using points.
   c. The Measure command will cut any object into chucks with user specified distance using points.
   d. The Divide command will cut any object to equally spaced intervals using points.

6. There are two methods to draw a spline in AutoCAD, Fit Points, or Control Vertices.
   a. True
   b. False

7. The Boundary command can create either _____ or _____.

CHAPTER REVIEW ANSWERS

1. c
3. a
5. b
7. Polyline, Region

CHAPTER 12

ADVANCED PRACTICES – PART I

In This Chapter

- Advanced features of Offset, Trim, and Extend commands
- How to utilize Cut/Copy/Paste when you open more than one file
- How to bring in AutoCAD objects from other software
- Hyperlink Purging
- Views and Viewports commands

12.1 OFFSET COMMAND – ADVANCED OPTIONS

- People who use AutoCAD on a daily basis are stuck with the default options, ignoring some powerful options, which may help significantly reduce production time. This chapter discusses some of the advanced options of AutoCAD. When you start the Offset command, you will see the following prompts:

  ```
  Current    settings:   Erase    source=No   Layer=Source
  OFFSETGAPTYPE=0
  Specify   offset   distance   or   [Through/Erase/Layer]
  <Through>:
  ```

- The first line is a message listing the current values for the different settings of AutoCAD. The message says: Erase source = No, Layer = Source, OFFSETGAPTYPE = 0. These settings and how to change them are discussed in the following sections.

12.1.1 Erase Source Option

- By default AutoCAD will keep both the source and the offset object. This option will keep the offset object but erases the source object. You will see the following prompt:

```
Erase source object after offsetting? [Yes/No] <No>:
```

- Input Yes and AutoCAD will get rid of the source object.

12.1.2 Layer Option

- When you use any command in AutoCAD that produces a copy of the original object, the copy will reside always in the same layer of the source object. Using this option, you can ask AutoCAD to send the generated object to the current layer instead. AutoCAD then shows the following prompts:

```
Enter layer option for offset objects [Current / Source] <Source>:
```

- Input Current to tell AutoCAD you want the offset object in the current layer.

12.1.3 System Variable: offsetgaptype

- This is not an option inside the Offset command but rather a system variable, which should be invoked before the command, using the command window. This system variable decides the outcome of the shape: normal (value = 0), filleted (value = 1), or chamfered (value = 2).

12.2 TRIM AND EXTEND – EDGE OPTION

- You cannot extend an object unless there is a real intersecting point between the boundary edge and objects to be extended. Also, you cannot trim an object unless there is a real intersecting point between the cutting edge and objects to be trimmed. See the following illustration:

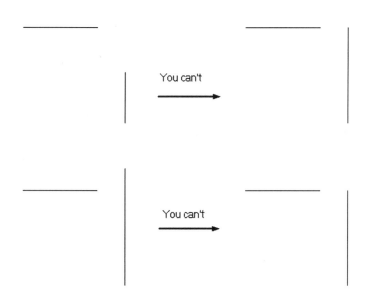

- The **Edge** option allows you to trim objects based on extended cutting edges and extend objects based on extended boundary edges. Selecting the Edge option invokes the following prompt:

```
Enter an implied edge extension mode [Extend/No extend]
<No extend>:
```

- The default option is **No extend** but you can select the **Extend** option to trim and extend based on the extended cutting and boundary edges. Note that all of the previous settings in Offset, Trim, and Extend will affect files from the time the change is implemented.

PRACTICE 12-1

Using Advanced Options in Offset, Trim, and Extend

1. Start AutoCAD 2015.
2. Open the file, Practice 12-1.dwg.
3. Using the Offset command, offset the top shape making sure that the new object will reside in the current layer, the original object will be deleted, and the offset will result in chamfered shape, using distance = 1.

4. Explode the newly created polyline.

5. Using the Edge option in both Trim and Extend, trim the upper horizontal line using the two vertical lines and extend the lower horizontal lines based on the same vertical lines.

6. Draw small vertical lines to complete the base.

7. You should get the following shape:

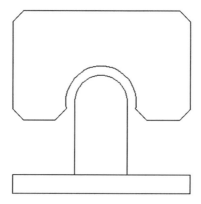

8. Save and close the file.

12.3 USING MATCH PROPERTIES

- This command allows you to match the correct properties of an object to the incorrect properties of other objects. Objects include everything in AutoCAD such as lines, arcs, circles, polylines, splines, ellipses, text, hatches, dimensions, viewport, and table. To issue this command, go to the **Home** tab, locate the **Properties** panel, then select the **Match Properties** button:

- You will see the following prompt:

  `Select source object:`

- AutoCAD is asking you to select the object which holds the correct properties. When done, you will see the following prompts:

  `Select destination object(s) or [Settings]:`

- With this prompt, you will see the cursor change to the following:

- Select objects that hold the incorrect properties, which will be matched with the source object. You can use the **Settings** option to specify the basic and advanced properties to be effected. You will see the following dialog box:

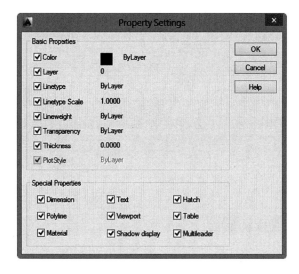

12.4 COPY/PASTE OBJECTS AND MATCH PROPERTIES ACROSS FILES

- In AutoCAD, you can open more than one DWG file at the same time, using a simple technique, which is holding the [Ctrl] key while selecting the names of the desired files in the **Open** file dialog box. But the

question is why would anyone want to open more than one file at the same time? The answer would be one or all of the following:
- To copy objects from one file to another
- To match properties across files

- To tile the opened files, go to the **View** tab, locate the **Interface** panel, then use one of the following two buttons:

- Use the normal Copy/Paste sequence in order to copy objects from one file to another (this Copy is different comparing to Copy command in Modify panel, as this command will copy objects from one file to another). Complete the following steps:
  - Without issuing any command, select the desired object(s).
  - Right-click and select the Clipboard option then select one of the two copying commands available.
  - Go to the other file. Right-click and select the **Clipboard** option, then select one of the three pasting commands available.
- These are the **Clipboard** options:

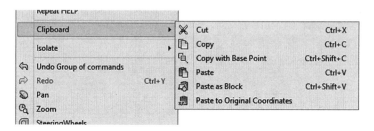

12.4.1 Copying Objects

- AutoCAD allows you to copy objects from one file to another using two techniques. They are:
  - The Copy option allows you to copy objects without specifying a base point.
  - The Copy with Base Point option allows you to copy objects by specifying a base point.
- You will see the following prompt:

```
Specify base point:
```

12.4.2 Pasting Objects

- There are three methods to paste objects across files. They are:
  - The Paste option allows you to paste the contents of the clipboard.
  - The Paste as Block option allows you to paste objects, as a block with an arbitrary name. You can use the Rename command to give the block the correct name.
  - The Paste to Original Coordinates option allows you to paste objects to the same coordinates used in the original file.

12.4.3 Drag-and-Drop Method

- If you do not like this method, you can use the Drag-and-Drop method, which includes selecting desired objects in the source file, then clicking and holding the left mouse button, going to the destination file, and dropping the objects there. You can also do the same but using the right mouse button instead. With this method, when you drop the objects in the destination file, a menu will appear like the following:

- This menu includes self-explanatory options.
- While you are copying objects across files, AutoCAD will create all the needed things to make the copying process successful such as creating layers, text styles, and dimension styles, etc.

12.4.4 Match Properties Across Files

- You can use the same command to match properties across files, selecting the source object in the current file (the file you will issue the command from) and then matching the objects holding the incorrect properties in the destination file. AutoCAD will also create the necessary layers, text styles, dimension styles, etc., in the destination file using this method.

PRACTICE 12-2

Using Match Properties, Copy/Paste Across Files

1. Start AutoCAD 2015.
2. Open the files, Practice 12-2.dwg and Ground Floor.dwg.

3. Tile them vertically.

4. Check layers in the file, Practice 12-2.dwg. Take a note of the existing layers.

5. Check the dimension styles in the file Practice 12-2.dwg and take a note of the existing styles.

6. From Ground Floor.dwg, Using Copy/Paste, copy the bathroom 9'x9' using base point to the suitable place in Practice 12-2.dwg.

7. Using the Match Properties select any dimension in Ground Floor.dwg and match it with all dimensions in Practice 12-2.dwg.

8. Close Ground Floor.dwg, and maximize Practice 12-2.dwg.

9. In Practice 12-2.dwg, match the properties of the copied text with all the other texts

10. Start the Match Properties command and select the hatch of the kitchen as the source object, then select the hatch of the two toilets.

11. Check the layers again. How many layers were added?

12. Check the dimension styles again. How many new styles were added?

13. If you have time, put all objects in the right layer.

14. Save and close the file.

12.5 SHARING EXCEL AND WORD CONTENT IN AUTOCAD

- As an application running under Windows OS, AutoCAD can take and give objects to and from any other Windows OS application, especially MS Office software such as MS Word and Excel, the most commonly used software today. AutoCAD uses OLE (**O**bject **L**inking & **E**mbedding) to copy the contents from and to AutoCAD. The Paste command used in AutoCAD dictates the type of object brought to AutoCAD. Sharing between Windows users is practical and easy to do.

12.5.1 Sharing Data Coming from MS Word

- Copy the contents from Word using Copy or Cut, then go to the **Home** tab, locate the **Clipboard** panel (if this panel is not shown by default, make sure that you are at the **Home** tab, right-click any panel, select **Show Panels**, then select **Clipboard**):

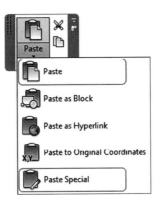

- You can use either Paste or Paste Special options. See the following:

  Word Content → Using Paste → OLE object embedded
  Word Content → Using Paste Special / Paste / Text → MTEXT
  Word Content → Using Paste Special / Paste / UniCode Text → MTEXT
  Word Content → Using Paste Special / Paste / AutoCAD Entities → Text
  Word Content → Using Paste Special / Paste Link → OLE object linked

- Once you issue the **Paste Special** command, you will see the following dialog box:

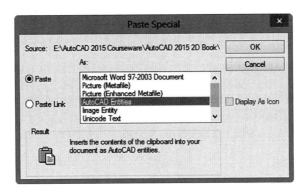

- As you can see that the first option (using Paste only) and the last option (using the Paste link) will bring in an OLE object. The first one will not be updated when the source changes but the last one will be. Using the first method, you will see the following dialog box:

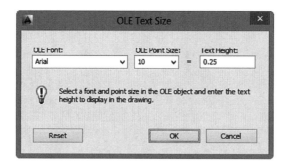

- The OLE Text Size dialog box will show the font name and font size used in the MS Word. It then asks you to specify the Text Height in AutoCAD units. At any time you can click Reset button to get the original value. Once you are done, click **OK**.
- After pasting the OLE object, AutoCAD allows you to edit it. To edit an OLE object, select it and right-click; you will see a menu in which you should select the OLE option. You then see the following:

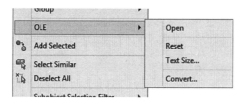

- The available choices are as follows:
  - The Open option allows you to open the source application.
  - The Reset option allows you to retain the font and font size of the original.
  - The Text Size option allows you to change the text size.
  - The Convert option allows you to change the nature of the OLE object.

12.5.2 Sharing Data Coming from MS Excel

- Copy the contents from Excel using either Copy or Cut, then go to the **Home** tab, locate the **Clipboard** panel, and choose either the Paste or

Paste Special options, like in the MS Word case. In Excel, you have the following choices:

Excel Content → Using Paste → OLE object embedded
Excel Content → Using Paste Special / Paste / Text → MTEXT
Excel Content → Using Paste Special / Paste / UniCode Text → MTEXT
Excel Content → Using Paste Special / Paste / AutoCAD Entities → Table Embedded
Excel Content → Using Paste Special / Paste Link → OLE object linked
Excel Content → Using Paste Special/Paste Link/AutoCAD objects → Table Linked

- You will see the same dialog boxes as previously discussed.
- The last choice allows you to paste a linked table in AutoCAD and is discussed in the next section.

12.5.3 Pasting a Linked Table from Excel

- This option allows you to paste Excel content into your current drawing, linking it to the original Excel sheet. With this option, when you update in Excel, it will affect the table in AutoCAD. When you update the file in AutoCAD, it will affect the file in Excel. If you hover over the table in AutoCAD, you will see the following image:

- This means the table inside AutoCAD is *locked* and *linked*.
- When you change anything in the Excel sheet and save, these changes will also be reflected in the AutoCAD table. At the lower-right corner of the AutoCAD window, you will see a chain called **Data Link**:

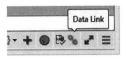

- If you right-click this icon, you will see a menu just like the following:

- Select the **Update All Data Links** option to get the latest copy of your Excel sheet. If the AutoCAD file is not open, the next time you open it, you will get the newest copy of the Excel sheet.
- Updating the Excel content from the AutoCAD table is a little bit harder and has a strict procedure to follow:
    - Go to the **Insert** tab, locate the **Linking & Extraction** panel, then select the **Data Link** button:

    - The following dialog box comes up:

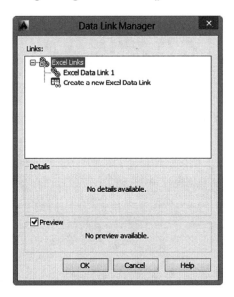

    - Under **Links** you will see there is a link called **Excel Data Link1**, which is created automatically when you paste link the Excel table. This name is temporary. To rename the link, click it, right-click, then select the Rename option:

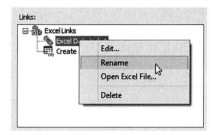

- There will be a **Preview** check box turned on to help you preview the link in case there is more than one link in the same file.
- Double-clicking the link or selecting the **Edit** option from the right-click menu opens the following dialog box:

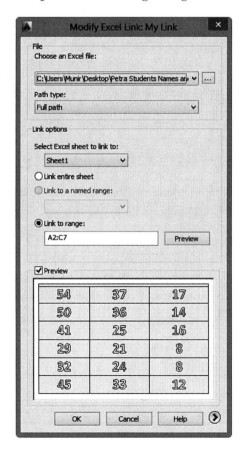

- Clicking the arrow at the lower-right part of the dialog box expands it to show more options:

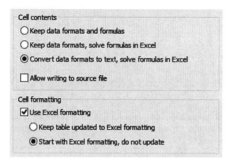

- Make sure that **Allow writing to source file** check box is turned on because without this the Excel sheet will not stay updated.
- Click **OK** several times to close all dialog boxes.
- You are ready to make the edits in the table pasted in AutoCAD.
- Using crossing, select the desired cells in the table.
- Right-click and a menu will appear; select the **Locking** option then **Unlocked** option:

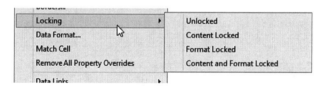

- Note that the locking symbol disappeared but the chain symbol is still there. Change the desired value in the unlocked cells.
- To upload these changes to the original Excel sheet, go to the **Insert** tab, locate the **Linking & Extraction** panel, then select the **Upload to Source** button:

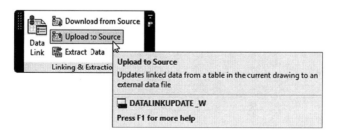

- AutoCAD will ask you to select objects; click the desired table from one of its outside borders.
- The following prompt then appears:

```
1 object(s) found.
1 data link(s) written out successfully.
```

- You will see the following bubble appear at the lower-right corner of AutoCAD window:

PRACTICE 12-3

Sharing Excel and Word Content in AutoCAD

1. Start AutoCAD 2015.
2. Open the file, Practice 12-3.dwg.
3. Using MS Word open the file House General Notes.doc and copy all its contents.
4. Go to the Full Plan layout.
5. Make layer Text current.
6. Using the Paste option, paste the text in the lower-left corner of the layout below the viewport.
7. Make the text size = 5.
8. Open the file Door Cost Schedule.xls and copy its contents.
9. Using Paste Special / Paste Link, select AutoCAD Entities.
10. Close the Excel sheet.
11. Change the quantity of Type 02 to be 21 and update the Excel sheet.
12. Update the AutoCAD file with the new value to recalculate the new values.

13. Make sure that the Excel file has changed.

14. Save and close the files.

12.6 HYPERLINKING AUTOCAD OBJECTS

- This command allows you to hyperlink any AutoCAD object(s) to a website, AutoCAD drawing, Excel sheet, or Power Point presentation, etc. To issue this command, go to the **Insert** tab, locate the **Data** panel, then select the **Hyperlink** button:

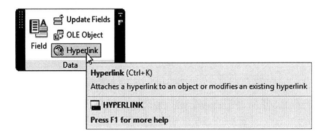

- You will see the following prompt:

```
Select objects:
```

- Select the desired objects and then press [Enter] when done. You will see the following dialog box:

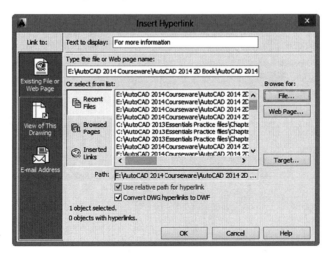

- Fill in the **Text to display**, which is like a help note and appears when you get close to the object holding the hyperlink. Then, input **Type the file or Web page name**, which is a website address or file path. Finally, click **OK** to end the command.
- Using the **Home** tab, locate the **Clipboard** panel and you will see an option called **Paste as Hyperlink**, which will do the job in reverse order and paste the contents as hyperlink to an object.

PRACTICE 12-4

Hyperlinking AutoCAD Objects

1. Start AutoCAD 2015.
2. Open the file, Practice 12-4.dwg.
3. Hyperlink the 3D shape to a file called Part Detail Dimension.dwg.
4. Test the hyperlink by holding the [Ctrl] key and clicking the 3D shape.
5. Save and close the files.

12.7 PURGING ITEMS

- Normally, you create a lot of content inside AutoCAD drawings such as layers, blocks, dimension styles, text styles, and multileader styles, etc. You use some of them whereas others are left without being used. The Purge command helps you get rid of these unused content. To use this command, go to the Application menu, select **Drawing Utilities**, then select the **Purge** command:

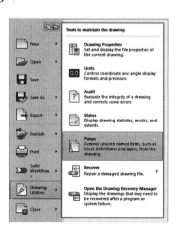

- You will see the following dialog box:

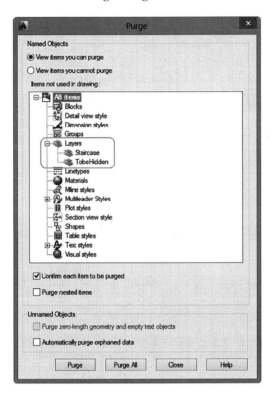

- AutoCAD will list all the unused items. In the previous example, you can see there are two unused layers and one unused multileader style. In the following, you see three checkboxes.
    - Confirm each item to be purged: if this turned on, you have to answer to purge or not. Each time you want to purge an item, you will see something like the following:

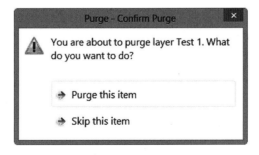

- Purge nested items: this option helps you purge a nested block. A nested block is a block that contains blocks. If this option is turned off, then you will purge only the big block and the nested blocks will remain there. But, if you turn it on, AutoCAD will purge the big block and all the nested blocks, if they are not being used in the current file.
- Purge zero-length geometry and empty text objects: this option removes all lines, arcs, or polylines that have a length of zero, and all MTEXT and Text that contains only spaces.
- Finally, you can purge one item at a time or purge all items at once; select either button, then click the **Close** button to end the command.
- You cannot purge the current layer, text style, or dimension style, etc.
- The Purge command can downsize the file size significantly.

PRACTICE 12-5

Purging Items

1. Start AutoCAD 2015.
2. Open the file, Practice 12-5.dwg.
3. Start the Purge command and type the name of the layers to be purged: _____, _____, _____, _____.
4. Click the plus sign beside the Layers category, select the Staircase, and then click the Purge button. When the dialog box comes up, select Purge this item.
5. Repeat the same for the other layers.
6. Click Purge All to purge all the other items and select Purge All Items option.
7. Check if there are any plus signs beside any of the items; if not, click the Close button.
8. Save and close the files.

12.8 USING VIEWS AND VIEWPORTS

12.8.1 Creating Views

- A view in AutoCAD is any rectangular shaped portion of the drawing, which will be saved under a name. There are two ways to define it. You can zoom to the part of the drawing you want to save and then issue the command or you can do so inside the View command. To issue this command, go to the **View** tab, locate the **Views** panel, then select the **View Manager** button (this panel may be hidden; to unhide it, right-click on any panel, select **Show Panels**, then select the name of the panel):

- You will see the following dialog box:

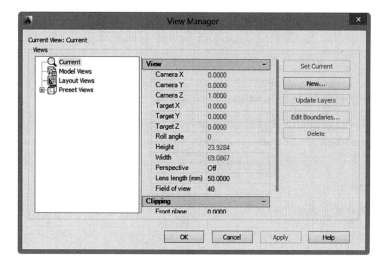

- To create a new view, click the **New** button. The following dialog box will appear:

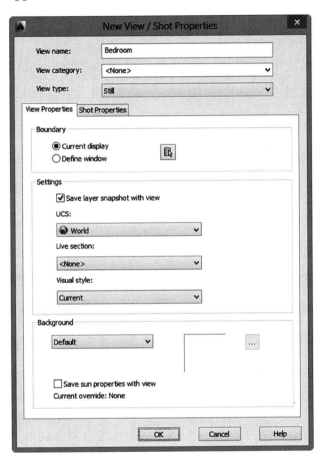

- As a first step, input the **View name**. Under **Boundary** select whether your view is the **Current display** or whether you want to **Define window** (click the small button at the right to zoom to the desired area). Using **Settings**, select whether you want to **Save layer snapshot with view** or not. Layer the snapshot and check the status of layers (on/off, Thaw/Frozen, Unlock/Lock, etc.). When done, click **OK** to save the view.

- There are multiple ways to retrieve the saved views. They are:
  - Using the In-canvas View control at the upper-left corner of the screen, select the **Custom Model Views** as shown:

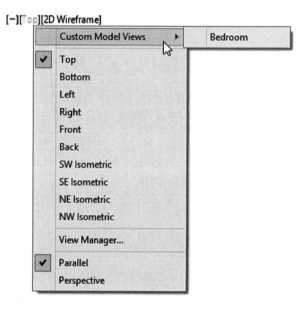

  - Using the **View** tab, locate the **Views** panel and click the name of the desired view, as shown:

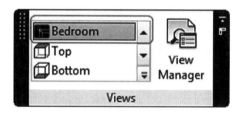

12.8.2 Using Views in Viewports

- You can show the saved views in viewports using the **Viewport** dialog box. Start the **Viewport** dialog box (if you are in one of the layouts, you have to use Vports or Viewports commands). Select one of the arrangements using Preview (the following example is Two: Vertical). If you click on one of the viewports it will be current so you can select one of the saved views to show inside it:

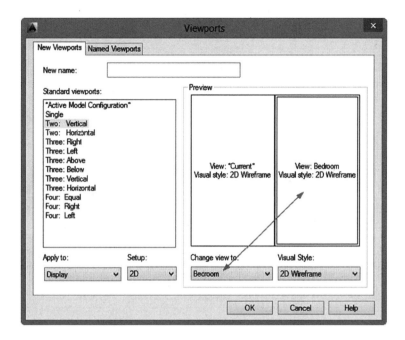

12.8.3 Creating Named Viewport Arrangement – Method (I)

- If you do not like the viewport arrangement, AutoCAD provides and you are thinking about creating your own arrangement then this part is for you. Make the following steps:
  - Make sure you are at Model Space.
  - Go to the **View** tab and locate the **Model Viewports** panel, then select **Viewport Configuration** button to see something like the following:

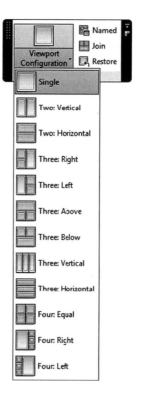

- Or you can reach the same command using the In-canvas viewport control.

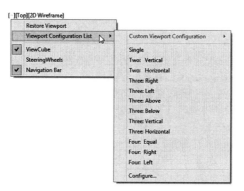

- Pick one of the existing arrangements to begin with.
- To break a viewport to even smaller view ports, select one of the Model viewport, Start the Viewport command again, and pick an arrangement.
- Make sure that at the lower-left corner of the dialog box, under Apply to, you select Current Viewport then click **OK**.

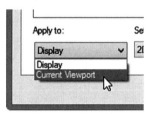

- This should cut your small viewport to even smaller areas.
- Do that to different viewports.
- Go to the **View** tab, locate the **Viewports** panel, and select the **Join** button. This will enable you to join the adjacent viewports (the condition here to form a rectangular shape).

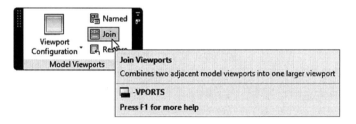

- Once you are done joining viewports, you will end up with a new arrangement, all of you have to do is to save it. Start **Named** command and type the name of the new arrangement as shown:

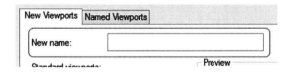

- To retrieve it in layout, go to the desired layout.

- Go to the **View** tab, locate the **Viewports** panel, and click the **Named** button:

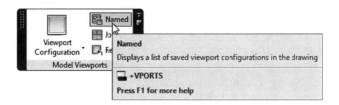

- Pick the desired named viewport arrangement and insert it as you do with the existing arrangements.

12.8.4 Creating Named Viewport Arrangement – Method (II)

- Another way to do the previous is the following:
  - Make sure you are at Model Space.
  - Go to the **View** tab, and locate the **Model Viewports** panel, then select the **Viewport Configuration** button to see something like the following:

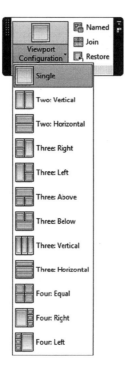

- Or you can reach the same command using the In-canvas viewport control.

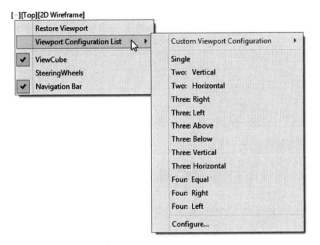

- Once you do this, click inside one of the viewports and you will see a thick blue line as a boundary.
- Go to one of the sides, the shape of the mouse will change to double arrows to resize the viewport horizontally or vertically.
- To resize in both directions in one step, go to one of the corners, the shape of the mouse will change to four arrows, click hold and move.
- Notice a small (+) in the thick black lines separating any two viewports, if you click it and move it (horizontally it will add a horizontal viewport, and if you move it vertically, you will add a vertical viewport). You see a green line, which will add a new viewport.
- To join viewports together, go to the **View** tab, locate the **Viewports** panel, and select the **Join** button. This will enable you to join adjacent viewports (the condition here is to form a rectangular shape).

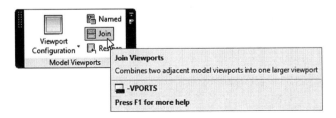

- Once you are done joining viewports, you will end up with a new arrangement, all of you have to do is to save it. Start the **Named** command and type the name of the new arrangement as shown:

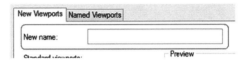

- To retrieve it in layout, go to the desired layout.
- Go to the **View** tab, locate the **Viewports** panel, and click the **Named** button:

- Pick the desired named viewport arrangement and insert it as you do with the existing arrangements.

PRACTICE 12-6

Using Views and Viewports

1. Start AutoCAD 2015.
2. Open the file, Practice 12-6.dwg.
3. Create four views making sure that you will save a *layer snap shot* with the view. The four views are:
   a. Master Bedroom
   b. Master Bedroom Bathroom
   c. WIC
   d. Kitchen

4. Thaw layer Dimension.

5. Retrieve one of the four views. What happens to the layer Dimension? Why?

---

6. Insert Four: Equal viewports in Detail 1 layout showing a view in each viewport.

7. Using Method (I) or Method (II) using Model Space create Four: Equal.

8. For the upper left, split it into Two: Vertical. For the lower-right, split it into Two: Vertical.

9. For the upper ones, join the small to big ones and at the lower ones join the smaller to the bigger one, then save this new arrangement under the name **Four Unequal**.

10. Go to Detail 2 layout, insert the new arrangement. Going to each viewport, zoom to different part of the floor plan.

11. Save and close the files.

NOTES:

CHAPTER REVIEW

1. One of the following statements is not correct.

    a. Using the Paste option, the content will be always OLE object in AutoCAD.

    b. You can insert a table in AutoCAD from the Excel sheet and the editing will be updated both ways.

    c. When you paste content from Word, it will always be MTEXT.

    d. To make the content linked, you have to use Paste Special.

2. Use _____ in the Offset command to create filleted or chamfered edges while offsetting.

3. All of the following is true about View and Viewport except:

    a. You can save a layer snapshot with the view.

    b. You can create a new viewport arrangement in Model and Paper spaces.

    c. You can insert a new viewport arrangement in Paper space.

    d. You can show in each viewport a saved view.

4. You can match properties across files.

    a. True

    b. False

5. While copying objects between files, you can use only drag-and-drop using the left button.

    a. True

    b. False

6. _____ option will trim objects based on extended cutting edges.

7. Using the Right-click menu and Clipboard option, you can use the Paste Special option.

    a. True

    b. False

CHAPTER REVIEW ANSWERS

1. c
3. b
5. b
7. b

CHAPTER 13

ADVANCED PRACTICES – PART II

In This Chapter

- How to use Content Explorer
- How to deal with Fields
- Quick Select, Select Similar, and Add Selected
- Partial Open and Partial Load
- Object Visibility

13.1 USING AUTODESK CONTENT EXPLORER

- Using the Autodesk Content Explorer, you can search drivers and folders containing thousands of AutoCAD files for anything you want, and Autodesk Content Explorer will find it and display it for you. Autodesk Content Explorer is Google-like index that allows you to search and find desired content quickly. To issue this command, go to the **Add-Ins** tab, locate the **Content** panel, then click the **Explore** button:

- You will see the following palette:

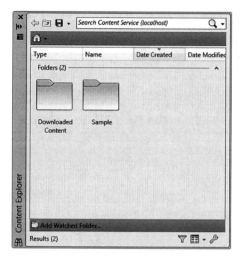

- By default, AutoCAD will look in two folders and one online site. The two folders are the Sample folder, which comes with AutoCAD, and the Downloaded Content folder. The online site is Autodesk Seek. In order to add more folders, click the **Configure Settings** button:

- You will see the following dialog box:

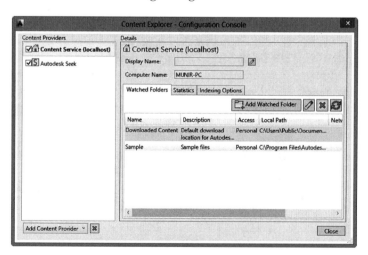

- Click the **Add Watched Folder** button to add more folders to watch when you issue the search process.
- To start a search, simply type your word(s) in the **Search** field, like the following:

- Then, click the magnifying glass and the results will be listed instantly:

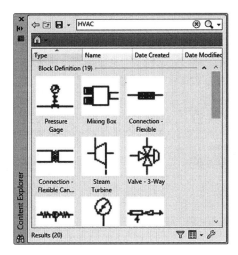

- AutoCAD is listing 19 files that contain the word, HVAC. When you double-click any of these blocks, AutoCAD will open the file and zoom to the block.
- The two buttons at the bottom will help you to **Toggle icon size**. You will see the following menu:

- The last button is **Configure filtering**, which allows you to filter the things that AutoCAD will search for when you input the search phrase. You will see the following menu:

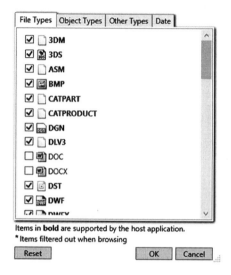

- Specify the File Types. Clicking the Object Types tab will lead to:

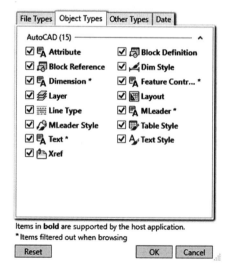

- Specify what objects AutoCAD should look for. Clicking Other Types tab will lead to the following:

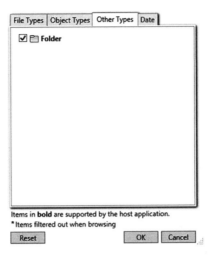

- Click the Date tab to see the following:

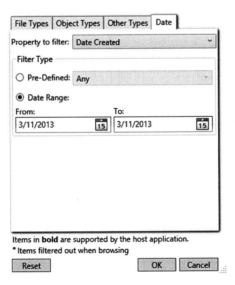

- Specify whether AutoCAD should search for all files created at any time or specify a date range for searching.

PRACTICE 13-1

Using Autodesk Content Explorer

1. Start AutoCAD 2015.

2. Open the file, Practice 13-1.dwg.

3. Start the Autodesk Content Explorer.

4. Make sure that the Sample folder, which comes with AutoCAD is included. If not, add it as Watched Folder.

5. Search for the word **duct**.

6. The result should be two files, containing four leaders, and two pieces of multiline text.

7. Double-click the Mtext, which starts with the word "NOTES DUCT-WORK: ...."

8. Copy it and paste it in your file.

9. Save and close both files.

13.2 USING QUICK SELECT

- The Quick Select command will select objects in the current drawing based on their properties, which is handy in dense and complicated drawings that contain hundreds of thousands of different types of objects. There are multiple ways to issue the Quick Select command. They are:
  - Go to the **Home** tab, locate the **Utilities** panel, then select the **Quick Select** button:

- Without issuing any command, right-click, then select the Quick Select option from the menu:

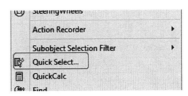

- You will see the following dialog box:

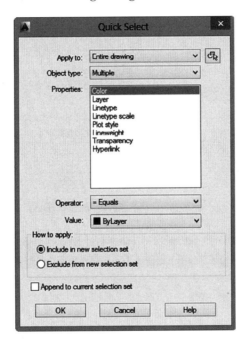

- This method is similar to SQL (Structured Query Language), which will apply a filter to find information based properties.
- First, select the **Apply to:** part, either the Entire drawing or click the button beside it to strict the filter by an area of your choice. Select the **Object type** to search for. AutoCAD will list objects found in the current file and you will see the following picture:

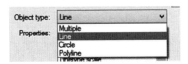

- Select the desired object and then select the **Properties** of the selected object. If you select **Multiple** in **Object type**, then you will see only general properties. If you select an object type, then you will see both general and specific properties to pick from. The following is a list of properties of a circle:

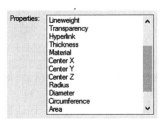

- The next step is to select the desired **Operator**. You will see something like the following:

- Specify the proper **Value** for the selected operator.
- Finally, select **How to apply** the filter in the drawing. You can create a fresh new selection set from it. If you selected objects prior to the Quick Select command, you can exclude the objects from it or you can append it to already selected objects.
- As a final note, you can access Quick Select while you are at the Properties palette, just like the following:

PRACTICE 13-2

Using Quick Select

1. Start AutoCAD 2015.

2. Open the file, Practice 13-2.dwg.

3. Since the drawing says that we have 12 x R 0.06, we want to make sure that this is right. So using the Quick Select, select all circles with R = 0.06. How many circles are selected? _____

4. Using Quick Select and Properties, change the radius of all the other eleven circles to have R = 0.06.

5. Using Quick Select and Properties, move all circles of R = 0.06 to layer Holes.

6. Using Quick Select and Properties, move all dimensions to layer Dimensions. (Hint: linear dimensions are called Rotated dimensions, and because you have two types rotated and radial, you should use append or do this step twice.)

7. Save and close the file.

13.3 USING SELECT SIMILAR AND ADD SELECTED

- The Select Similar command allows you to select an object, then select all similar objects which holds the same properties following certain settings. Adding the Selected command will allow you to select an object then initiate the command, drawing the same object with the same properties.

13.3.1 Select Similar Command

- There are two ways to issue this command. The first method involves the following steps:
  - Select the desired object.
  - Right-click then choose **Select Similar** option:

  - Based on the current settings, AutoCAD will select the similar objects.

- The second method involves the following steps:
  - Type SELECTSIMILAR in the command window and the following prompt will appear:

  ```
  Select objects or [SEttings]:
  ```

  - Right-click and select the **Settings** option. You will see the following dialog box:

- As previously shown, AutoCAD will select based on layer and name. Other things to pick are Color, Linetype, Linetype scale, Lineweight, Plot style, or Object style (anything has a style like dimension, text, etc.). If you select more than one object (arc and circle), then right-click and chose Select Similar, AutoCAD will select similar objects to both selected objects.

13.3.2 Add Selected Command

- Let us assume you have a command lies in layer Centerline, the color is yellow, and the linetype is dashdot. You want to create a new line that lies in the same layer and holds the same properties. This command allows you to accomplish your mission easily. Select the desired object, then right-click and choose **Add Selected** option:

- You will see AutoCAD started the command you want and you are ready to start drafting. The new object will reside in the same layer and will hold the same properties of the selected object.

13.4 WHAT IS OBJECT VISIBILITY IN AUTOCAD?

- The normal practice in AutoCAD to hide an object is to turn the layer off or freeze it. But this means all the other objects in the same layer will be hidden as well. This command will hide an object or group of objects without hiding all the other objects in the same layer. To do this, select the desired object (or as many objects as you wish), right-click, and then select the **Isolate** option. You will see the following:

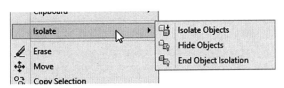

- As you can see, there are three commands. They are: Isolate Objects, Hide Objects, and End Object Isolation. Here is a discussion of each one of these commands:
    - Isolate Objects allows you to select objects and will hide all other objects.
    - Hide Objects allows you to hide the selected objects and show all other objects.
    - End Object Isolation allows you to cancel the first command, meaning it will show all objects.
- You can see a button containing a circle, rectangle, and triangle at the lower-right hand of the screen; it is either on or off. If you click it, you will see the same menu so you can do everything from there as well.

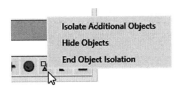

PRACTICE 13-3

Using Select Similar and Add Selected

1. Start AutoCAD 2015.
2. Open the file, Practice 13-3.dwg.
3. Type the SELECTSIMILAR command and select the Settings option, clear all checkboxes, then select one of the red circles, and press [Enter].
4. You will notice that AutoCAD selected all circles in the drawing.
5. Right-click and select Isolate/Isolate Objects.
6. Make the layer Centerlines current.
7. Start the Line command then using OTRACK and Center point of one of the two large circles take 1.2 from the center to the left to start the line. Draw a 2.4 horizontal line that passes through the center of the two large circles.
8. End Object Isolation.
9. Make layer 0 current.
10. Select one of the green centerlines at the right or at the left, right-click, select Add Selected, and draw a vertical centerline for the large circles.
11. Does it look different than the horizontal line? _____ Why? (The linetype scale of the two lines at the right and the left is 0.5; the new line holds all the properties of the original line.)
12. Select one of the two linear dimensions, then right-click and choose Select Similar. Two dimension blocks will be selected, right-click, select Precision and then select 0.000.
13. Select the four centerlines, right-click, and then select Isolate/Hide Objects.
14. Save and close the file.

13.5 ADVANCED LAYER COMMANDS

- This section discusses some advanced commands related to layers. Most of these commands depend on selecting a tool to perform a certain task related to the layer of the object selected. These commands will make your life easier and decrease the time it takes to complete a drawing. To use these commands, go to the **Home** tab and locate the **Layers** panel:

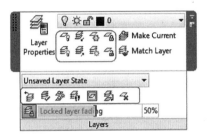

13.5.1 Isolate and Unisolate Commands

- Use the following two buttons:

- The Isolate command allows you to select object(s). The layer(s) of these objects will be shown, whereas the other layers will be turned off (or locked). The Unisolate command allows you to cancel the effects of the Isolate command.

13.5.2 Using Freeze and Off Commands

- Use the following two buttons:

- The Freeze and Off commands allow you to select object(s) and the layer(s) of these objects are then turned off or frozen.

13.5.3 Using Turn All Layers On and Thaw All Layers Commands

- Use the following two buttons:

- These two commands allow you to turn all layers on then thaws them. It is very helpful to complete this process in one shot.

13.5.4 Lock and Unlock Commands

- Use the following two buttons:

- The Lock command allows you to select object(s) and then the layer(s) of these object(s) can be locked. You can lock a layer at a time. The Unlock command allows you to select object(s) then the layer(s) of these objects will be unlocked. You can unlock a layer at a time.

13.5.5 Change to Current Layer Command

- Use the following button:

- This command allows you to select object(s) then changes the layer of these object(s) to be the current layer.

13.5.6 Copy Objects to New Layer Command

- Use the following button:

- This command allows you to select object(s) and then copy them to a new location in the drawing. You can then change the layer of the new objects to a new layer, either by selecting an object resides in the desired layer or by typing its name.

13.5.7 Layer Walk Command

- Use the following button:

- This command allows you to see a dialog box listing all layers in the current file. To show the contents of a layer, click its name in the list (by default all layers are selected). A checkbox at the bottom says "Restore on Exit," which means whenever you close this dialog box, all layers are restored to their previous status. If a layer is turned off, this command will keep its effects:

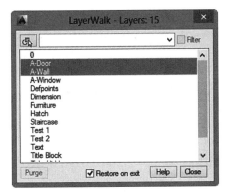

13.5.8 Isolate to Current Viewport Command

- Use the following button:

- This command allows you to select the object and then freeze the layer of this object in all viewports except the current viewport.

13.5.9 Merge Command

- Use the following button:

- This command allows you to merge a layer or more in a target layer. First, AutoCAD asks you to select an object (AutoCAD will list the name of the layer of the selected object). AutoCAD will keep asking you to select objects until you press [Enter] and are done. The last step is to select an object in the target layer; AutoCAD then will report the deleted layers.

13.5.10 Delete Command

- Use the following button:

- This is a very helpful command because you know that AutoCAD will not delete a layer unless it is empty. This command allows you to delete (purge) a layer by selecting an objects that resides in it, except the current layer.

13.6 LAYER'S TRANSPARENCY

- You can set the visibility of a layer. The default value for all layers is 0 (zero) and can be as much as 90. If you are making a test plot for a drawing with lots of solid hatching, set the visibility for the minimum so you do not lose plotter ink. You can use the same color for many layers then control the visibility of the layer in order to give each layer a different tone of the color. When you start the Layer Properties Manager, you will see a column called Transparency, and if you are at a layout, you will see another column called VP Transparency:

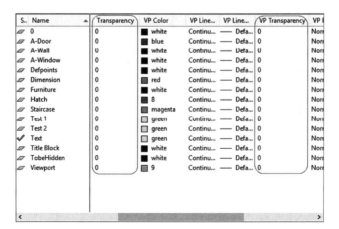

- You can control the visibility of new objects. To do that, go to the **Home** tab and locate the **Properties** panel:

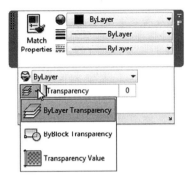

- Select either to set the **Transparency** to ByLayer, ByBlock, or Transparency Value (which will set it to 0 (zero)) or move the slider to any desired value.
- Also, set the value for object(s) using the Properties palette, as follows:

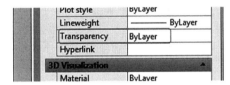

PRACTICE 13-4

Using Advanced Layer Commands

1. Start AutoCAD 2015.

2. Open the file, Practice 13-4.dwg.

3. Using Freeze command, select one of the dimension blocks. What happens to the Dimension layer? _____

4. Start the Isolate command and select a line of the walls. What happens to other objects in other layers? _____

5. Unisolate.

6. Start the Layer Walk command and you will notice that all layers are selected.

7. Click any of the layer names, select A-Door layer, then hold [Ctrl].and select A-Wall and A-Window. Uncheck the Remove on exit checkbox and click Close. At the warning message, click Continue.

8. Start the Layer Properties Manager palette. What is the state of layers you did not select in the previous step? _____

9. Start the Layer Walk again and while holding the [Ctrl] key, select the Furniture and Toilet Furniture layers. Close the Layer Walk command.

10. Merge layer Toilet Furniture to layer Furniture.

11. What happens to the layer Toilet Furniture after the merging process?

12. Save and close the file.

13.7 USING FIELDS IN AUTOCAD

- AutoCAD stores a lot of data in the drawing database. Some of it is constant and some of it is variable. You can utilize these types of data by inserting them in the drawing to benefit from the updating feature that AutoCAD will perform on the variable data. This approach is better than writing text using the Text and Mtext commands, which need to be updated manually. There are four methods to insert a field in your drawing:
    - Using the Field command. To issue this command, go to the **Insert** tab, locate the **Data** panel, then select the **Field** button:

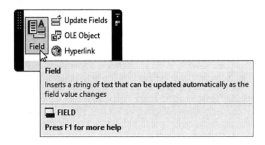

- Using the Text command, after you specify the starting point, height, and rotation but before you start writing, right-click, and select the Insert Field option:

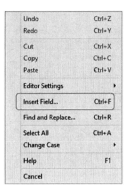

- Using the Mtext command, after you specify the text area, you will find the **Text Editor** context tab and the **Insert** panel. Select the **Field** button:

- Also, you can insert fields in the Table and use the Attribute commands.
- Regardless of the method used to reach the **Field** command, the following dialog box comes up:

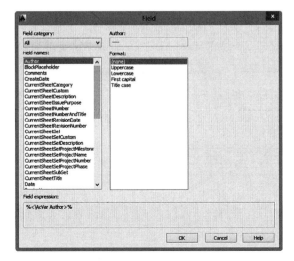

- The first thing to control is the Field category, which will help you find the desired data quickly. Clicking the pop-up list will show the following list:

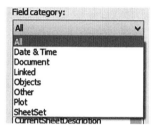

- There are seven field categories and each one will show related field names. For instance, Objects category will show four field names. They are:

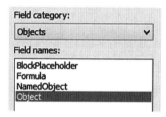

- Based on the field category and field name, you will find at the right side of the dialog box things to help you control the appearance of the field. For instance, if you select the Field category Document and the Field name is File name, then you will see the following at the right:

- As you can see, you will select the Format and whether you want to show the Filename, Path only, or Path and filename. Finally, you can control whether or not to show the file extension. If you use the Filed category Objects and Field name Object, you will need to select an object from the drawing:

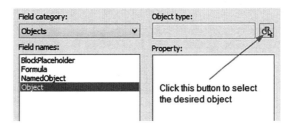

- When you click this button, the dialog box will disappear temporarily to let you select the desired object. When done, you will see something like the following:

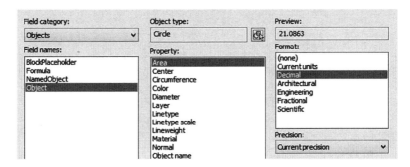

- As you can see, we selected a circle and we chose to use the Area with decimal format.
- When you insert a field in the drawing, the default settings will display it with a background just like the following:

- To remove the background, go to the Application menu and select the **Options** button. Select the **User Preferences** tab, then locate **Fields**, and uncheck **Display background of fields** checkbox, as in the following:

- Click the checkbox **Display background of fields** off if you do not want to display the background. Click the **Field Update Settings** to see the following dialog box:

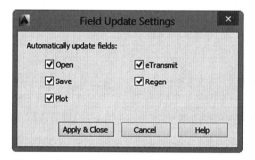

- Control to update the field automatically each time you save, open, plot, eTransmit, or Regen. If you need to update manually, then go to the **Insert** tab, locate the **Data** panel, then select the **Update Fields** button:

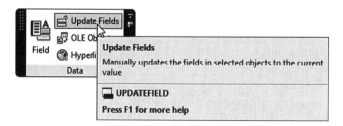

- You need to select the desired fields to be updated.

PRACTICE 13-5

Using Fields in AutoCAD

1. Start AutoCAD 2015.
2. Open the file, Practice 13-5.dwg.
3. Make sure that current layer is Polyline.
4. Using the Boundary command, click inside the master bedroom to add a polyline.
5. Make layer Text current.
6. Go to the Annotate tab and make sure that Room Titles are the current text style.
7. Start the Single Line text command and specify a point almost at the middle of the room, with rotation angle = 0. Type "Area =" then right-click, select the Insert Field option, and insert the area of the polyline showing 266 SQ. FT.
8. Go to Full Plan layout.
9. Zoom to the lower-right portion of the title block.
10. Using the Application Menu, go to the Drawing Utilities, then select the Drawing Properties. Go to the Summary tab and at the Title input, Munir Hamad Villa. At Author, input your name then click OK to end the command.
11. Make Standard the current text style.
12. Using the Field command and under the title Project Name, insert the Title field using uppercase letters. Under Designed By, insert the Author, using an initial capital letter.
13. Under Date insert today's date using the MMMM d, yyyy format.
14. Under Filename insert the filename, only without the extension using lowercase letters.
15. Remove the background for fields.
16. Save and close the file.

13.8 USING PARTIALLY OPENED FILES

- You may deal with huge drawings, containing plenty of views and layers. Big files tend to take a long time to open. To eliminate this problem, you can use the Partial Open command. Later, you can use the Partial Load command to add more contents to the partially opened file.

13.8.1 How to Open a File Partially

- Use the normal Open command to select the desired file but DO NOT click Open as you always do. Instead, click the small arrow at its right to see the following list of options. Select Partial Open:

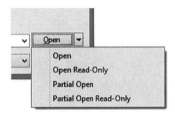

- You will see the following dialog box:

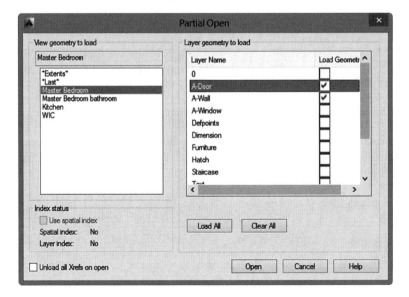

- You can do any of the following:
  - Select the desired layers to open.
  - Select the desired views to open, or you can use Extents, or Last views.
  - Then click Open to open the file with the settings you selected.

13.8.2 Using Partial Load

- This command is not applicable on normal files; instead, you can use it only on the files that are partially opened. To use this command, make sure the menu bar is shown, then select **File / Partial Load**, or you can type **PARTIALLOAD** in the command window. You will see the following dialog box:

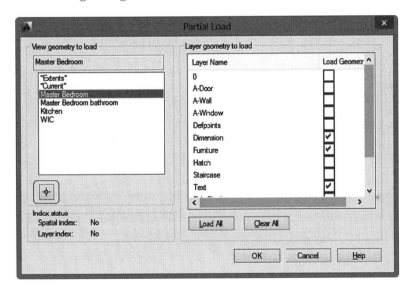

- As you can see, this dialog box matches the Partial Open dialog box with the exception of the small button at the lower-left portion of the dialog box. This button will specify a window in the drawing to specify the extents of objects to be loaded.

- Saving the partially loaded file means AutoCAD will keep these settings working until you change them. When you try to open the drawing again, the following dialog box will be shown:

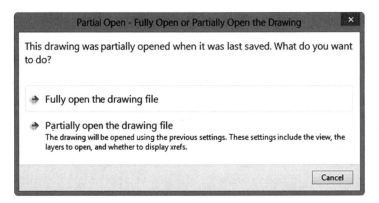

PRACTICE 13-6

Using Partially Opened Files

1. Start AutoCAD 2015.
2. Using the Partial Open command, open the file, Practice 13-6.dwg partially, using Master Bedroom view and A-Door, A-Wall, and A-Window layers.
3. Using Partial Load, select layer Furniture.
4. Save and close the file.
5. Open the drawing again using the normal Open command; there will be a message asking you either to open it fully or partially, select partially.
6. Zoom out to see that all of the furniture is shown.
7. Use the Partial Load again, select Hatch layer, and click the Pick a window button. Select the area of the kitchen and then click OK.
8. Save and close the file.

CHAPTER REVIEW

1. The Content Explorer:

    a. Has a Google-like index, which allows it to search and find the desired content quickly.

    b. Adds watched folders.

    c. Searches for a word in a drawing.

    d. All of the above.

2. The _____ command will work only on partially opened files.

3. Using the Select Similar command:

    a. You can select all circles.

    b. You can select all circles in the same layer.

    c. You can select all circles with the same color.

    d. All of the above.

4. One of the following statements is not true.

    a. You can isolate objects and isolate layers.

    b. You can control field update settings.

    c. You need to empty the layer in order to delete using Delete in Layers panel.

    d. You can hide objects rather than hiding layers.

5. You can insert the area of a polyline as a field without a background.

    a. True

    b. False

6. _____ will show a dialog box listing all layers in the current file. To show the contents of a layer, click its name in the list.

CHAPTER REVIEW ANSWERS

1. d
3. d
5. a

CHAPTER 14

USING BLOCK TOOLS AND BLOCK EDITING

In This Chapter

- How to use Automatic Scaling
- How to use Design Center
- How to use and customize Tool Palette
- How to edit a block

14.1 AUTOMATIC SCALING FEATURE

- When you create a block, you input the Block Unit, as if you are telling AutoCAD that each unit you used in this block will represent a certain unit. In order for this circle to be completed, you have to control the drawing file unit. Go to the **Application menu**, select the **Drawing Utilities**, and then select **Units**. You will see the following dialog box:

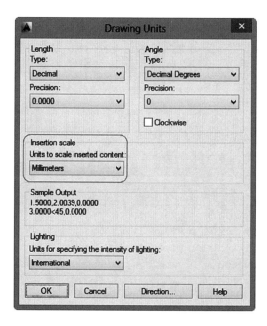

- Under **Units to scale inserted content,** select the desired unit that represents your drawing file unit. AutoCAD will convert the block unit to the drawing unit, and will show this in the **Insert** dialog box under the title **Block unit**, which is read-only.

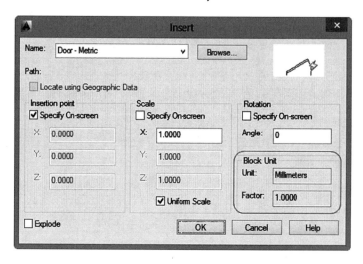

14.2 DESIGN CENTER

- Before AutoCAD 2000, there was no direct method to share blocks, layers, and other things. In AutoCAD 2000, Design Center was introduced as the ultimate solution for this problem. While in Design Center, you can share blocks, layers, dimension styles, text styles, and table styles, etc., that are created in other files and put them in yours.
- To issue this command, go to the **Insert** tab, locate the **Content** panel, then click the **Design Center** button:

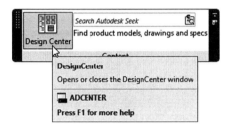

- You will see the following palette:

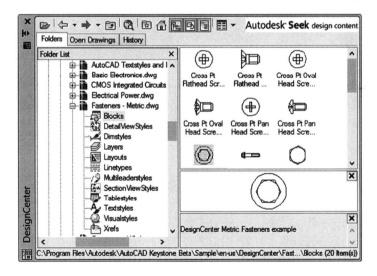

- The Design Center left pane is like My Computer in Windows OS, containing a list of all of your drives, folders, and files. This will enable you to locate the desired file, which contains the needed blocks, and layers, etc. When you locate the file, click the plus sign at the left of the file name, a list will come up. See the following:

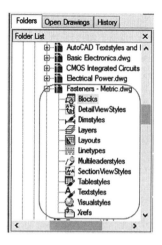

- Select the desired content. If you select Blocks, you will see the shapes and the names in the right pane. There are three methods to copy blocks using the Design Center to your current drawing.
  - Drag-and-Drop using left button to insert the block in the current file.
  - Drag-and-Drop using right button. When you release it, you will see the following menu (self-explanatory).

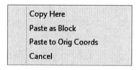

  - Double-click and the Insert dialog box will open.

NOTE *You can download free-of-charge AutoCAD 2D/3D blocks from Autodesk Seek. All you have to do is type the word of what you want to search for in the same panel and click the Autodesk Seek button, as in the following:*

USING BLOCK TOOLS AND BLOCK EDITING • **421**

- The browser will open on the Autodesk Seek website, along with the search results, as in the following:

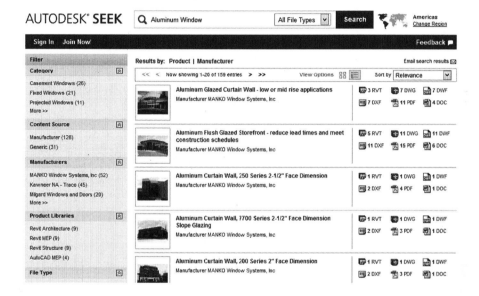

- Click on DWG file symbol at the right of the desired block and select it to download to your local hard disk.

PRACTICE 14-1

Using Design Center

1. Start AutoCAD 2015.

2. Open the file, Practice 14-1.dwg.

3. Make layer Furniture the current layer.

4. Start Design Center.

5. Locate where your AutoCAD 2015 folder is, then go to the following path: \Sample\en-us\Design Center (en-us folder is assumed for the English language AutoCAD; this folder may change depending on the language chosen).

6. Locate the file Home-Space Planner.dwg and choose Blocks to put the furniture shown.

7. Make layer Toilet the current layer.

8. Locate the file House Designer.dwg and choose Blocks to put the toilet furniture as shown.

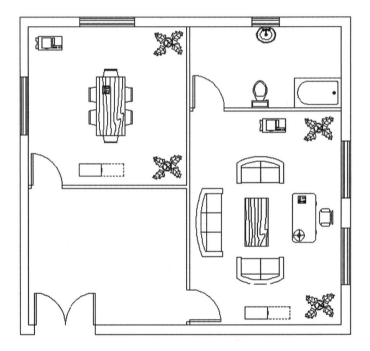

9. Save and close the file.

14.3 TOOL PALETTES

- Using the Design Center to share data between files was a blessing. However, there remain some hardships to deal with. You have to be careful of the current layer, scale, rotation angle, and agonizing idea of

searching for your desired file each time you need to copy something from it. We waited until AutoCAD 2004, when AutoCAD introduced us to Tool Palettes which will take blocks to the next level.
- Tool Palettes allow you to store any type of objects and then retrieve them in any opened file. You can also control the object's property so next time you drag it in your file, you will not worry about layers, scale, or rotation angles. You can keep several copies of the same block; each will hold different properties.
- To start the command, go to the **View** tab, locate the **Palettes** panel, then click the **Tool Palettes** button:

- You will see something like the following:

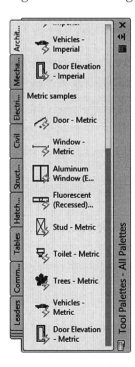

- By default, there are some tool palettes come with AutoCAD, the possibility to create your own palette is there.

14.3.1 How to Create a Tool Palette from Scratch

- This method will create an empty palette, so you will be able to fill it using different methods. To do this, right-click over the name of any existing tool palette. You will see the following menu:

- Pick the **New Palette** option, a new empty palette will be created, allowing you to name it right a way. Type in the name of the new palette as shown:

14.3.2 How to Fill the New Palette with Content

- You can fill the new empty palette with content using the drag-and-drop method from the different drawings to the palette. Say you opened one of your colleague drawing files, and you discovered that she created several new blocks, simply click the block, avoiding the grips, hold, drag it to the palette and here you go. You can do the same with normal objects (lines, arcs, circle, polylines, hatches, tables, dimensions, etc.).

14.3.3 How to Create a Palette from Design Center Blocks

- AutoCAD allows users to create a palette from all blocks in a file using the Design Center. In order for this to work, make sure that both the Design Center and Tool Palettes are displayed in front of you. Go to your desired file in the left pane of the Design Center and right-click. You will see something like the following:

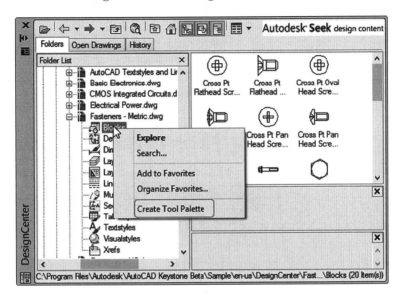

- Select the **Create Tool Palette** option; a new tool palette holding the same name of the file will be created containing all the blocks.

NOTE *You have the ability to make a drag-and-drop for any block in any file from the Design Center to any tool palette.*

14.3.4 How to Customize Tools Properties

- You can create several copies of blocks and hatches in the tool palettes using the normal Copy/Paste procedure. Once you have several copies of the same block/hatch, you can change the properties of these copies according to your needs. Follow these steps:

- Right-click on the copied block/hatch. You will see the following menu:

- Pick the **Properties** option. You will see something like the below dialog box:

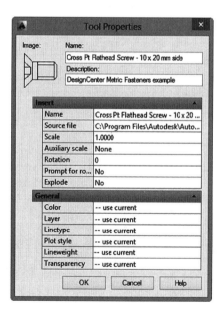

- Type in a new **Name** and new **Description**. Properties of block/hatch are cut into two categories:
  - **Insert** type of properties.
  - **General** type of properties.
- By default, the **General** properties are all **use current**. For a block or a hatch, you can specify that whenever you drag from the tool palette

it will reside in a certain layer (regardless of the current layer) and it will hold a certain color, linetype, or lineweight, etc.
- With this, you can create your own tool palettes, which hold all the needed blocks and hatches, customized according to the company standard so it will become a simple drag-and-drop process. If we know that 30 to 40% of any drawing are blocks and 10 to 20% are hatches, this means the time you save in these two areas will be enormous, if you use tool palettes effectively.

14.4 HATCH AND TOOL PALETTE

- Just like blocks, you can store hatch in tool palettes. Storing hatch in tool palettes is not important as storing hatches with altered properties such as layer, scale, color, etc., ensures a simple and fast hatch insertion in the drawing. Issue the Tool Palettes command and select the Hatches tab as in the following:

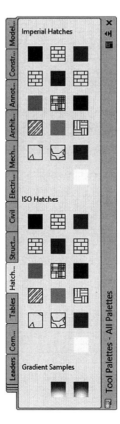

- Use the Drag-and-drop technique from the drawing and to the drawing. Inside the tool palette right-click and pick the **Properties** option to change the properties of a saved hatch.

PRACTICE 14-2

Using Tool Palettes

1. Start AutoCAD 2015.
2. Open the file, Practice 14-2.dwg.
3. Open the Design Center, and locate /Sample/en-us/Design Center/Fasteners-US.dwg.
4. Locate the Block underneath it, right-click, then select the Create Tool Palette option.
5. If the Tool Palettes is not displayed, it will be displayed with a new palette called Fasteners-US. Close the Design Center.
6. Make sure that the current layer is 0.
7. Using the newly created tool palette, locate Hex Bolt ½ inch-side, right-click, and select Properties. Change the layer to be Bolts.
8. Create another copy from it and make the Rotation angle = 270.
9. Using drag-and-drop, drag the two blocks to the proper places as shown.
10. Using the newly created tool palette, locate the Square nut ½ inch – top and set its layer to be Nut.
11. Create a copy of Square nut ½ inch – top and set the scale to 1.5.
12. Using drag-and-drop, drag the new block to the proper places as shown.
13. Erase the lines to get the following shape:

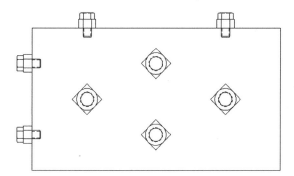

14. Close tool palettes. Save and close the file.

14.5 CUSTOMIZING TOOL PALETTES

- You can customize the looks of tool palettes to satisfy their needs. Set the transparency, set view options, and so on. All of these functions will be available at the right-click menu.

14.5.1 Allow Docking

- You can be allowed to dock the tool palette to right or left of the screen to make the tool palette permanent. The Default mode is to ensure the tool palette is floating. To do this, make sure that the tool palette is shown then right-click any empty space within the tool palette (avoid icons). You will see the following menu. Select Allow Docking (if you see (✓) at its left, this means is on):

- You can drag the tool palette to the right or left of the screen.

14.5.2 Transparency

- You can set the transparency value of the tool palette. The default value is 100% opacity. To do this, make sure that the tool palette is shown. Right-click any empty space within the tool palette (avoid icons). You will see the following menu. Select the Transparency option:

- You will see the following dialog box:

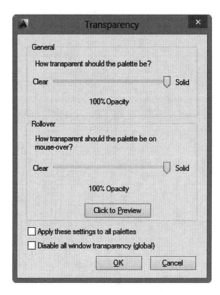

- You can change the value of the transparency for the palette (use the slider under General) and for Rollover (when your mouse is over the palette). The value of the transparency of Rollover should be always equal or more than the general value.
- Also, select whether these settings affect the current palette or all palettes, and whether to disable all window transparency or not.

14.5.3 View Options

- You can set how the icons will appear inside the palette. To do this, make sure that tool palette is shown, right-click any empty space within the tool palette (avoid icons). You will see the following menu and should select the View Options option:

- You then see the following dialog box:

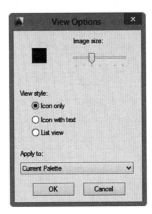

- Using this dialog box, you can:
  - Set the size of the image.
  - Set the view style: Icon only, Icon with text, or List view.
  - Choose whether to apply these changes to the current palette or all palettes.

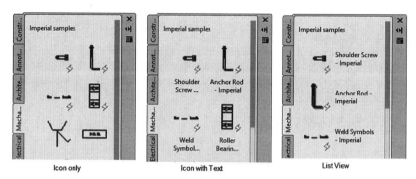

14.5.4 Add Text and Add Separator

- You can add text to create internal grouping for each palette and also add some lines that will work as a separator. To do this, make sure that the tool palette is shown and right-click any empty space within the tool palette (avoid icons). You will see the following menu, and should select the Add Text and Add Separator options:

- This is what you will get:

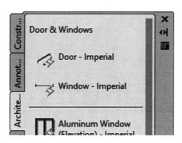

14.5.5 New / Delete / Rename Palette

- These three commands will create a new palette, delete an existing palette, or rename an existing palette. To do this, make sure that the tool palette is shown and right-click any empty space within the tool palette (avoid icons). You will see the following menu and should select the New Palette, Delete Palette, and Rename Palette options:

14.5.6 Customize Palettes

- You can create palette groups to organize palettes. You can then set one of the groups to be the current palette group. To do this, make sure that the tool palette is shown and right-click any empty space within the tool palette (avoid icons). You will see the following menu and should select the Customize Palettes option:

- You will see the following dialog box:

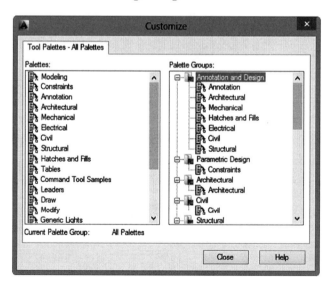

- To the right part, you will see the pre-defined palette groups along with the palettes belonging to them. To the left, you will see the defined palettes. In order to create a new group, at the right-part right click anywhere. You will see the following menu:

- Select the New option and type the name of the new group. You may move the group from any place to any place using the drag-and-drop technique.
- To fill this group with palettes, use the drag-and-drop technique again from the left part to the right part.
- To make this group the current group, right-click it then select Set Current option:

- Using this menu, you can Rename a group, Delete a group, Export one group, Export all groups, and finally Import groups. The Palette group file extension is °.XPG.

PRACTICE 14-3

Customizing Tool Palettes

1. Start AutoCAD 2015.
2. Open the file, Practice 14-3.dwg.
3. Start the Tool Palette command.

4. Using the tool palette created from Practice 14-2 rearrange the tools in Fasteners to see all the nuts, screws, and bolts together.

5. Add separators between the three types and add text as the title for each type.

6. Change the transparency of the tool palette to be 50% for general.

7. Using the View Options, change the size of the tool to be less than maximum with one degree, showing Icons with text.

8. Using the Customize Palettes create a new group and call it My Group.

9. Drag-and-drop Fasteners – US to it.

10. Make My Group the current group (you can see that this group contains only one tool palette).

11. Export My Group to be My Group.xpg (you can go to another computer and try to import the same).

12. Save and close the file.

14.6 EDITING BLOCKS

- You can edit the original block in Block Editor, which is used to create dynamic features of a block (this is an advanced feature of AutoCAD). The Block Editor will redefine the block by adding/removing/modifying the existing objects.
- To issue the command, go to the **Insert** tab, locate the **Block Definition** panel, then click the **Block Editor** button:

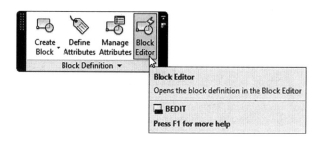

- You will see the following dialog box:

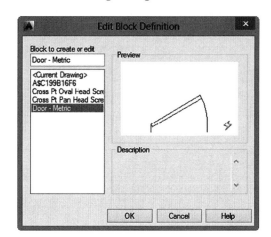

- Select the name of the desired block to edit and then click **OK** to start editing. Another easy way is to double-click one of the block incidences.
- When the Block Editor opens, you will see many things take place such as the background color will be different; the new context tab "Block Editor" will appear; and numerous new panels will appear as well. Ignore all of this and start adding, removing, or modifying objects of your block. Once you are done, click the **Save Block** button in the **Open/Save** panel (you can find it at the left) and then click **Close Block Editor** button in the **Close** panel to end the command.

PRACTICE 14-4

Editing Blocks

1. Start AutoCAD 2015.
2. Open the file, Practice 14-4.dwg.
3. Double-click of the incidences of block Single Door.
4. Change the arc linetype to be Dashed2.
5. Save the block with the new changes.
6. What happened to the other incidences of block Single Door?

7. You should have the following shape:

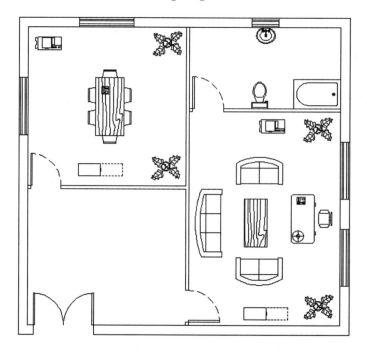

8. Save and close the file.

NOTES:

CHAPTER REVIEW

1. Tool Palette can store commands such as line and polyline, etc.

    a. True

    b. False

2. _____ is the tool to allow users to share blocks, layers, etc.

    a. Insert Center

    b. Design Office

    c. Design Center

    d. The Internet

3. Using Tool Palettes, you can customize all tools by changing their properties.

    a. True

    b. False

4. Double-clicking a block will show _____ dialog box.

5. If both Design Center and Tool Palettes are open, you can drag any block from Design Center to a Tool Palette.

    a. True

    b. False

6. Using _____ of the right-click menu will give you Icon only, Icon with text, and List view.

    a. Transparency

    b. View Options

    c. Auto Hide

    d. Allow Docking

CHAPTER REVIEW ANSWERS

1. a
3. a
5. a

CHAPTER 15

CREATING TEXT, TABLE STYLES, AND FORMULAS IN TABLES

In This Chapter

- How to create a Text Style
- How to create a Table Style
- Using the Table command to insert a Table
- How to use Formulas in Tables
- How to utilize the cell functions in Tables

15.1 STEPS TO CREATE TEXT AND TABLES

- Writing text in AutoCAD involves two simple steps:
  - Create a text style (normally will be created once) that includes the size and the shape of the text
  - Write text using either the Single Line Text or Multiline Text
- Normally creation of any style is tedious and lengthy and usually falls on the shoulders of the CAD managers, who should always think about standardization, Text styles is also part of this process.
- The same thing applies to table, because users will perform two steps:
  - Create a table style.
  - Insert and fill a table.
- Text Style and Table Style and all the other styles should be part of Template, which should hold the company standards. If you do not

have a template, you still can share these styles using the Design Center discussed in Chapter 14.

15.2 HOW TO CREATE A TEXT STYLE

- The first step in writing text in AutoCAD is to create a text style. Text style is where you define the characteristics of your text. To start the Text Style command, go to the **Annotate** tab, locate the **Text** panel, and then click the small arrow at the lower right part:

- The following dialog box will appear:

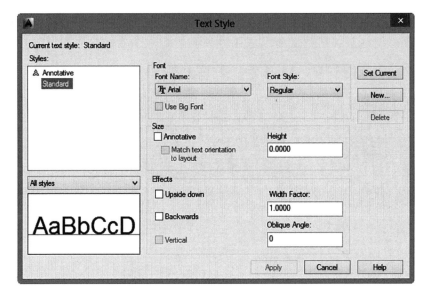

- There will be two pre-defined text styles, one called **Standard** and the other called **Annotative**. Virtually they are the same, except the latter uses the Annotative feature (discussed in Chapter 9). Both styles use Arial as font. Professional users will be advised not to use these two but to create their own. To create a new text style, click the **New** button and you will see the following dialog box:

- Input the name of the new text style and when you are done click **OK**. Then, select the **Font**. There are two types of fonts which you can use in AutoCAD:
    - Shape files (*.shx) is the very old method of fonts (out-of-date).
    - True Type fonts (*.ttf).
- See the following illustration to identify that TTF files are more accurate and look better:

True Type Font   Shape file font

- Next, select the **Font Style** (applicable if true type font) and choose one of the following:
    - Regular
    - Bold
    - Bold Italic
    - Italic
- Keep **Annotative** off for now (discussed later in this chapter). You have to specify the **Height** of the text, which is the height of the

capital letters; small letters will be 2/3 of the height. See the following illustration:

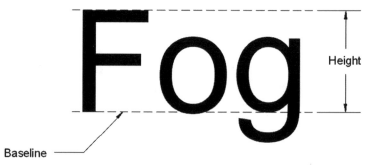

Text height is the Capital letter height

- How do you set the height of the text?
  - Setting the height to 0 (zero) means it will be variable (you have to input the value each time you use this style).
  - Setting the height to a value greater than 0 (zero) means the height is fixed.
- Finally, set the effects. You have five of them:
  - Upside down writes text upside down.
  - Backwards writes text from right to left.
  - Vertical to write from top-to-bottom. Good for Chinese words, but only for Shape files.
  - Width Factor to set the relationship between the width of the letter and its length. If the value>1.0, text is wide. If the value is <1.0, text is long.
  - Oblique Angle sets the angle to italicize either to the right (positive) or to the left (negative).
- Whenever you are done, click the **Apply** button then the **Close** button.
- You can show **All styles** or show **Styles in use**. See the pop-up list at the left as shown:

PRACTICE 15-1

Creating Text Style and Single Line Text

1. Start AutoCAD 2015.
2. Open the file, Practice 15-1.dwg.
3. Create a text style with the following specs:
   a. Name = Room Names
   b. Font = Tahoma
   c. Annotative = off
   d. Height = 0.3
   e. Leave the rest to default values
4. Make Room Names text style current.
5. Make layer Text current.
6. Type the room names as shown.
7. Make the Standard text style current (this text style has the height = 0 so you should set it every time you want to use this style).
8. Make layer Centerlines current.
9. Zoom to the upper-left centerline and check the letter A is missing.
10. Start Single Line text, right-click, and select the Justify option from the list. Pick MC (Middle Center), select the center of the circle as the Start point, and set the height to be 0.25, rotation angle = 0. Type A then press [Enter] twice.

11. You should get the following results:

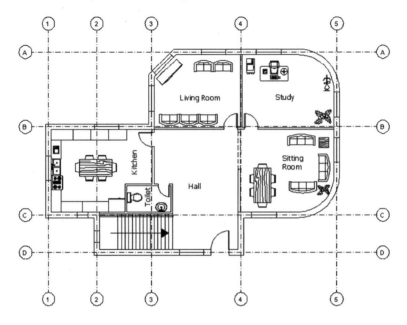

12. Save and close the file.

15.3 CREATING TABLE STYLE

- In order to create a professional table, you should create a Table Style, which holds the features of the table specifying the Title, Header, and Data rows. Using this style, you can insert as many tables as you wish. Table Style can be shared using the Design Center.
- To issue this command, go to the **Annotate** tab, locate the **Tables** panel, then select the **Table Style** button:

- The following dialog box will be displayed:

- Just like text style, there is a pre-defined table style called Standard. Click the **New** button to create a new table style, you will see the following dialog box:

- Input the name of the new table style and click **Continue**.

- You will see the following dialog box:

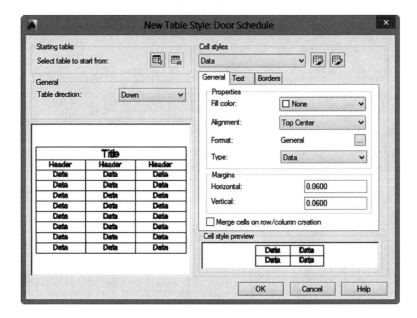

- AutoCAD will allow you to start a new table style based on an existing table in your current drawing. Click the button shown:

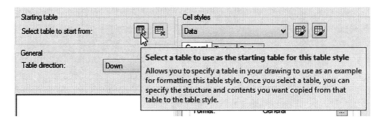

- If not, you need to specify the characteristics of your table by selecting the **Table direction**, whether **Down** or **Up**. See the following illustration:

- AutoCAD has three portions of any table: Title, Header, and Data. Using the Table Style command, you can set these three parts by selecting the desired portion, then by setting the features using the **General** tab, **Text** tab, and **Border** tab.

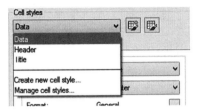

15.3.1 General Tab

- This the General tab:

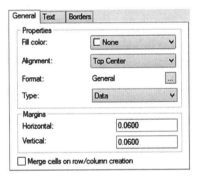

- Edit the following features:
  - Change the **Fill Color** of the cells (by default it is None).
  - Change the **Alignment** to setup the text related to the cell borders. For instance, if you say Top Left. This means the text will reside in the top left part of the cell. See the following illustration:

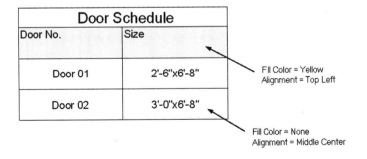

- Change the **Format**. You will see the following dialog box, which allows you to set the format of the cell, whether currency, percentage, date, etc.

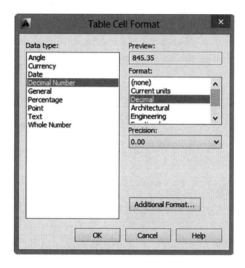

- Change the **Type** of the contents of the cell whether **Data** or **Label**. This portion is very important, as some of the cells may hold numbers, but these numbers should not be included in a mathematical formula. Therefore, the type of the data will be Label, and not Data.
- Change the **Margins** left for the text away from the border of the cell, the **Horizontal** margin and **Vertical** margin.

15.3.2 Text Tab

- This is the Text tab:

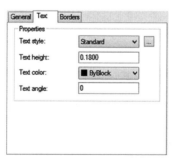

- Edit the following features:
  - Change the **Text style** of the text filling the cells.
  - Change the **Text height** of the text filling the cells (noting that if the used Text style has a height > 0.0, the number here is meaningless).
  - Change the **Text color**.
  - Change the **Text angle**. See the following illustration:

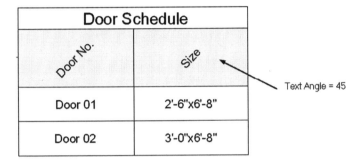

15.3.3 Borders Tab

- This is the Borders tab:

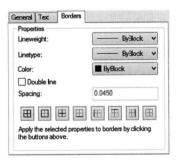

- Edit the following features:
  - Change the **Lineweight**, **Linetype**, and **Color** of the border lines.
  - Change the border to Double line instead of Single line (default value). Also choose the Spacing between the lines.
  - Change whether you want lines representing column separators, row separators, or not.

15.4 INSERTING A TABLE IN THE DRAWING

- This command allows you to insert a table in the current drawing. To issue this command, go to the **Annotate** tab, locate the **Tables** panel, then select the **Table** button:

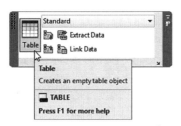

- You will see the following dialog box:

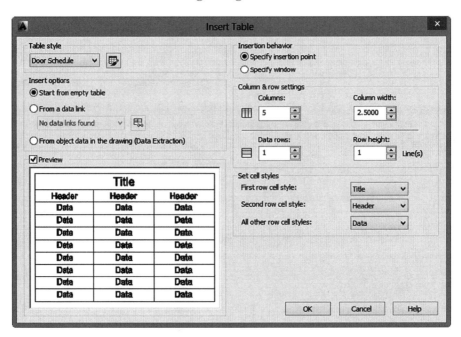

- The first step is to select the desired pre-defined **Table style** name from the available list. If you did not create your table style yet, click the small button beside the list to go and define it right now.
- Select the proper **Insert options**. To insert the table in your drawing, pick the following three choices:

- Start from empty table.
- From a data link: this is an advanced feature of AutoCAD.
- From object data in the drawing (Data Extraction): this is also an advanced feature of AutoCAD.

■ Most likely, you will click the first choice, which is Start from empty table. Now select the **Insertion behavior** and select one of the two available choices:
- Specify insertion point
- Specify window

15.4.1 Specify Insertion Point Option

■ The insertion point meant here is the upper-left corner of the table. You will hold the table from this point. To fulfill the rest of information, fill in the following data:
- Columns (which means the number of columns)
- Column width
- Data rows (which means the number of rows, but without Title and Heads)
- Row height (in lines)

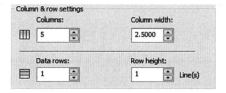

■ Click **OK** and AutoCAD will show the following prompt:

```
Specify insertion point:
```

■ Specify the location of the table, and start filling the cells. You can use arrows on the keyboard or the [Tab] key to jump from one cell to another. Use the [Shift] + [Tab] to go backwards.

15.4.2 Specify Window Option

■ Using this option you will specify a window, which means you will give AutoCAD the total length and the total width of the table. In order to fulfill the rest of the information, fill in the following:

- Input the number of **Columns** and AutoCAD will calculate the column width. Or input the **Column width** and AutoCAD will calculate the number of columns.
- Input the number of **Data rows** and AutoCAD will calculate the row height. Or input the **Row height** and AutoCAD will calculate the number of rows.

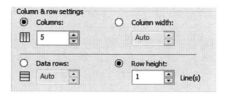

- Click **OK** and AutoCAD will show the following prompt:

```
Specify first corner:
Specify second corner:
```

- Specify the two opposite corners of the table, then start inputting the cell contents using the arrow, [Tab], and [Shift] + [Tab] keys.

PRACTICE 15-2

Creating Table Style and Inserting a Table in the Current Drawing

1. Start AutoCAD 2015.
2. Open the file, Practice 15-2.dwg.
3. Create a new Table Style with the following features:
   a. Name = Door Schedule
   b. Title Text style = Notes
   c. Header Fill color = Yellow
   d. Header Text style = Notes
   e. Data Alignment = Middle Left
   f. Data horizontal margin = 10
   g. Data Text style = Notes
   h. Data Text color = Blue

4. Create a table using the previous Table style and the frame drawn. Use the Window option to insert the table.

Door Schedule	
Door No.	Size
Door 01	2'-6" x 6'-8"
Door 02	3'-0" x 6'-8"

5. After finishing the table input, erase the frame.

6. Save and close the file.

15.5 USING FORMULAS IN TABLE CELLS

- This section discusses how to create formulas in AutoCAD table cells, along with more functions such as merging cells and using premade functions. If you know how to use Excel, you can skip this part. If not, read it because you will need it in the coming parts:
  - We call the intersection of column and row in a table a **Cell**.
  - The cell address comes from the column letter and row number. Example B15 means this cell is in column B and row 15.
  - To make AutoCAD know that you will be writing a formula, start with the equal sign. Along with the cell addresses, you can use arithmetic functions such as add, subtract, multiply, and divide. For example, =(B3/2)+A9.
  - AutoCAD provides some premade functions like SUM, COUNT, and AVERAGE.
  - To copy formulas from one cell to another, AutoCAD opted to make the cell address relative. For example, say you have a formula in cell C3 that says =A3+B3. If you copy it to C4, what is the result? AutoCAD will change the formula to =A4+B4. AutoCAD provides a tool called

Autofill grip in order to copy formulas (or any cell content) click and drag it upward or downward.

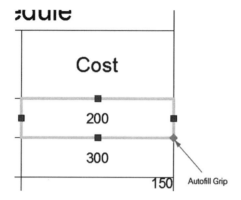

- Adding a dollar sign $ will attach the formula either to a column, row, or both. For example, $F12, means column F is the column to use while copying formulas with variable rows. If you type F$12, this means row 12 is the row to use while copying with variable columns. F12 means a fixed address.

15.6 USING TABLE CELL FUNCTIONS

- There are three different modes for table editing. They are:
  - If you click the outer frame, you can edit the whole table by adjusting the total width/height for the table or columns and rows just like the following:

  - If you click a cell or group of cells (to select a group of cells, click and drag) a new context tab called Table Cell will be added to control the cells.
  - If you double-click inside a cell, you will be able to input data or edit existing data. The Text Editor context tab will appear to help you finish your job.

- The following section covers the second case, which is the Table Cell context tab and its contents.

15.6.1 Using Rows Panel

- You will use this panel:

- You can do any of the following:
  - **Insert** a new row **Above** the selected cell(s).
  - **Insert** a new row **Below** the selected cell(s).
  - **Delete** the selected **Row(s)**.

15.6.2 Using Columns Panel

- You will use this panel:

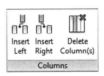

- You can do any of the following:
  - **Insert** a new column to the **Left** of the selected cell(s).
  - **Insert** a new column to the **Right** of the selected cell(s).
  - **Delete** the selected **Column(s)**.

15.6.3 Using Merge Panel

- You will use this panel:

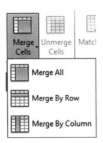

- You can do any of the following:
  - Merge all selected cells in a single cell.
  - Merge selected cells using rows.
  - Merge selected cells using columns.
  - If you have cells containing different types of data (numbers, text, dates, etc.) which one will be used? The answer is the content of the first cell. You will see the following message:

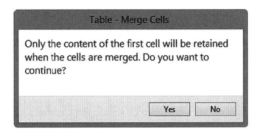

  - Unmerge cells to retain them back to the previous condition before merging.

15.6.4 Using Cell Styles Panel

- You will use this panel:

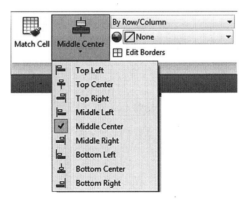

- You can do any of the following:
  - **Match cell** matches the properties of a selected cell with other cells.
  - Select the **Alignment** using one of the nine available alignments for the content relative to the cell borders.

CREATING TEXT, TABLE STYLES, AND FORMULAS IN TABLES • **461**

- Select one of the existing cell style for the selected cells:

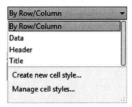

- Select one color to be the background for the selected cells:

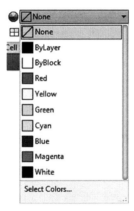

- Click the **Edit Borders** button and the following dialog box will come up:

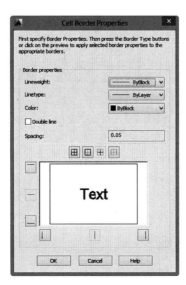

- You can control all properties of border such as lineweight, linetype, color, double line, and spacing. Show border lines in different locations.

15.6.5 Using Cell Format Panel

- You will use this panel:

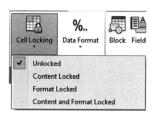

- You can do all or any of the following:
  - Select whether to lock the contents or format (or both) the selected cells. You can unlock the locked cells as well.
  - Select Data Format of the selected cells by choosing one of the following:

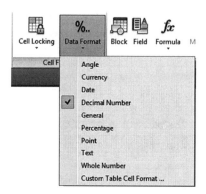

- You can also customize the data format to something else. Select the Custom Table Cell Format option and you will see the following dialog box:

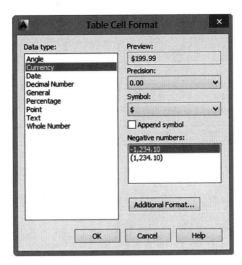

- Select the Data type you want to change and then make the necessary changes. You can click the Additional Format button to see the following dialog box then specify what to use for the Decimal and Thousands separator, and whether or not to suppress leading and trailing zeros:

15.6.6 Using Insert Panel

- You will use this panel:

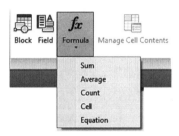

- You can do any of the following:
  - Click the **Block** button and the following dialog box will come up:

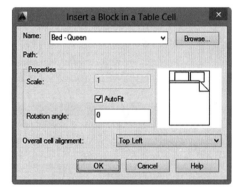

  - Select block name or click the Browse button to bring in any block or file, then specify the block's **Scale** or **AutoFit** the **Rotation angle**. Finally, specify the Overall cell alignment of the block reference to the cell border.
  - Click the **Field** button to insert a field in the cell.
  - Click the **Formula** button to insert a premade formula.

15.6.7 Using Data Panel

- You will use this panel:

- You can any of the following:
  - Click the **Link Cell** button and the following dialog box will come up:

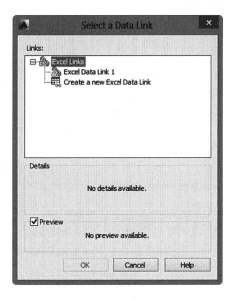

  - This is the same dialog box, which we dealt with while linking data with Excel. Create the link that you want. Then use the **Download from Source** button to bring in the data from this link.

PRACTICE 15-3

Formulas and Table Cell Functions

1. Start AutoCAD 2015.

2. Open the file, Practice 15-3.dwg.

3. Insert the block in the first column using the second column as your guide (for the first two blocks use AutoFit and for the rest use Sale = 0.01).

4. Select all Data cells and make them Middle Center.

5. Select one of the Cost column cells.

6. Insert a new column to its left and name it Quantity.

7. Input the following values in it from top to bottom: 4, 1, 1, 6, 4.

8. Select all the Data cells of Cost column.

9. Select the Custom Table Cell Format (from the Cell Format panel) and change the currency format to show a number like the following: $999.99.

10. Select one of the Cost column cells.

11. Insert a new column to its right and call it Total Cost.

12. Input a formula: Quantity * Cost.

13. Using the AutoFill grip, copy it to the other cells.

14. Select all the Data cells of Total Cost and make sure to add a comma as the thousands separator.

15. Select any cell in the lower row and add a new row below it.

16. Merge by row all the cells, except the rightmost cell. Type in this cell Total Cost and make it Middle Right.

17. Add the new cell Sum function using the range of cells containing the Total Cost.

18. Select the Title cell and make sure that the border lines at the top, right, and left are removed.

19. Set the background color of the Header to be Cyan.

20. Save and close the file.

NOTES:

CHAPTER REVIEW

1. In _____ you define the background color of the table cell.
2. There are two methods to insert a table in a drawing.
   a. True
   b. False
3. By default, the first row cell style is _____.
4. Height in Text style is for:
   a. Small letters
   b. Capital letters and small letters
   c. Capital letters only
   d. Everything above the baseline
5. While inserting a table using a window, specifying number of columns is enough. You do not need to specify also the column width.
   a. True
   b. False
6. In Table style, you can specify different text styles for the header and title.
   a. True
   b. False

CHAPTER REVIEW ANSWERS

1. Table Style
3. Title
5. a

CHAPTER 16

DIMENSION AND MULTILEADER STYLES

In This Chapter

- How to create a Dimension Style
- More dimension advanced commands
- How to create Multileader Style
- How to insert Multileader in the drawing

16.1 WHAT IS DIMENSIONING?

- Dimensioning in AutoCAD is just like text and tables. As the first step, you should prepare dimension style and then you can use it to insert dimensions. The Dimension Style controls the overall outcome of the dimension block generated by the different types of the dimension commands.
- To insert a dimension, depending on the type of the dimension, you should specify points, or select objects, and then a dimension block is added to the drawing. For example, in order to add a linear dimension,

you should select two points representing the distance to be measured, and a third point will be the location of the dimension block. See the following illustration:

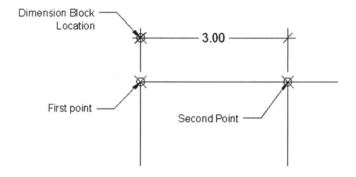

- The generated block consists of three portions. They are:
  - Dimension line
  - Extension lines
  - Dimension Text
- See the following illustration:

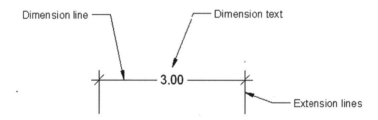

16.2 HOW TO CREATE A NEW DIMENSION STYLE

- This command allows you to create a new dimension style or modify an existing one. To issue this command, go to the **Annotate** tab, locate the **Dimensions** panel, then select the **Dimension Style** button:

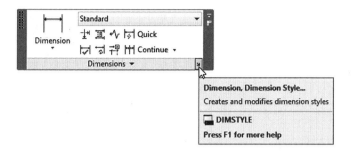

- The following dialog box appears:

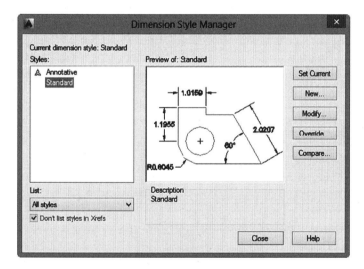

- As you can see, you will find two pre-defined dimension styles, Standard and Annotative. Click the **New** button to create a new style and the following dialog box appears:

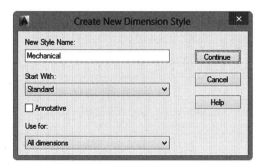

- Input the name of the new style to be made. Under **Start With** select the existing style, which is your starting point. Leave the Annotative checkbox off (discussed in Chapter 9) but do make sure that Use for All dimensions is selected. (We cover this later in this chapter.) To start creating the new style, click the **Continue** button.

16.3 DIMENSION STYLE: LINES TAB

- As a rule-of-thumb, and while we are discussing the different tabs of the dimension style command, we will leave Color, Linetype, and Lineweight to their default settings because we want these things to be controlled by layers rather than the individual dimension block.
- The first tab in the dimension style dialog box is Lines and it allows you control dimension lines and extension lines:

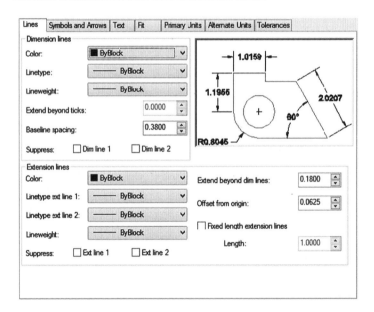

- Under **Dimension lines**, change any of the following settings:
  - Control **Extended beyond ticks** as the following illustrates (this option works only with **Arrowhead** equals to **Architectural tick** or **Oblique**):

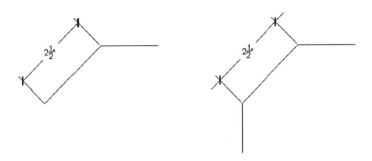

Extended beyond ticks = 0   Extended beyond ticks = 3/8"

- As we will see in this chapter, when you add Baseline dimension you do not control the spacing between a dimension and another control **Baseline spacing**, as illustrated:

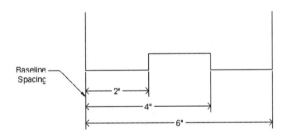

NOTE *From now on, "First" means nearest to the first point picked, hence "Second" means nearest to the second point picked.*

- Choose whether to **Suppress Dim line 1, Dim line 2**, or leave them as is. See the following picture:

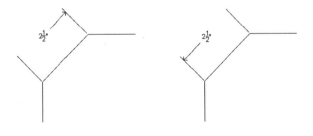

Suppress Dim Line 1   Suppress Dim Line 2

- Under **Extension lines** change any of the following settings:
  - Choose whether to **Suppress Ext line 1, Ext line 2,** or leave them as is. See the following illustration:

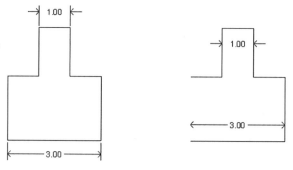

  - Input Extend beyond dim lines and Offset from origin. See the following illustration:

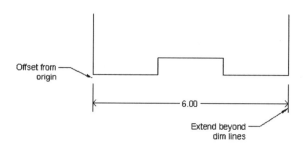

  - Choose whether to fix the length of the extension lines or not. If yes, what is the length? See the following example:

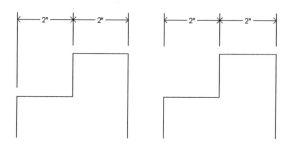

16.4 DIMENSION STYLE: SYMBOLS AND ARROWS TAB

- This tab allows you to control the arrowheads and related features. This is the **Symbols and Arrows** tab:

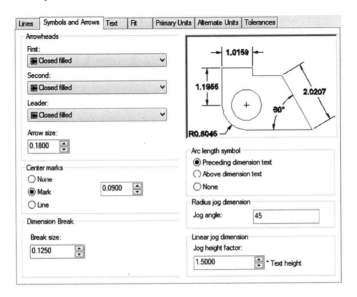

- Under **Arrowheads**, change any of the following:
  - The shape of the **First** arrowhead. When you set the shape of the first, the **Second** arrowheads automatically changes. If you want them to be different, change the second.
  - The shape of the arrowhead to be used in the **Leader** (Radius and Diameter are not leaders).
  - The **Size** of the arrowhead.
- Under **Center marks**, choose whether to show or hide the Center mark of arcs and circles as shown, then set the **Size** of the center mark:

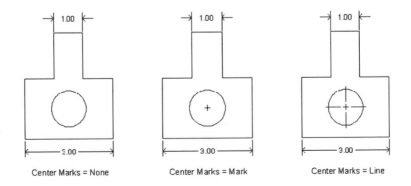

- Under **Dimension Break**, input the **Break size**. Break size is defined as the distance of void left between two broken lines of dimension. See the following illustration:

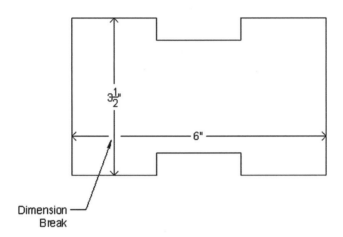

- Under the **Arc length symbol**, choose whether to show (as shown in the illustration below) or hide the arc length symbol:

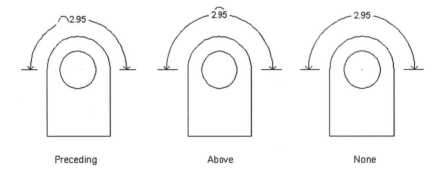

- Under the **Radius dimension jog**, input the **Jog angle** as shown in the following illustration:

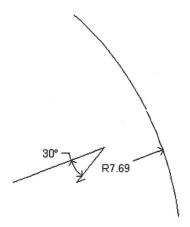

Jog Angle

- Under the **Linear Jog dimension**, input the **Jog height factor** as shown in the following illustration. The Jog height factor is defined as a factor to multiply the height of the text used in dimension:

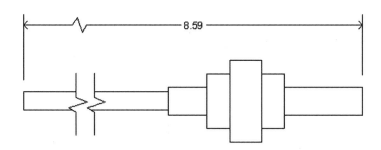

Jog height factor = 2.0

16.5 DIMENSION STYLE: TEXT TAB

- This tab allows you to control the text appearing in the dimension block. This is the **Text** tab:

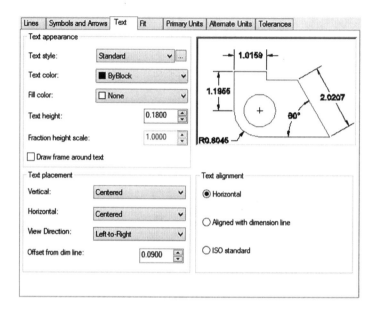

- Under **Text appearance**, change any of the following:
  - Select the desired Text Style or create a new one.
  - Specify the **Text color** and the **Fill color** (text background color).

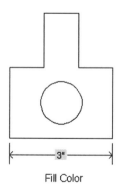

Fill Color

- If your text style has a text height = 0 (zero) then input the **Text height**.
- Depending on the primary units (discussed next), set the **Fraction height scale**.

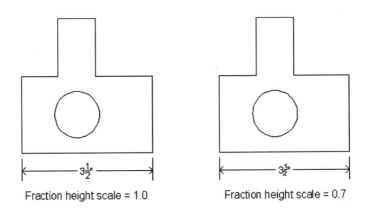

Fraction height scale = 1.0   Fraction height scale = 0.7

- Select whether your text will be with frame or without frame, see the following illustration:

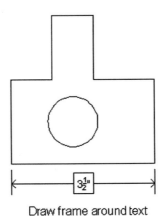

Draw frame around text

- Under **Text placement** change any of the following:
  - Choose the **Vertical** placement of your text related to the dimension line. There are five available choices: Centered, Above, Outside, JIS (Japan Industrial Standard), and Below. See the following picture:

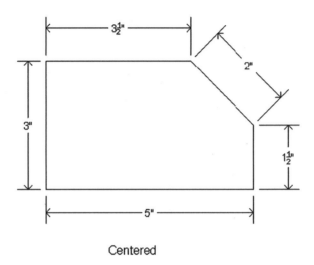

Centered

Above

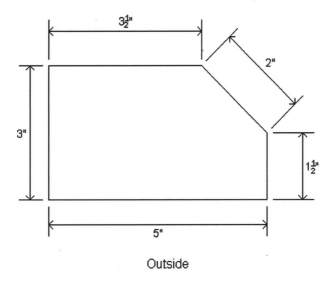

Outside

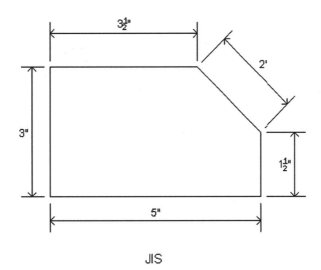

JIS

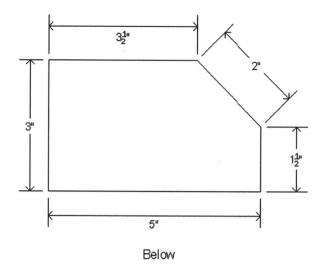

Below

- Choose the **Horizontal** placement; you have five choices to pick from. See the following illustration:

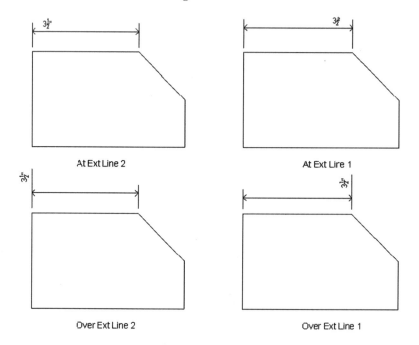

At Ext Line 2   At Ext Line 1

Over Ext Line 2   Over Ext Line 1

- Choose the View Direction of the dimension text Left-to-Right or Right-to-Left. See the following illustration:

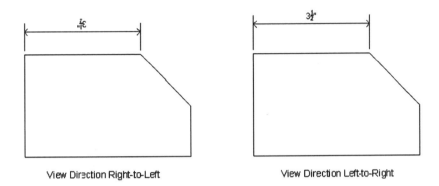

View Direction Right-to-Left   View Direction Left-to-Right

- Input the **Offset from dim line** as shown in the following:

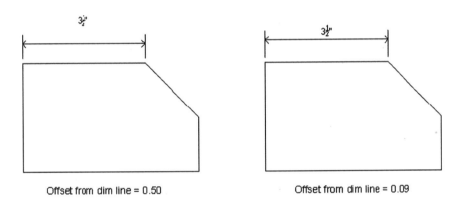

Offset from dim line = 0.50   Offset from dim line = 0.09

- Under **Text alignment**, control the alignment of the text related to the dimension line, whether always horizontal regardless of the alignment of the dimension line or aligned with the dimension line. ISO will influence only the Radius and Diameter dimension, and all of the dimension types will be aligned except for these two.

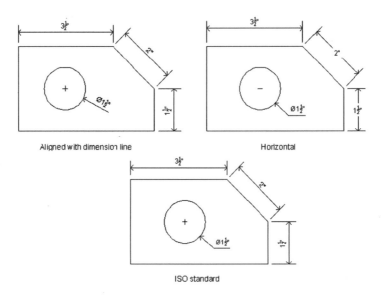

16.6 DIMENSION STYLE: FIT TAB

- This tab allows you to control the relationship between the dimension block components. This is the **Fit** tab:

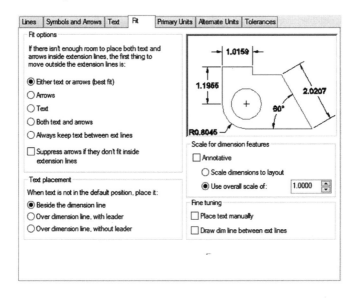

- Under **Fit options**, change any of the following:
  - If there is no room for the text and/or the arrowheads inside the extension lines, what do you want AutoCAD to do? Select the desired option.
  - If there is no room for arrows to be inside the extension lines, do you want AutoCAD to suppress them or not?
- Under **Text placement**, when the text is not in the default position, select one of the following options:

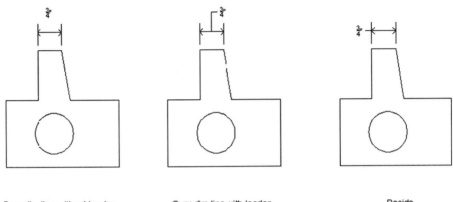

Over dim line without leader   Over dim line with leader   Beside

- Under **Scale for dimension features**, you can control the size of the text features (length, sizes, etc.). Will it be scaled automatically if it was annotative? Or will it follow the viewport scale if input in the layout? If you want to input it in the Model space, you can set the scaling factor.
- Under **Fine tuning**, select whether you want to place your text manually rather than leaving it to AutoCAD. Also, select whether to always force text inside extension lines.

16.7 DIMENSION STYLE: PRIMARY UNITS TAB

- This tab allows you to control everything related to the numbers, which appear in the dimension block. This the **Primary Units** tab:

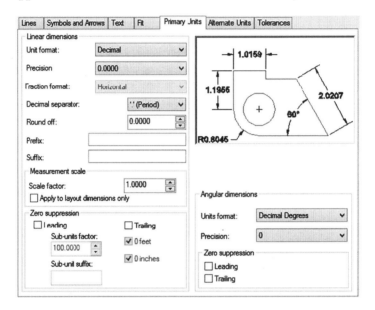

- Under **Linear dimensions**, change any of the following settings:
  - Select the desired **Unit format**, then select its **Precision**.
  - If your selection was **Architectural** or **Fractional**, then choose the desired **Fraction format**. You will have three choices to pick from. They are **Horizontal**, **Diagonal**, and **Not Stacked**.

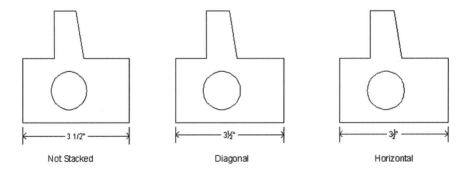

Not Stacked   Diagonal   Horizontal

- If your selection was **Decimal**, then choose the **Decimal Separator**. You have these three choices, **Period**, **Comma**, and **Space**.
- Input the **Round off** number.
- Input the **Prefix** and/or the **Suffix**, as in the following:

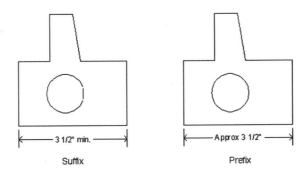

- Under the **Measurement scale**, change any of the following:
  - By default, AutoCAD will measure the distance between the two points specified by you (if linear) and input the text in the format set by you. What if you want to show a different value than the measured value? You simply input the **Scale factor**.
  - Then choose whether this scale will affect only the dimension input in the layout (discussed in Chapter 9).
- Under **Zero suppression**, choose whether to suppress the **Leading** and/or the **Trailing** zeros as shown:

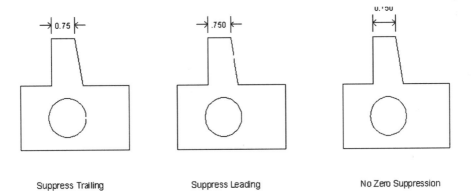

- A sub-unit is when you have meters as your unit and the measured value is less then one. Select the sub-unit factor and the suffix for it (in this example it is cm).
- Under **Angular dimensions**, choose the Unit format and the Precision. Control **Zero suppression** for angles as well.

16.8 DIMENSION STYLE: ALTERNATE UNITS TAB

- This tab shows two numbers in the same dimension block, one showing primary units and the other alternate units. This is the **Alternate Units** tab:

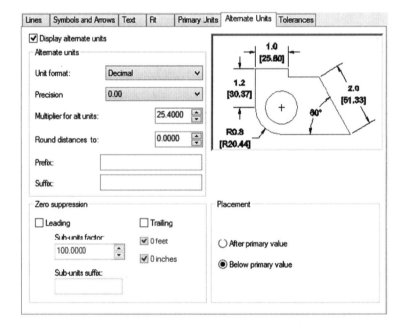

- Click on the **Display alternate units** option, then change the following:
  - Choose the Alternate **Unit format** and its **Precision**.
  - Input the Multiplier for all units value.
  - Input the Round distance.
  - Input the **Prefix** and the **Suffix**.
  - Input the **Zero suppression** method.

- Choose the method of displaying alternate units, whether **After primary value** or **Below primary value**. See the following:

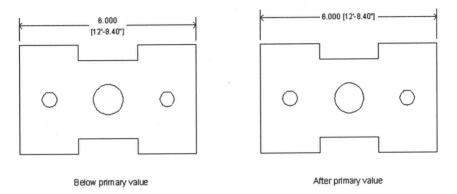

Below primary value　　　　　　　After primary value

16.9 DIMENSION STYLE: TOLERANCES TAB

- This tab allows you to control whether or not to show tolerances and the method used. This is the **Tolerances** tab:

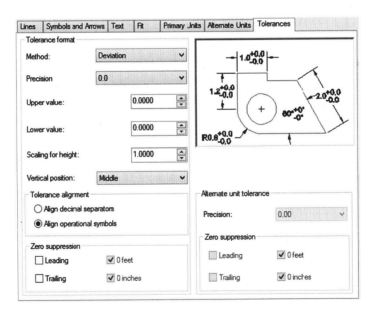

- There are four methods to show the tolerance. They are:
  - Symmetrical
  - Deviation
  - Limits
  - Basic
- The following illustrates each of the four choices:

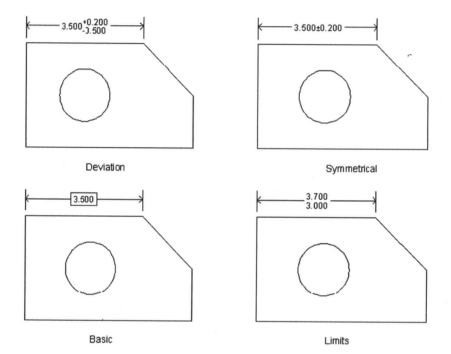

- Under **Tolerance format**, change a any of the following:
  - Select the proper **Method** and then select its **Precision**.
  - Depending on the method, specify the **Upper value** and **Lower value**.
  - Input **Scaling for height** for the tolerance values if desired.
  - Choose the **Bottom**, **Middle**, or **Top** vertical position for the dimension text with reference to the tolerance values. See the following illustration:

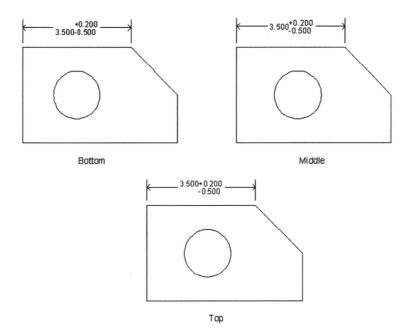

- If either **Deviation** method or **Limits** method is selected, then you have to choose whether to **Align decimal separators** or **Align operational symbols** as shown:

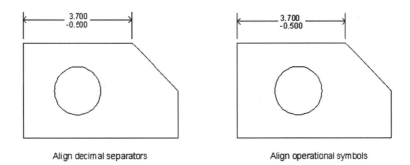

- Under Alternate units tolerances, and if **Alternate units** option is turned on, then specify the **Precision** of the numbers. Consequently, choose the **Zero suppression** for both the **Primary units** tolerance and the **Alternate units** tolerance.

16.10 CREATING A SUB DIMENSION STYLE

- By default, the dimension style created affects all types of dimension. If you want the dimension style to affect only a certain type of dimension and not the others, you have to create a sub style. Do the following steps:
    - Select an existing dimension style.
    - Use the **New** button to create a new style and you will see the following dialog box:

- Do not input anything. Go to **Use for** and select the type of dimension (in the following Diameter is selected). The dialog box will change to:

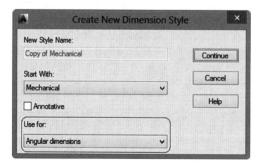

- Click the **Continue** button and make the changes you want. These changes affect diameter dimensions only.
- The picture of the **Dimension Style** dialog box allow you to differentiate between the style and the sub-style, and its features.

Dimension and Multileader Styles • 493

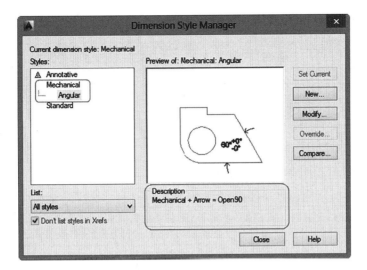

PRACTICE 16-1

Creating Dimension Style

1. Start AutoCAD 2015.

2. Open the file, Practice 16-1.dwg.

3. Create a new dimension style based on Standard, using the following information:

   a. Name: Part

   b. Extend beyond dim line = 0.3

   c. Offset from origin = 0.15

   d. Arrowhead = Right angle

   e. Arrow size = 0.25

   f. Center mark = Line

   g. Arc length symbol = Above dimension text

   h. Jog angle = 30

   i. Text placement vertical = Above

   j. Offset from dim line = 0.2

k. Text Alignment = ISO Standard

l. Text placement = Over dimension line with leader

m. Primary unit format = Fractional

n. Primary unit precision = 0 ¼

o. Fraction format = Diagonal

4. Click OK to end the creation process.
5. Select Part and create a sub style for Radius.

    a. Arrowhead = Closed filled

    b. Arrow size = 0.15

6. Make Dimension the current layer.
7. Make Part the current dimension style and add the dimensions so the shape looks like the following:

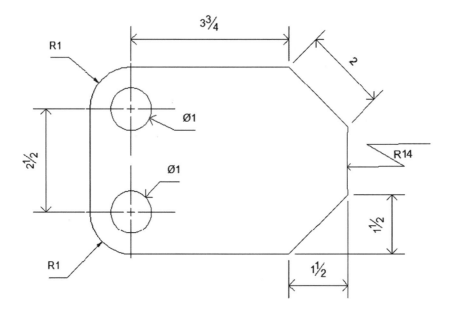

8. Save and close the file.

16.11 MORE DIMENSION FUNCTIONS

- This section discusses other dimension functions that help you produce a better look for the final drawing. These functions are:
  - Dimension Break
  - Dimension Adjust Space
  - Dimension Jog Line
  - Dimension Center Mark
  - Dimension Oblique
  - Dimension Text Angle
  - Dimension Justify
  - Dimension Override

16.11.1 Dimension Break

- If two or more dimension blocks intersect in one or more points, this command allows you to break one of the blocks at the intersection point. To issue this command, go to the **Annotate** tab, locate the **Dimensions** panel, then select the **Break** button:

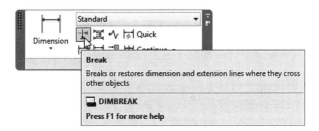

- You will see the following prompts:

```
Select dimension to add/remove break or [Multiple]:
Select object to break dimension or [Auto/Manual/Remove] <Auto>:
```

- The first prompt asks you to select the dimension block that will be broken and the second prompt asks you to select the dimension block that will stay as is. These prompts remove a break if it exists. At the second prompt, right-click and select the Remove option. These two prompts will be repeated until you press [Enter] to end the command.

- The final result looks like the following:

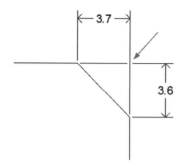

16.11.2 Dimension Adjust Space

- This command allows you to adjust the spaces between dimension blocks to be either aligned or have equal spaces between them. To issue this command, go to the **Annotate** tab, locate the **Dimensions** panel, then select the **Adjust Space** button:

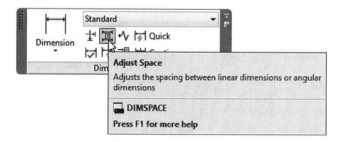

- You will see the following prompts:

```
Select base dimension:
Select dimensions to space:
Select dimensions to space:
Enter value or [Auto] <Auto>:
```

- The first step to select the base dimension block, which the other blocks will follow, and then select all the other dimension blocks. When done,

press [Enter] and AutoCAD will ask for the value. There are three options:
- Value = 0 (zero); all the other blocks will be aligned with the base dimension block.
- Value > 0: the distance that will separate the base dimension block from the nearest block and the others as well.
- Value = Auto: AutoCAD will try to figure out the best arrangement for the selected blocks.

- See the following example.

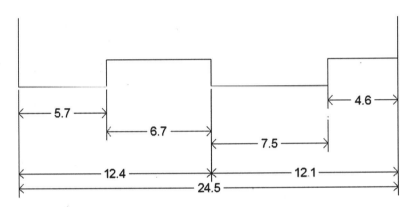

- As the first step, start the Adjust Space command and select the dimension block at the left reading 5.7 as your base dimension block. Then select the adjacent 6.7, 7.5, and 4.6. Set the value to 0:

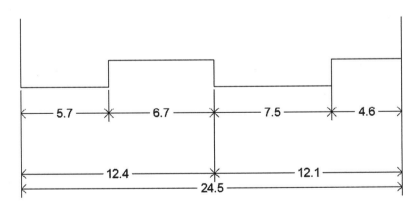

- Start the command again and select the dimension block reading 5.7 as your base dimension block. Then selected the other three dimension blocks reading 12.4, 12.1, and 24.5. Set the value to 1.0:

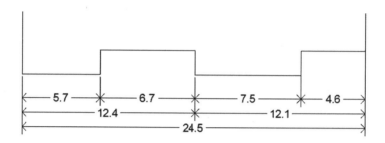

16.11.3 Dimension Jog Line

- This command allows you to add/remove a jog line to an existing linear or aligned dimension. To issue this command, go to the **Annotate** tab, locate the **Dimensions** panel, then select the **Jog line** button:

- You will see the following prompts:

```
Select dimension to add jog or [Remove]:
Specify jog location (or press ENTER):
```

- The first prompt asks you to select the desired linear or aligned dimension block. You can select the position of the jog, but you can also press [Enter] to let AutoCAD locate it automatically. You will see something like the following:

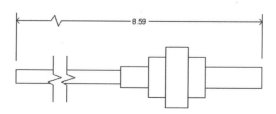

16.11.4 Dimension Center Mark

- This command allows you to add a center mark to an existing circle/arc. To issue this command, go to the **Annotate** tab, locate the **Dimensions** panel, then select the **Center Mark** button:

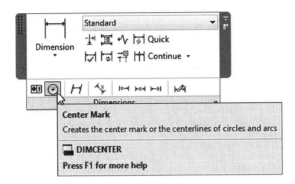

- You will see the following prompt:

```
Select arc or circle:
```

- Simply select the existing arc or circle and a center will be added automatically.

16.11.5 Dimension Oblique

- This command allows you to change the angle of the extension lines to any angle you want. To issue this command, go to the **Annotate** tab, locate the **Dimensions** panel, then select the **Oblique** button:

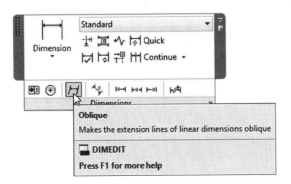

- You will see the following prompts:

```
Select objects:
Select objects:
Enter obliquing angle (press ENTER for none):
```

- Select the desired dimension block(s) and then press [Enter]. Next, input the oblique angle, which can be positive or negative. See the following example:

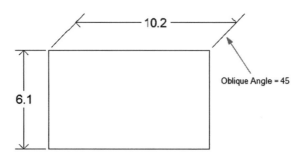

16.11.6 Dimension Text Angle

- This command allows you to change the angle of the dimension text to any angle you want. To issue this command, go to the **Annotate** tab, locate the **Dimensions** panel, then select the **Text Angle** button:

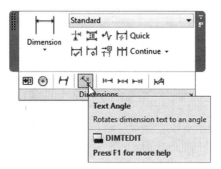

- You will see the following prompts:

```
Select dimension:
Specify angle for dimension text:
```

- Select the desired dimension block(s) then input the text angle, which can be positive or negative. See the following example:

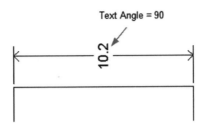

16.11.7 Dimension Justify

- This command allows you to change the horizontal position of the dimension text. To issue this command, go to the **Annotate** tab, locate the **Dimensions** panel, then click one of the following buttons:

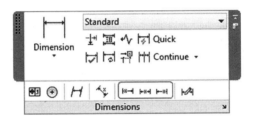

- Each one of the functions will move the dimension text either to the left, center, or to the right. In the following example, the text was moved the left.

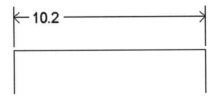

16.11.8 Dimension Override

- This command allows you to override a dimension system variable (you should memorize the system variable) or simply remove the override.

To issue this command, go to the **Annotate** tab, locate the **Dimensions** panel, then select the **Override** button:

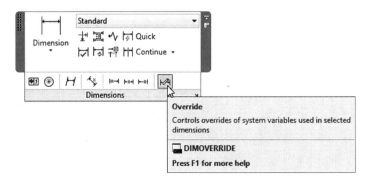

- You will see the following prompts:

```
Enter dimension variable name to override or [Clear
overrides]:
```

- This command is very useful if you want to remove (clear) all the overriding steps you make on a dimension block using the right-click menu. To answer to the previous prompt, either type the name of the dimension system variable or type C to clear the override. See the following illustration:

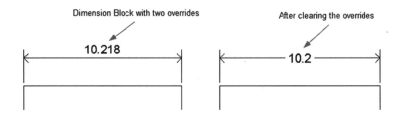

PRACTICE 16-2

More Dimension Functions

1. Start AutoCAD 2015.
2. Open the file, Practice 16-2.dwg.

3. Using Adjust Space, adjust the space between the six continuous dimensions to be all the same alignment to 2 ¾ dimension.

4. Using Adjust Space, adjust the space between the six continuous dimensions and the total single dimension to 1.5.

5. Make sure that the Dimension layer is the current layer; if not, make it current.

6. Add center marks to the two circles.

7. Using Dimension Break, break the horizontal 1 ½ using the vertical 1 ½.

8. Rotate the text of the vertical 1 ½ to be 45.

9. What is the upper horizontal dimension? _____

10. Using Dimension Override, clear the override. What does it read now? _____

11. Make the same dimension left justified.

12. You should have the following:

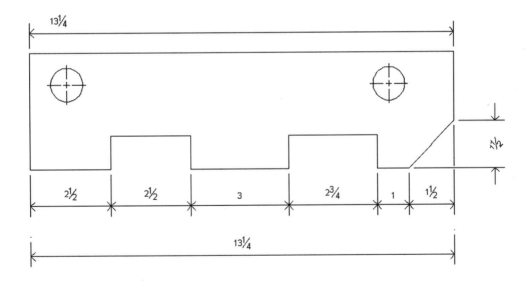

13. Save and close the file.

16.12 HOW TO CREATE A MULTILEADER STYLE?

- Multileader is a replacement for the normal leaders used to exist in AutoCAD.
- Leaders used to follow the current dimension style and they were always single. Multileader has its own style and a single leader can point to different location in the drawing.
- Multileader style allows you to set the characteristics of the Multileader block. To start this command, go to the **Annotate** tab, locate the **Leaders** panel, then select the **Multileader Style** button:

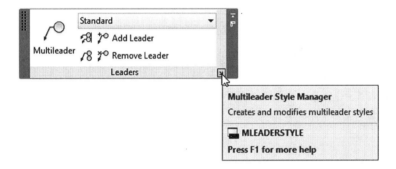

- You will see the following dialog box:

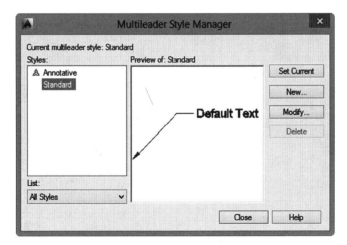

- There are two pre-defined styles, one called Standard (default) and the other called Annotative. Click the **New** button to create a new multileader style. You will see the following dialog box:

- Input the name of the new style, then click the **Continue** button. There are three tabs, each one will control part of the multileader block. They are:
  - Leader Format
  - Leader Structure
  - Content

16.12.1 Leader Format Tab

- Below is the picture of Leader Format tab:

- Change any of the following:
  - Edit the type of the leader by choosing one of the three options, Straight, Spline, or None. The following is an example of both straight and spline options:

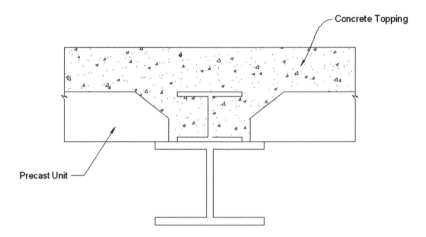

  - Edit the Color, Linetype, and Lineweight.
  - Select the arrowhead shape and its size.
  - Set the distance of the dimension break from any two blocks that will intersect.

16.12.2 Leader Structure Tab

- This is the Leader Structure tab:

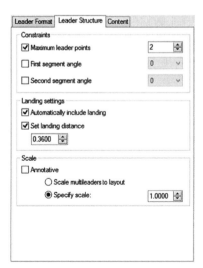

- You can change any of the following:
  - Specify whether you want to change the **Maximum leader points**. Then input the desired value. By default, this value is 2; that is, the first point points to the geometry and the second points is the end of the multileader.

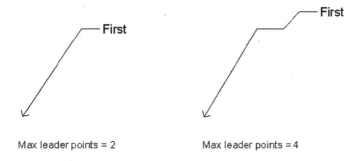

Max leader points = 2   Max leader points = 4

  - Specify whether you want to change the **First segment angle** and the **Second segment angle**. If yes, what are the angle values?
  - Specify whether you want AutoCAD to **Automatically include landing**. If yes, what is the **landing length**?
  - Specify whether the multileader will be Annotative. (Annotative is discussed in the coming chapters.)

16.12.3 Content Tab

- This is the Content tab:

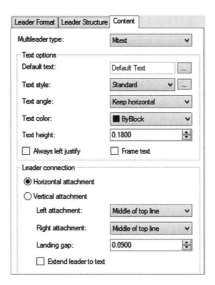

- In AutoCAD, there are two Multileader types:
  - Mtext
  - Block (either pre-defined or user-defined)
- The following displays the two types:

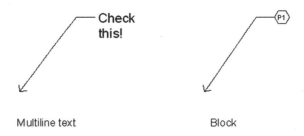

- If you select the **Mtext** option, you can change any of the following:
  - If there is Default text.
  - Select **Text style**, **Text angle**, **Text Color**, and **Text height** (if Text Style's height = 0).
  - Select whether the text is **Always left Justify** and with **Frame**.
  - Choose whether the leader connection is horizontal or vertical. If vertical, then edit the position of the text relative to the landing for both left and right leader lines. Next, control the gap distance between the end of landing and the text

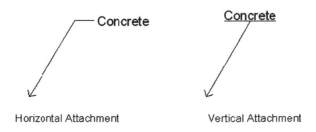

  - If you select the **Block** option, you can change any of the following:

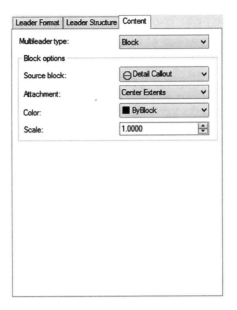

- Specify the **Source block**:

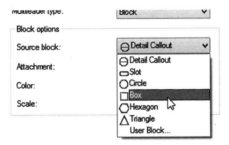

- Specify the **Attachment** position, **Color** of attachment, or **Scale** of the attachment.

16.13 INSERTING A MULTILEADER DIMENSION

- This group of commands allows you to add a single multileader then add a leader to an existing multileader; remove a leader form an existing multileader; or, align and group an existing multileader. You will always start with the **Multileader** command, which inserts a single leader. To start this command, go to the **Annotate** tab, locate the **Leaders** panel, then select the **Multileader** button:

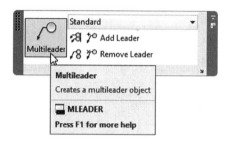

- You will see the following prompts:

```
Specify leader arrowhead location or [leader Landing
first/Content first/Options] <Options>:
Specify leader landing location:
```

- First, specify the leader arrowhead location and then specify the leader landing location. Next, type the text you want to appear beside the leader. To add a leader to an existing multileader, go to the **Annotate** tab, locate the **Leaders** panel, then select the **Add Leader** button to add more leaders:

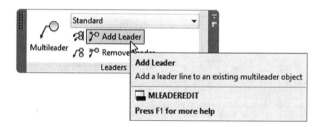

- You will see the following prompts:

```
Select a multileader:
1 found
Specify leader arrowhead location:
Specify leader arrowhead location:
```

- To remove a leader to an existing multileader, go to the **Annotate** tab, locate the **Leaders** panel, then select the **Remove Leader** button to remove leaders:

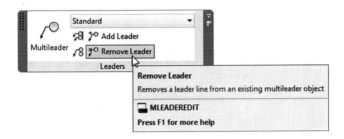

- You will see the following prompts:

```
Select a multileader:
1 found
Specify leaders to remove:
Specify leaders to remove:
```

- To align a group of multileaders, go to the **Annotate** tab, locate the **Leaders** panel, then select the **Align** button:

- You will see the following prompts:

```
Select multileaders: 1 found
Select multileaders: 1 found, 2 total
Select multileaders:
Current mode: Use current spacing
Select multileader to align to or [Options]:
Specify direction:
```

- To collect a group of similar multileaders to be a single leader, go to the **Annotate** tab, locate the **Leaders** panel, and then select the **Collect** button. This command works only with leaders containing blocks:

- You will see the following prompts:

```
Select multileaders:
Select multileaders:
Specify collected multileader location or [Vertical/
Horizontal/Wrap] <Horizontal>:
```

PRACTICE 16-3

Creating the Multileader Style and Inserting Multileaders

1. Start AutoCAD 2015.
2. Open the file, Practice 16-3.dwg.
3. Create a new multileader style based on Standard:
    a. Name = Texture and Painting
    b. Arrowhead symbol = Dot small
    c. Arrowhead size = 0.35
    d. First segment angle = 0
    e. Automatically include landing = Off

f. Multileader type = Block

   g. Source block = Circle

4. Create a new multileader style based on <u>Standard</u>:

   a. Name = Material

   b. Leader format = Spline

   c. Arrowhead symbol = Right angle

   d. Arrowhead size = 0.25

5. Make layer Dimension current.

6. Using both styles insert the following multileaders:

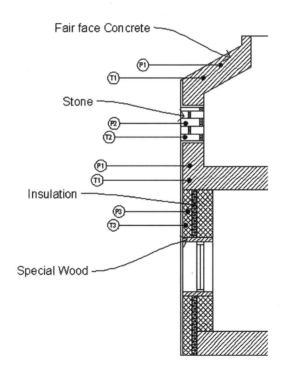

7. Using Add leader, Align, and Collect, try to get the following picture:

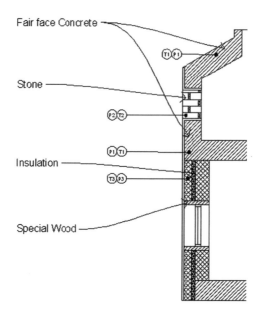

8. Save and close the file.

NOTES:

CHAPTER REVIEW

1. Symmetrical and Deviation are two types of _____ in dimension style.

2. There are two types of multileader blocks, multiline text or block.

    a. True

    b. False

3. When creating a new dimension style, which of the following is NOT correct:

    a. You can create dimension style affecting all types of dimensions.

    b. You have to select the existing dimension style to start with.

    c. You cannot create a dimension style affecting only one type of dimension.

    d. You can create a sub style.

4. You can show _____ units and _____ units in a dimension block.

5. Using the Multileader style, there should be always a landing in the block.

    a. True

    b. False

6. Collect and Align are _____ commands.

CHAPTER REVIEW ANSWERS

1. Tolerance

3. c

5. b

CHAPTER 17

PLOT STYLE, ANNOTATIVE, AND DWF

In This Chapter
- How to create and use the two types of Plot Styles
- Using the Annotative in AutoCAD
- Creating and viewing DWF files

17.1 PLOT STYLE TABLES – FIRST LOOK

- Plot styles are used to convert colors used in the drawing to the printed colors. The default setting is to keep the same color as the printer. Before AutoCAD 2000, there was only one type of conversion method and it depended on the colors used in the drawing. After AutoCAD 2000, a new concept called Plot Style was introduced. There are two types of plot styles:
  - Color-dependent Plot Style Table
  - Named Plot Style Table

17.2 COLOR-DEPENDENT PLOT STYLE TABLE

- This plot style table is simulation for the only method that existed before AutoCAD 2000. The essence of this method is simple: for each used color in your drawing, specify the color to be used in the printer. Auto-CAD will give the user the ability to set the lineweight and linetype, etc., for each color. The problem of this method is its limitation because you have only 255 colors to use.

- Each time you create a Color-dependent Plot Style Table, AutoCAD creates a file with extension *.**ctb**. You can create Plot Style tables from outside AutoCAD using the Control Panel of Windows or from inside AutoCAD using the menu bar.
- From outside AutoCAD, start the Control Panel of Windows, double-click **Autodesk Plot Style Manager** icon, or show the menu bar. Then select **Tools/Wizards/Add Plot Style Table** and you will see the following dialog box:

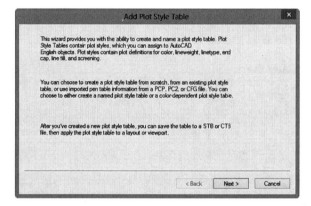

- The first screen is an introduction to plot styles. You should read it to understand the next steps. When done, click the **Next** button and you will see the following dialog box:

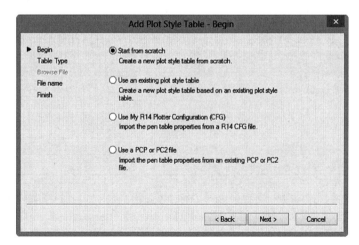

- AutoCAD will list the four possible choices:
  - Start from scratch.
  - Use an existing plot style.
  - Import the AutoCAD R14 CFG file and create a plot style from it.
  - Import PCP or PC2 file and create a plot style from it.
- Select **Start from scratch**, then click the **Next** button, and you will see the following dialog box:

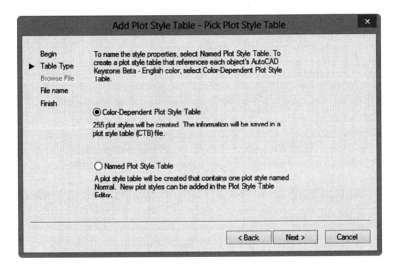

- Select **Color-Dependent Plot Style Table** and click the **Next** button. You will see the following dialog box:

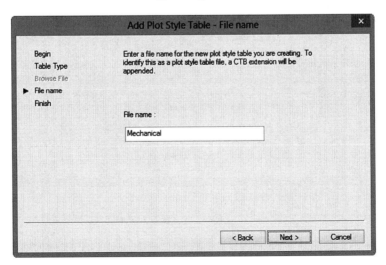

- Input the name of the new plot style, click the **Next** button, and you will see the following dialog box:

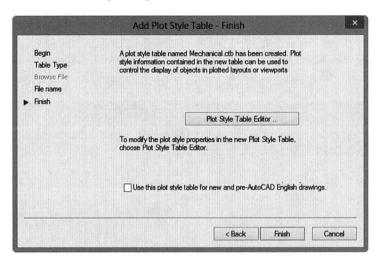

- Select the **Plot Style Table Editor** button and the following dialog box will be displayed:

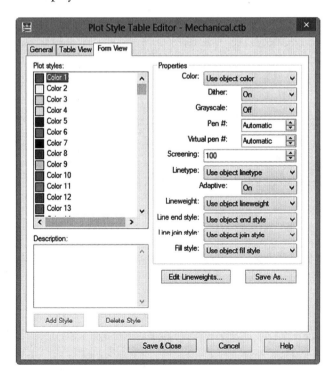

- From the left, select the color you used in your drawing file, and then at the right, change any of the following settings:
  - Change the **Color** to be used in the plotter.
  - Switch **Dither** on/off. This option will be dimmed if your printer or plotter does not support Dithering. Dither is a method to give the impression of using more colors than the 255 color used by AutoCAD. It is preferable to leave this option off but should be turned on, if you want **Screening** to work.
  - Change **Grayscale**. This method is good for laser printers.
  - Change **Pen #.** This option is valid for the old types of plotter such as pen plotters which are not used these days.
  - Change **Virtual pen #.** For non-pen plotters to simulate pen plotters by assigning a virtual pen for each color; leave it **Automatic**.
  - Change **Screening**. This option reduces the intensity of the shading and fill hatches, reducing the amount of ink used. It also depends on **Dithering**.
  - Change **Linetype**. Set a different linetype for the color or leave it to the object's linetype.
  - Change **Adaptive**. This option changes the linetype scale of all objects using the color to start with a segment and end with a segment. Turn this option off if the linetype scale is important for your drawing.
  - Change **Lineweight**. This option changes the lineweight for the color selected.
  - Change **Line end style**. This option allows the user to select the end style for lines, pick one of the following: Butt, Square, Round, and Diamond.
  - Change **Line join style** to select the line join shape. The available choices are: Miter, Bevel, Round, and Diamond.
  - Change **Fill style**. This option sets the fill style for filled area in the drawing (good for trial printing).
- Click **Save & Close** and then click **Finish**.
- Your last step should be linking your plot style with a layout. Do the following steps:
  - Select the desired layout, then start the **Page Setup Manager**.
  - Select the name of the current Page Setup and click **Modify**.

- At the upper-right part of the dialog box and under the **Plot style table (pen assignment)**, select the desired plot style table:

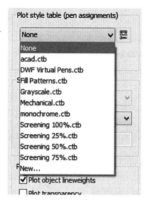

- Click the Display plot styles checkbox on.
- You can assign one **ctb** file for each layout. In order to see the linetype and lineweight of the objects, you have to switch the **Show/Hide Lineweight** button in the status bar on, as shown:

17.3 NAMED PLOT STYLE TABLE

- This method does not depend on colors. The created plot style tables are linked later on with layers. You may have two layers holding the same color, but they will print with different colors, linetypes, and lineweights.
- The Named Plot Style Table has a file extension of *****.stb**. The creation procedure of the Named Plot Style is identical to the Color-dependent Plot Style, except for the last step, which is configuring the **Plot Style Table Editor**. You can create it from outside AutoCAD, using the Control Panel, then double-click the **Autodesk Plot Style Manager** icon or you can show the menu bar, then select **Tools/Wizards/Add Plot Style Table**. You will see the same screen you saw while creating the Color-dependent plot style table.

- Follow the same steps until you reach the **Plot Style Table Editor** button. Select it and you will see the following dialog box. Click the **Add Style** button:

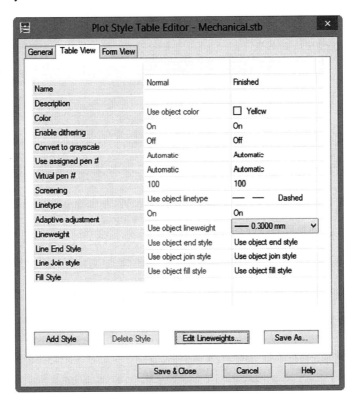

- As you can see, you can change any of the following:
  - Input the Name of the style and brief Description.
  - Change the Color to be used in the plotter.
  - The rest of features in the Color-dependent Plot Style Table were explained in the previous section.
- You can add as many styles as you wish in the same Named Plot Style. Click **Save & Close** and then click **Finish**.

- Linking the Named Plot Style Table with any drawing is a bit more complicated than linking the color-dependent plot style table. Do the following steps:
  - The first step is a precaution. You may want to print a drawing, and then discover it takes only ctb files. To solve this problem, you have to convert one of the **ctb** files to a **stb** file. In the command window, type **convertctb** and a dialog box listing all the ctb files will appear. Select one of them, keeping the same name or giving it a new name, then click **OK**. You will see the following dialog box:

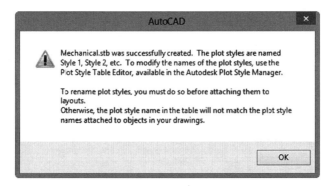

  - Convert the drawing from Color-dependent Plot Style to Named Plot Style. In the command window, type **convertpstyles** and you will see the following warning message:

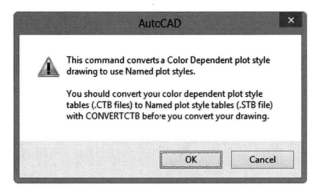

- Click **OK** and you will see the following dialog box:

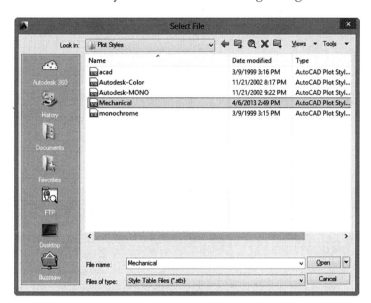

- Select the newly created Named Plot Style Table, then click **Open**, and you will see the following prompt:

```
Drawing converted from Color Dependent mode to Named
plot style mode.
```

- The second step is to select the desired layout. Start the **Page Setup Manager** and in the upper-right part of the dialog box, under **Plot style table (pen assignment)**, select the name of the newly created named plot style table. Click the **Display plot styles** checkbox on and end the Page Setup Manager command:

- Select the **Layer Properties Manager** and for the desired layer(s) click the name of the plot style under the **VP Plot Style** column:

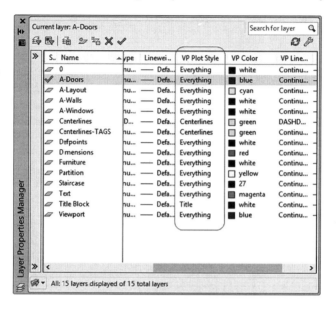

- You will see the following dialog box:

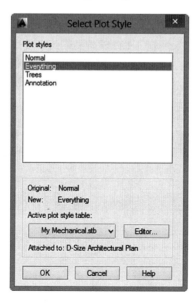

- Select the desired plot style. When done, click **OK**.
- Repeat the same for other layers.
- You may need to type in the command window **regenall** (which means regenerate all viewports) in order to see the effect of what you did.

If you create a new drawing using acad.dwt, this means your drawing will accept only the Color-dependent plot style table. Use acad-Named Plot Styles.dwt to create a new drawing, which will accept only Named plot style table.

PRACTICE 17-1

Color-Dependent Plot Style Table

1. Start AutoCAD 2015.
2. Open the file, Practice 17-1.dwg.
3. Create a new Color-dependent plot style table and call it **Architectural** using the following table:

Color	1	2	3	4	5	6	7	27
Plot color	Black	Black	Use object color	Black	Black	Black	Black	Black
Lineweight	0.3	0.3	0.3	0.3	0.7	0.3	0.3	0.3

4. Switch to D-Size Architectural Plan layout.
5. Link this layout to Architectural plot style.
6. Check that the Show/Hide Lineweight button at the status bar is turned on.
7. Check the doors comparing to the other objects and you will find it thicker because color (5) used for both the title block and doors is using 0.7 lineweight.
8. Save and close the file.

PRACTICE 17-2

Named Plot Style Table

1. Start AutoCAD 2015.
2. Open Practice the file, 17-2.dwg.
3. Create a new Named plot style table and name it **Architectural** using the following table:

Style name	Color	Lineweight
Everything	Black	0.3
Centerlines	Green	0.5
Title	Black	0.7

4. Link the newly created table with the D-Size Architectural Plan layout
5. Using the Layer Properties Manager make the following:
    a. Layer = Centerlines and Centerlines-TAGS linked to Centerlines
    b. Title Block linked to Title
    c. The rest of layers linked to Everything
6. Zoom in to compare the Centerline lineweight to the other objects.
7. Save and close.

17.4 WHAT IS THE ANNOTATIVE FEATURE?

- Because you will always print from layout, you need to use viewports. Because you use viewports, you have to set the viewport scale for each viewport. Viewport scale affects all the objects in the Model space. If you hatch type text, insert dimensions, multileader, or a block containing text in Model space, all of these objects will be scaled. If they were scaled down, then text and dimension will be unreadable, and hatch will look like solid hatching.
- What is needed is a feature that scales everything except the annotation objects (namely, hatch, text, and dimension). The feature is **Annotative**.
- The annotative feature can be found in different places:

- In Text style and under Size:

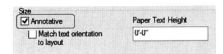

- In dimension style, go to the Fit tab:

- In the multileader style, go to the Leader Structure tab:

- In the Hatch context tab, locate the Options panel:

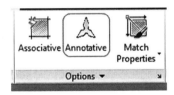

- In the Block creation command, under Behavior:

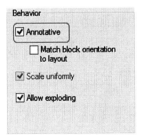

- How do you recognize whether you are dealing with a style supporting annotative feature? You will see this special symbol, as shown:

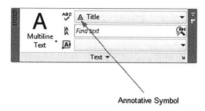

Annotative Symbol

- How do you recognize if an object was inserted using a style or a command supporting annotative feature? Simply hover over it and if you see something like the following, you know it was inserted with annotative feature on:

This is my first Annotative text

- To work with annotative annotation objects, follow these simple steps:
  - Create your drawing in Model space without any annotation objects (namely, hatch, text, dimensions, multileader, and blocks with text).
  - Select the desired layout, add viewports, and then scale them. This scale will be for both the annotation objects and the viewport.
  - Double-click inside the viewport to make it active.
  - Insert all the annotation objects you need.
  - Once you insert annotation objects in the viewport, you can see them in this viewport and any other viewport holding the same scale value.
  - Changing the scale of the viewport will result in loosing the annotation object.
  - To show the annotation object in more than one viewport holding different scale values, right-click on **Annotation Visibility** (a button on the right side of the status bar) and you will see the following menu:

- Select to Show Annotation Objects for Current Scale Only or Show Annotation Objects for All Scales.

- Control how to add a scale values to the viewport; is it Automatic or manual? Right-click on the **Add Scale** button (a button at the right portion of the status bar) and you will see the following menu:

- If you used the zooming command inside the viewport, this ruins the current viewport scale. To retain it, back click the Synchronize button (on the right side of the status bar) and you will see the following message (you can also use the lock in the status bar to lock the viewport scale to avoid this problem):

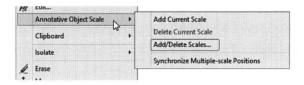

- Clicking this message restores the viewport scale.
- To make an annotation object appear in viewports holding different scales, select it and right-click. Select the **Annotative Object Scale** option and you then see the following sub-menu:

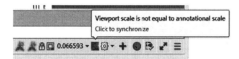

- Choose the **Add/Delete Scales** option and the following dialog box will appear:

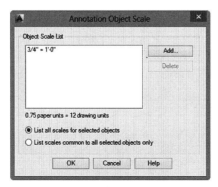

- Select the **Add** button and the following dialog box will appear:

- Select the desired scale value and click **OK** twice.

PRACTICE 17-3

Annotative Feature

1. Start AutoCAD 2015.
2. Open the file, Practice 17-3.dwg.
3. Make layer Dimensions current.
4. Switch to the Details layout.
5. There are three viewports, the first one at the left is scaled to be 1:20 and the other two at the right are scaled to be 1:10.
6. Go to the Annotate tab, locate the Dimensions panel, and check the available dimension styles. You will see only one with a distinctive symbol at its left, which is mechanical. This dimension style is annotative.
7. Double-click inside the big viewport.
8. Start the Radius command and select the big magenta circle, then add the dimension block. Check that it does not appear in the upper-right viewport.
9. Add another Radius block to the small magenta circle.

10. Add two linear dimensions for the total width and the total height.
11. While you are still inside the same viewport, make layer Text current.
12. Make text style "Annotative" current.
13. Using Multiline text add beneath the shape the word "Bearing."
14. Make layer Dimensions current again.
15. Click inside the upper-right viewport to make it current and add a Radius dimension to one of the small circles.
16. Make layer Hatch current.
17. Click inside the lower-right viewport.
18. Start the Hatch command, make sure that Annotative is on, set the scale = 10, and hatch the area between the two dashed lines, then finish the Hatch command.
19. Using the Add/Delete scales add scale 1:20 so the two hatches will appear in the big viewport.
20. You should have the following picture:

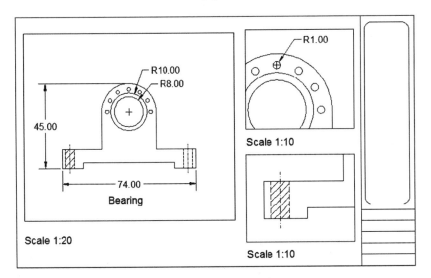

21. Save and close the file.

17.5 DESIGN WEB FORMAT (DWF) FILE

- Sharing files and data is vital to most everyone in today's world. But if you send DWG files to others (joint venture companies, third-party companies, clients, vendors, etc.), you are taking a risk that your design could be stolen. Moreover, DWG files are always bulky (one file may reach to more than 50 to 100 MB, depending on if it is 2D or 3D) and very difficult to send by email. Add to the previous points, that you need AutoCAD to open DWF files and not everyone has AutoCAD installed. For all these reasons and more, AutoCAD offers to plot to the **D**esign **W**eb **F**ormat (DWF) file. The DWF file cannot be modified, so there is no worry that your design will be stolen. Moreover, the DWF size is relatively small in comparison to DWG file size. To view DWF, you can use free software called Autodesk Design Review, which comes in the same DVD as AutoCAD or can be installed from Autodesk website. This software is not for viewing and printing only; you can use it to measure and red-line DWF files as well.
- There is also another version of DWF called DWFx, which can be viewed with both Windows Vista and Windows 7 using an Internet browser and without Autodesk Design Review software.

17.6 EXPORTING DWF, DWFX, AND PDF FILES

- This command allows you to export your current DWG file to DWF, DWFx, and PDF. To start this group of commands, go to the **Output** tab, locate the **Export to DWF/PDF** panel, then select the **Export** button:

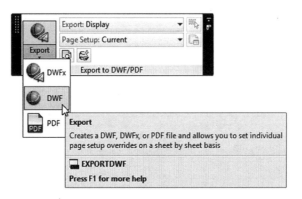

- You will see the following dialog box:

- At the top-right part of the dialog box, under **Current Settings**, AutoCAD lists the current settings:

- To edit these settings, click the **Options** button. The following dialog box will appear:

- Change any of the following settings:
  - Indicate the location in the hard disk where you want to save the DWF file.
  - By default, the file is always Multiple sheet.
  - Indicate the Override precision. You can pick from Manufacturing and Architectural, etc. This setting allows the user to pick the dpi (dot per inch) precision, which allows DWF users to measure distances accurately.
  - Input the name of the DWF file or let AutoCAD prompt you after executing the command.
  - Indicate whether to include Layer information. This allows you to turn off layers in the Autodesk Design Review software.
  - Indicate whether to include a Password.
  - To finish, click **OK**.
- Under **Output Controls**, you will see the following:

- Change any of the following:
  - Whether to open DWF using Autodesk Design Review automatically after executing the command.
  - Whether to include a plot stamp.
  - Indicate what to export. If you execute the command from Model space, then select Display, Extents, or Window. On the other hand, if you are exporting while you are at one of the layouts, then the available selections are Current Layout or All layouts.
  - Select Page Setup.
  - To finish, click the **Save** button.

17.7 USING THE BATCH PLOT COMMAND

- This command allows you to produce DWF files containing multiple layouts from the current drawing and from other drawings. To issue this command, go to the **Output** tab, locate the **Plot** panel, then select the **Batch Plot** button:

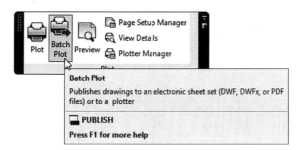

- You will see the following dialog box:

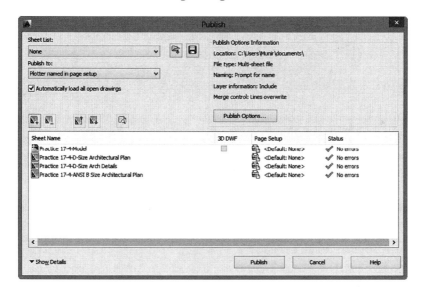

- You will see a table that contains Model Space and layouts. You can do any of the following:
  - Indicate **Publish to** either printer/plotter defined in the layout or DWF, DWFx, PDF.
  - Indicate whether to load all open drawings automatically.

- Using the five buttons above the table, you can add a sheet, remove a sheet, and move any sheet up or down to specify its order relative to the document sheets. Finally, you can preview the sheet:

- You can rename the sheets by clicking on any sheet name; you will see the name becomes editable.
- Click **Publish Options** and the following dialog box will appear, which is identical to what was discussed for the Exporting command:

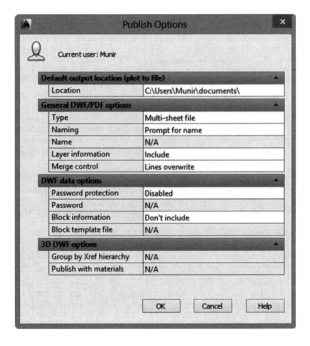

- When done, click **OK**.
- Click the **Show Details** button at the lower-left corner of the dialog box then set the following: Specify the number of copies. Select whether to include Plot Stamp. Indicate whether you want to publish DWF in the background. Indicate whether to open DWF using Autodesk Design Review. Set the precision for the DWF file (dpi precision) for measurement in Autodesk Design Review.

- To finish, click the **Publish** button. A final message will come up as a bubble similar to the following:

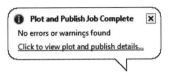

17.8 VIEWING DWF AND DWFX FILES

- AutoCAD comes with free software to view, print, and red-line DWF, and DWFx files called **Autodesk Design Review**. Once you locate the shortcut on your desktop, start it and open a DWF or DWFx file (you can open one file at a time). See the following picture:

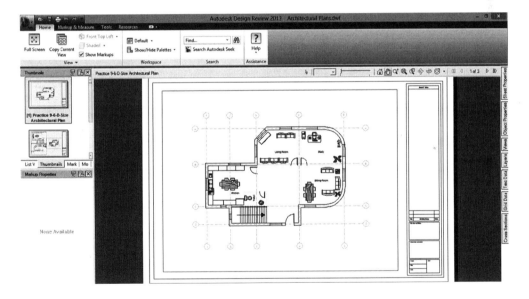

NOTE

Though this is AutoCAD 2015, the software being discussed is Autodesk Design Review 2013. This because Autodesk, Inc. stopped developing Autodesk Design Review because of the Autodesk 360 (Autodesk Cloud) application called Design Feed.

PRACTICE 17-4

Creating and Viewing DWF File

1. Start AutoCAD 2015.
2. Open the file, Practice 17-4.dwg.
3. Produce a multi-sheet DWF file including all layouts except the Model space sheet. Be sure to include layers in the DWF file. Save the file with the name "Architectural Plans.dwf." Open it using Autodesk Design Review.
4. If you have Windows Vista or Windows 7, produce a DWFx file, and open it using the Internet browser.
5. Save and close the file.

NOTES:

CHAPTER REVIEW

1. Color-dependent plot styles are similar to those used before AutoCAD 2000.

    a. True

    b. False

2. The best practice is to insert dimensions, text, and hatches in _____, using _____ feature.

3. _____ is better than Color-dependent plot style.

4. There is no difference between DWF and DWFx.

    a. True

    b. False

5. In order to insert an annotative object, you have to be inside the viewport.

    a. True

    b. False

CHAPTER REVIEW ANSWERS

1. a

3. Named plot style table

5. a

CHAPTER 18

HOW TO CREATE A TEMPLATE FILE AND INTERFACE CUSTOMIZATION

In This Chapter

- How to create a template file
- How to use the CUI command to customize the interface of AutoCAD

18.1 WHAT IS A TEMPLATE FILE AND HOW DO YOU CREATE ONE?

- Any company using AutoCAD should think of standardizing their work in order to shorten the production time. Templates can answer these two issues in a simple way. Standardization includes using the same layer naming, colors, linetype, and lineweight. It also includes standard text, tables, standard dimension and leaders, and standard layouts. Companies will use always the same shapes for the blocks.
- To create a good template, you should include the following:
  - Drawing units
  - Drawing limits
  - Grid and Snap settings
  - Layers
  - Linetypes

- Text Styles
- Table Styles
- Dimension Styles
- Multileader Styles
- Layouts (including Border blocks, and Viewports)
- Page Setups
- Plot Style tables

■ Follow these steps to create a template file:
  - Create a new file using the template file **acad.dwt** or **acadiso.dwt**.
  - Prepare the settings discussed in the previous section.
  - Create inside this file all the same settings as discussed.
  - From the Application menu, Save As/AutoCAD Drawing Template.

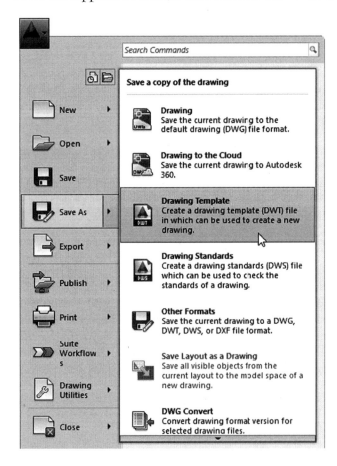

How to Create a Template File and Interface Customization • 545

- You will see the following dialog box:

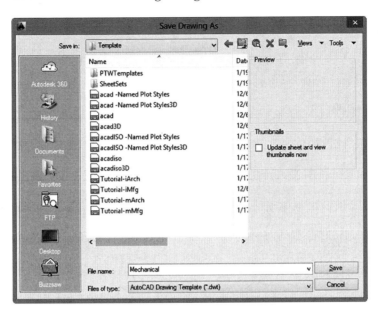

- Input the name of the template file. By default, you will be directed to the same folder that AutoCAD save its default template files in. You can save your template there or you can create your own folder, which is highly recommended. But using this method will mean that you have to specify the new folder for AutoCAD. You can do this using the Options dialog box. See the following illustration:

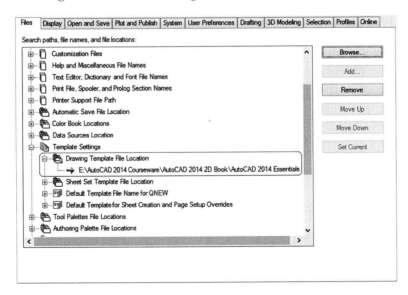

18.2 EDITING A TEMPLATE FILE

- To edit an existing template, do the following:
  - Use the normal Open file command.
  - You will see the following dialog box. Using **Files of type**, pick the **Drawing Template (*.dwt)**.

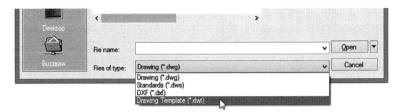

- It will take you to the default Template folder. Select and open the desired template, then make your edits.
- Save it using the same name or use a new name.

PRACTICE 18-1

Creating and Editing a Template File

1. Start AutoCAD 2015.
2. Open Practice the file, 18-1.dwg.
3. Go to Model and erase all objects.
4. Delete D-Size Arch Details layout.
5. Go to ANSI B Size Architectural Plan layout and delete the circular viewport.
6. Rename ANSI B Size Architectural Plan to be ANSI B Size.
7. Go to D-Size Architectural Plan layout and delete the viewport.
8. Rename D-Size Architectural Plan to ANSI D Size.

9. Go to the Annotate tab and check the existing dimension style and text style.
10. Go to the Home tab and check the existing layers.
11. Go to the Application menu and start the Units command. Set the precision for the length to be 0'-0 ½". Click OK to end the command.
12. Save the file as a template file under the name My Company.dwt (it will be saved in Template folder where AutoCAD saves all templates). You can save the template file in a different folder, but this should be specified in the Options dialog box in the Files tab.
13. Close the file.
14. Start a new file using My Company.dwt template file. You will find that all of your settings are in their in the new file so you can start working without the need to set anything.
15. Close the new file without saving.

18.3 CUSTOMIZING THE INTERFACE – INTRODUCTION

- Customizing the interface means to change the ribbons and panels to fit your own needs. You can add new workspaces, tabs, panels, and even new commands by using a single command called the CUI (Customization User interface).
- To issue this command, go to the **Manage** Panel, locate the **Customization** panel, then select the **User Interface**:

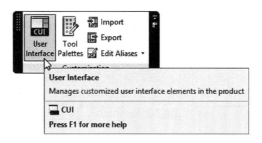

- You will see the following dialog box:

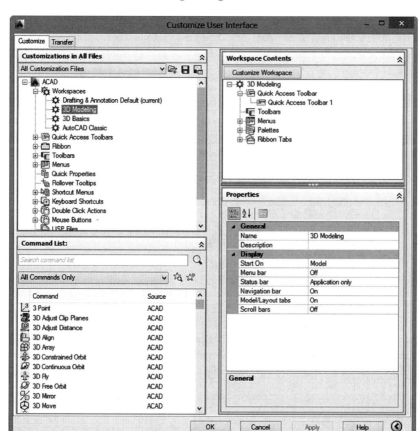

- This dialog box is cut into four parts:
  - This Customization part allows you to specify what you want to customize: Workspace, Tab, Panel, Toolbar, etc. This part also allows you to create new, delete existing, or rename, etc.
  - This Command list contains all AutoCAD commands.
  - Depending on what you choose from the left, the right part will change to Content and Properties.

18.4 HOW TO CREATE A NEW PANEL

- Panels are part of tabs. To create a new panel, in the upper left-part of the CUI dialog box locate Ribbons, then expand it. You will find beneath

How to Create a Template File and Interface Customization • 549

Panel it. With or without expanding, right-click Panels and select **New Panel** option as shown:

- Type the name of the new panel and press [Enter]. You will see something similar to the following:

- To the right, you will see Panel Preview with a small rectangle just like the following:

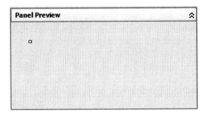

- Beneath it, you will see the Properties part, as shown:

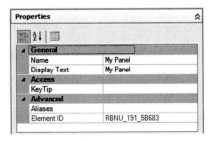

- When you create a new panel, three things will be automatically added. They are as follows:

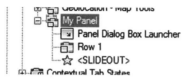

- Panel Dialog Box Launcher
- Row 1 (To add commands to this row or any other row, use the Command List part, a drag-and-drop the desired command.)
- SLIDEOUT
- To understand these three things, see the following illustration:

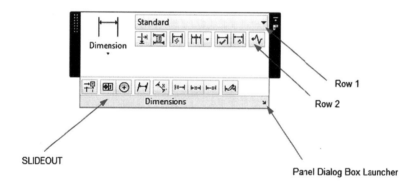

- To create a panel, decide on the following:
  - How many rows you will have.
  - How many sub-panels you will have.
  - The commands to be included and where (in rows or sub-panels)
  - The command presented as the large icon on the left
  - If it will have a Panel Dialog Box Launcher, and what is it (normally it is style or palette).
  - Will it have SLIDEOUT or not, and the commands in it.

- Answer all the six questions and write them on a paper before you start working with AutoCAD CUI.
- By default all buttons will be small. To make a button large, do the following:
  - Drag it to the panel.
  - From Panel Preview, click the desired button.
  - In the Properties panel, check the Button Style and select Large (either with Text horizontal, vertical, or without text) as in the following:

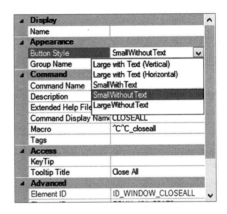

- To add a new row in a panel, do the following:
  - Select the desired panel.
  - Right-click the name and select New Row option.

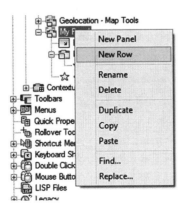

- Sub-panels as the name suggest can cut the panel into smaller parts to control. Each sub-panel will have row1, row2, etc. To add a new sub-panel do the following steps:
  - Select the desired row
  - Right-click and select New Sub-Panel:

- To add a vertical separator between buttons, do the following steps:
  - Select the desired row.
  - Right-click and select the Add Separator option.

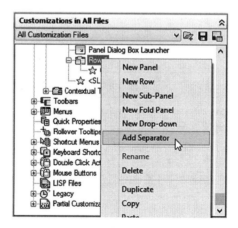

NOTE *Any row located below the slideout will be shown in it.*

18.5 HOW TO CREATE A NEW TAB

- Tabs consist of panels. You can include panels with that are included with AutoCAD, or create your own panels, or mix and match.
- To create a new tab, start the CUI command, at the upper-left part, expand Ribbon, then right-click Tabs, and select the New Tab option:

- Type the name of the new tab.
- You should fill the tab with panels and use one of the following approaches:
    - Drag-and-drop. This method is not practical, as we have a huge list of panels and tabs.
    - Go to the desired panel, right-click, and select Copy, then go to the desired tab, right-click and select Paste.

18.6 HOW TO CREATE A QUICK ACCESS TOOLBAR

- The Quick Access appears at the top left part of AutoCAD window to the right of Application menu. You can specify which commands to be included in this toolbar. To add a new Quick Access Toolbar, start the CUI command in the upper left, locate the Quick Access Toolbar, right-click and select the New Quick Access Toolbar option:

- Type the name of the new Quick Access Toolbar. By default, six commands will be included: New, Open, Save, Undo, Redo, Plot. You can drag-and-drop any new command from the Commands List.

18.7 HOW TO CREATE A NEW WORKSPACE

- Workspace is the set of tabs (panels) that appear together along with palette(s), menu(s), toolbar(s), and Quick Access toolbars. To create a new workspace, start the CUI command. At the upper left part, locate Workspaces, right-click, and select New Workspace option:

- Type the name of the new workspace. Click the newly created workspace and in the upper-right part, you will see something like the following:

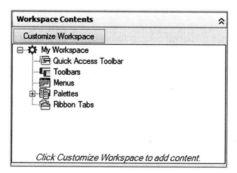

- There are five different things to be included inside a workspace. They are: Quick Access Toolbar, Toolbars, Menus, Palettes, and Ribbon Tabs. In order to add/remove the contents of the workspace click the button at the top of the window, titled "Customize Workspace." The text will change to blue in this part and at the upper-left part, you will see everything with a checkbox to its left:

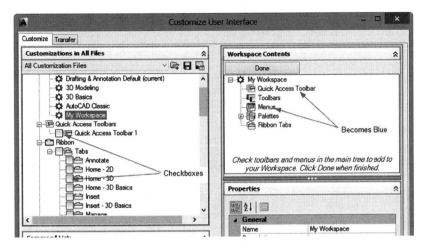

- If you want to add anything to your workspace, simply click the checkbox to copy it to it. When you are through adding, click the "Done" button to finish the process of customizing your workspace.

PRACTICE 18-2

Customizing Interface

1. Start AutoCAD 2015.
2. Start a new file.
3. Start the CUI command.
4. Create a new workspace and call it My Workspace.
5. Go to Ribbon, and expand it, then locate Tabs.
6. Create a new tab and call it My Tab, then collapse the list.

7. Locate Panels then create a new panel and call it My Panel. Do the necessary steps to look like the following:

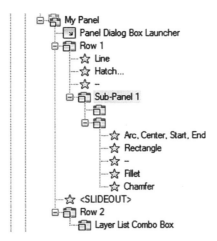

8. The Panel Dialog Box Launcher = Tool Palette command.

9. Copy My Panel to My Tab.

10. Add the following premade panels to My Tab:

    a. Annotate – Dimensions

    b. View – Viewports

    c. Home – Properties

11. Select My Workspace then at the right click Customize the Workspace button.

12. At the left click My Tab to be the only tab included in this workspace (you can include other premade tabs if you like).

13. Collapse Tabs.

14. Click the Quick Access Toolbars and click Quick Access Toolbar 1.

15. Click OK to end the CUI command.

16. Go to the top of the screen and select My Workspace to make it current.

17. Test the panel you created.

18. Restore Drafting and Annotation workspace.

19. Close the file without saving.

18.8 HOW TO CREATE YOUR OWN COMMAND

- In this part, we will learn how to create a new command by creating a macro using AutoCAD commands with certain values, then assigning this macro to a button, and putting it in a panel.
- Use these characters in the macro:
  - ^C^C is equal to pressing [Esc] twice. Previously AutoCAD used [Ctrl] + C to cancel commands, then later on (AutoCAD R12) changed it to pressing [Esc] instead. You have to start each macro with ^C^C in order to make sure that macro will cancel all running commands before it starts.
  - Use ";" to simulate [Enter].
  - Use "\" to wait for the user input.
  - Macros cannot deal with commands which will initiate palettes or dialog boxes like LAYER or INSERT. You will use the command prompt version of these types by adding a hyphen before the command like -LAYER and INSERT.
- Before you start writing a macro, test the command prompts using AutoCAD prompts, which appear in the command window. Then, do the following:
  - Type "circle" in the command window, then press [Enter].
  - See the prompts, what is the default option? The answer is "Center."
  - You have to wait for the user to input the center (use \).
  - The next prompt will be to input either the radius of diameter. With radius as the default, input value = 2.
- The macro will look like the following:

  ```
  ^C^Ccircle;\2;
  ```

- To add a new command, do the following steps:
  - Start the CUI command.
  - Using the lower-left part (namely the Command List), click Create a new command button:

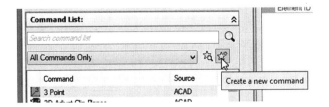

- A new command called command1 will be added.
- A Properties window will open at the right part.
- Type the new name of the command (in this case, it is called CircleR2).
- Type the macro as you produce it on the paper.
- You will see the following:

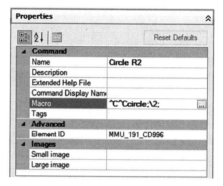

- At the top right, you will see a part called Button Image. You can select an existing image and use it as is or select it and make some modification. If so, click the Edit button and you will see the following dialog box:

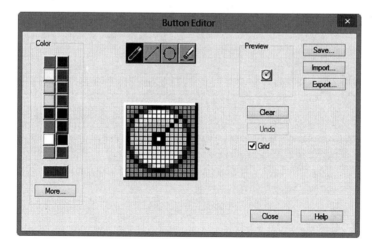

- All of the preceding are very familiar image changing functions like adding a line, circle, or erasing. Use Color part to set the current color to be used, Grid to show rows, and columns to make it easier for you to draw.

- When done, click Save and give the new image a name.
- The new image will be used as representation of the new macro you just created.
- Copy it to an existing panel to be used.

PRACTICE 18-3

Creating New Commands

1. Start AutoCAD 2015.
2. Start a new file using My Company.dwt.
3. Start the CUI command.
4. Create a new command with the following specifications:
    a. Name = Circle R2
    b. Description = This command will draw a circle with R = 2 units.
    c. Macro = ^C^Ccircle;\2;
    d. The image should show a circle with number 2 inside it.
5. Copy the command to My Panel, under Sub-Panel 1, under Row 1, beside Rectangular Array button.
6. Make another command with the following specifications:
    a. Name = CL 0
    b. Description = This command will make 0 the current layer.
    c. Macro = ^C^C-layer;m;0;;
    d. The image should show couple of layers and then 0 (zero).
7. Copy the command to My Panel, under Sub-Panel 1, under Row 2, beside Chamfer button.
8. Click OK to end the CUI command.
9. Make My Workspace the current workspace and test the two new commands.
10. Close the file without saving.

NOTES:

CHAPTER REVIEW

1. Creating template file involves:

    a. Creating text styles.

    b. Setting up units and limits.

    c. Creating layers.

    d. All of the above.

2. _____ is the command to customize the interface.

3. You cannot save your template file except in the Template folder designated by AutoCAD.

    a. True

    b. False

4. While customizing the AutoCAD interface:

    a. You can create new workspace.

    b. You can create new panel.

    c. You can create new commands using macros.

    d. All of the above.

5. Macros cannot deal with commands that show dialog boxes.

    a. True

    b. False

6. Which of the following simulate [Enter] in macro:

    a. Backslash (\)

    b. Semicolon (;)

    c. ^C

    d. ^C^C

CHAPTER REVIEW ANSWERS

1. d
3. b
5. a

CHAPTER 19

PARAMETRIC CONSTRAINTS

In This Chapter

- What are Parametric Constraints?
- Parametric constraints – Geometric
- Parametric constraints – Dimensional

19.1 WHAT ARE PARAMETRIC CONSTRAINTS?

- Design software such as Inventor has had the concept of parametric constraints for a long time. Now AutoCAD includes this concept so AutoCAD is considered a design tool as well as a drafting tool.
- Parametric constraints are Geometric and Dimensional. The first type creates a relationship between different objects like parallel, horizontal, and concentric, etc. Dimensional constraints impose a certain dimension to a line or a radius to an arc, or a circle, and can also create a relationship between two or more objects by writing a formula.
- With this, the designer has the ability to express his/her own design intent, without the fear that somebody alters the design in an accidental or intentional way because the design holds all the necessary measures to keep objects as they are.
- This is a huge step forward for AutoCAD, which makes it complete with other software in this specific category.

- This chapter discusses the two types of constraints and how can you apply them to objects. They are:
  - Geometric constraints
  - Dimensional constraints

19.2 USING GEOMETRIC CONSTRAINTS

- Geometric constraints allow you to set rules for the objects. You can decide to make a line horizontal always and perpendicular on another line. In addition, you can set two circles to always share the same center point and so on. In order to reach to this type of constraint, select the **Parametric** tab and locate the **Geometric** panel:

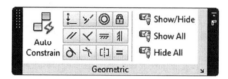

- This panel shows twelve different types of constraints and other buttons like the Auto Constraint button and Show and Hide buttons. The following sections discuss each one.

19.2.1 Using the Coincident Constraint

- Locate the **Geometric** panel, then select the **Coincident** button:

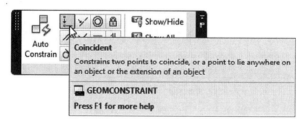

- According to AutoCAD, this will constrain two points to coincide or a point to lie anywhere on an object (or the extension of an object).

- You will see the following prompts:

  ```
  Select first point or [Object/Autoconstrain] <Object>:
  Select second point or [Object] <Object>:
  ```

- This command requires by its basic method to select two points on two existing objects. The first object will stay in its place but the second will move to connect with the first object. A small blue square then appears at the point connecting the two objects:

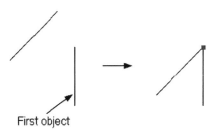

- Another variation of this command is to select an object rather than to select a point on an object. At the first prompt, right-click and select the Object option. You will see a prompt asking you to select the desired object. Once you select it, the following prompt appears:

  ```
  Select point or [Multiple]:
  ```

- If you select a point on another object, the first object will stay in its place and the other will link with the extension of the first object, as in the following:

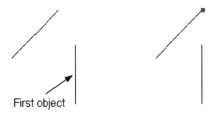

- If you use the Multiple option, you then see the following prompts:

  ```
  Select Point:
  Select Point:
  ```

- You are able to select multiple points to link them with the first object using the previous rule. The following example shows three objects coincident with the first object connecting their end points (you can select any other desired point):

19.2.2 Using the Collinear Constraint

- Locate the **Geometric** panel then select the **Collinear** button:

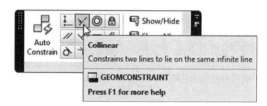

- According to AutoCAD, this will constraint two lines to lie on the same infinite line.
- You will see the following prompts:

  ```
  Select first object or [Multiple]:
  Select second object:
  ```

- The simplest method is to select the first line (which stays intact), then select the second line, which will move to be collinear. See the following illustration:

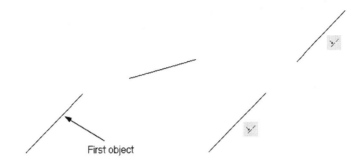

- If you select the **Multiple** option, this allows you to set several lines to be collinear.

19.2.3 Using the Concentric Constraint

- Locate the **Geometric** panel then select the **Concentric** button:

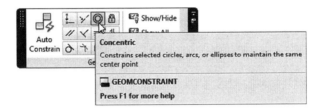

- According to AutoCAD, this will constraint selected circles, arcs, or ellipses to have the same center point.
- You will see the following prompts:

```
Select first object:
Select second object:
```

- The following is an example:

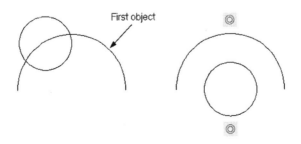

19.2.4 Using the Fix Constraint

- Locate the **Geometric** panel then select the **Fix** button:

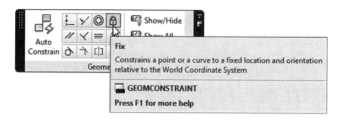

- According to AutoCAD, this will constraint a point or a curve to a fixed location and orientation relative to the World Coordinate System. You will see the following prompt:

```
Select point or [Object] <Object>:
```

- Either you will select a point on an object or select an object option. To be allowed to select the desired object, see the following illustration which shows the two cases:

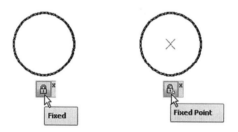

19.2.5 Using the Parallel Constraint

- Locate the **Geometric** panel then select the **Parallel** button:

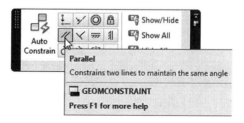

- According to AutoCAD, this will constraint two lines to be parallel. You will see the following prompts:

```
Select first object:
Select second object:
```

- The first object will keep its current angle but the second object will rotate to have the same angle of the first. See the following illustration:

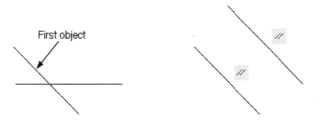

19.2.6 Using the Perpendicular Constraint

- Locate the **Geometric** panel then select the **Perpendicular** button:

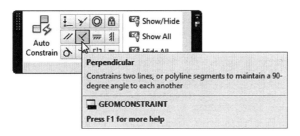

- According to AutoCAD, this will constraint two lines or polyline segments to maintain a 90-degree angle to each other. You will see the following prompts:

```
Select first object:
Select second object:
```

- The first object will keep its current angle; the second object will rotate it to make it 90° relative to the first object. See the following illustration:

19.2.7 Using the Horizontal Constraint

- Locate the **Geometric** panel then select the **Horizontal** button:

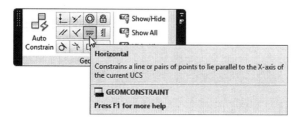

- According to AutoCAD, this will constraint a line or pair of points to lie parallel to the X-axis of the current UCS. You will see the following prompts:

```
Select an object or [2Points] <2Points>:
```

- If you select a line, it becomes horizontal and the command ends. The endpoint nearest to the selection point will remain but the other end will move.
- See the following illustration:

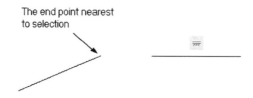

- If you select the **2 Points** option, you will see the following prompts:

```
Select first point:
Select second point:
```

- This option ensures that the two points selected will form a imaginary horizontal line. See the following example:

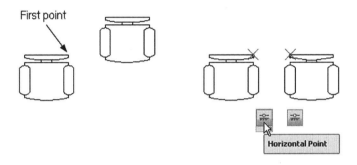

- Note that although polyline is considered a single object, if you use this constraint, AutoCAD will treat each object alone.

19.2.8 Using the Vertical Constraint

- Locate the **Geometric** panel then select the **Vertical** button:

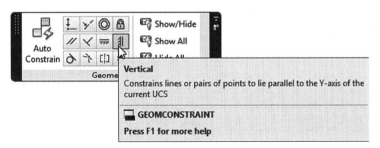

- This command is identical to the previous command, allowing objects or points to be vertical.

19.2.9 Using the Tangent Constraint

- Locate the **Geometric** panel then select the **Tangent** button:

- According to the AutoCAD, this will constraint two curves to maintain a point of tangency to each other or their extension. You then see the following prompts:

```
Select first object:
Select second object:
```

- The first object stays in its current place but the second object moves using the nearest tangent point close to its current position. See the following illustration:

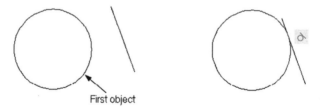

19.2.10 Using the Smooth Constraint

- Locate the **Geometric** panel then select the **Smooth** button:

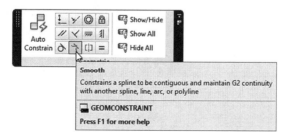

- According to AutoCAD, this will constrain a spline to be contiguous and maintain G2 continuity with another spline, line, arc, or polyline. You then see the following prompts:

```
Select first spline curve:
Select second curve:
```

- The first object should be a spline, but the second object can be a Spline, Line, Arc, or Polyline, etc. The following illustration demonstrates the continuity concept:

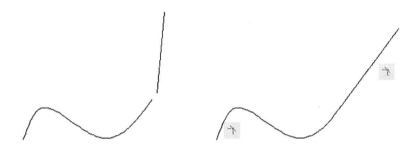

- Although you are selecting objects, you should be careful because the point that is highlighted while you are selecting is very important to the end result.

19.2.11 Using the Symmetric Constraint

- Locate the **Geometric** panel then select the **Symmetric** button:

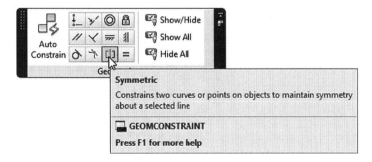

- According to AutoCAD, this will constraint two curves or points on objects to maintain symmetry within a selected line. You will see the following prompts:

```
Select first object or [2Points] <2Points>:
Select second object:
Select symmetry line:
```

- This is similar to the Horizontal command because it uses the same prompts. You will select either two objects or two points. The first object maintains its current angle but the second object rotates to make the mirror image of the first object around the symmetry line selected. See the following illustration:

19.2.12 Using the Equal Constraint

- Locate the **Geometric** panel then select the **Equal** button:

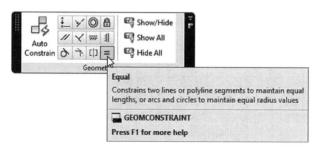

- According to AutoCAD, this will constraint two lines or polylines segments to maintain equal lengths, or arcs, and circles to maintain equal radius values. You will see the following prompts:

```
Select first object or [Multiple]:
Select second object:
```

- The first object will maintain its length (radius for arcs and circles) and the second object will change its length to match the first object. If you select the Multiple option, you can match the length of the first object with multiple lines rather than a single line. See the following illustration:

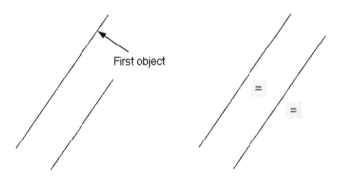

19.3 GEOMETRIC CONSTRAINTS SETTINGS

- To control the display of the Geometric Constraints, you should use the **Settings** in the dialog box. To issue this command, locate the **Geometric** panel then click the arrow at the lower-right corner of the panel:

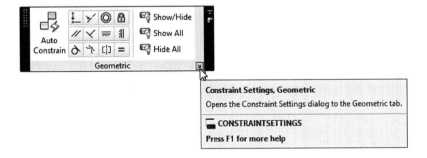

- You will see the following dialog box:

[Constraint Settings dialog box image]

- By default, all of the constraint will be displayed so you can select to hide any of the twelve constraints in the drawing. Also, control the following:
  - Whether to display only constraint bar for objects in the current plane.
  - Constraint bar transparency value.
  - Whether or not to show the constraint bar after applying constraint to selected objects.
  - Whether or not to show constraints bars when objects are selected.

19.4 WHAT IS THE INFER CONSTRAINT?

- In the previous dialog box, one checkbox was overlooked. This is the Infer geometric constraint checkbox, which is at the top-left of the dialog box. The Infer Constraint flips the process of constraining in AutoCAD because it sets the constraints while drafting rather than after drafting. Also, commands like Fillet and Chamfer will be affected.

- To activate this command, go to the Status bar and click the following button:

- After you switch this checkbox on, AutoCAD applies constraints to all objects as they are added to the drawing, using proper constraints depending on the object and the attached objects to it. Do not assume that this method is a replacement to the original method discussed at the beginning of this chapter, because you will be mistaken. This method helps you complete your work in fewer steps but that is all. The following is an example:
  - Switch on the Infer Constraint checkbox.
  - Start the Line command and draw a line a vertical line. Note that AutoCAD applies a Vertical constraint.
  - Continue by drawing a horizontal line. Note that AutoCAD applies a Perpendicular constraint to the new horizontal line.
  - Continue drawing another vertical line and close the rectangle. Notice that AutoCAD applies the Perpendicular constraint, along with the Coincident on all corner points.
  - Start the Fillet command and set the Radius to a suitable value then fillet one of the corners. You will notice that AutoCAD applies the Tangent constraint between the two lines and the added arc along with the Coincident constraint at the two ends of the arc.
  - You then see something like the following:

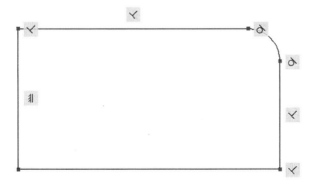

- If you right-click the button in the Status bar and select **Settings**, you see the same dialog box shown in the previous section, which allows you to control which constraint to be displayed on the screen.

19.5 WHAT IS AUTO CONSTRAIN?

- The Auto Constrain command applies multiple geometric constraints to the selected objects, based on the current relationship between these objects and the selected constraint to be applied in the Settings dialog box of the Auto Constraint.
- To issue this command, locate the **Geometric** panel then select the **Auto Constrain** button:

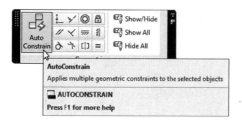

- You will see the following prompt:

  ```
  Select objects or [Settings]:
  ```

- As a first step, select the **Settings** option. You then see the following dialog box:

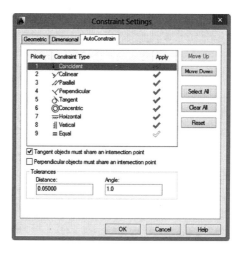

- Using this dialog box, you can do any of the following:
  - Move the constraints up and down to set the priority for each constraint.
  - Turn off any unwanted constraint by clicking the green (✓).
  - Use Select All, Clear All, and Reset buttons.
  - Tangent objects must share an intersection point or not.
  - Perpendicular objects must share an intersection point or not.
- As you can see, three of the normal twelve constraints will not be used. They are: Symmetrical, Fix, and Smooth.
- You should use the following shape as an example:

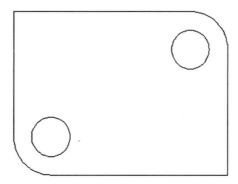

- Before you apply any constraint, select the whole object using grips and try to move one of the grips. Take note of this movement.
- Start the **Auto Constrain** command.
- Select all objects, then press [Enter], and the following message will appear:

```
16 constraint(s) applied to 8 object(s)
```

- The shape will change to:

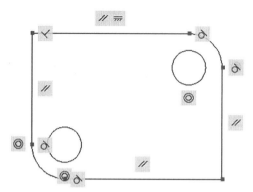

- Select the objects using grips and try to move one of the grips. You will feel that all the objects are moving together as a coherent set of objects that understand each other and keep the correct relationship at all times.
- As a final note for Auto Constrain, you should note that the first prompt of the Coincident constraint contains Autoconstrain:

```
Select first point or [Object/Autoconstrain] <Object>:
```

- If you select this option, then select the objects, all of them will hold a coincident constraint.

19.6 CONSTRAINT BAR AND SHOWING AND HIDING

19.6.1 Constraint Bar

- The Constraint Bar is the small bar, which appears beside the object after you apply a geometric constraint on it. Hover over the small bar and it will be highlighted along with the effected objects. See the following illustration:

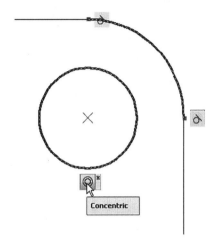

- On the other hand, if you hover over an object with geometric constraint applied to it, the Constraint Bars will be highlighted as well. Occasionally, one object may hold more than one constraint so the bar will show more buttons.
- Right-clicking the buttons in the Constraint Bar will show the following menu:

- The following is a list of these commands:
  - The Delete command deletes the selected geometric constraint. You can use normal [Del] on the keyboard.
  - Hide the current Constraint Bar (even if it contains more than one button).
  - Hide All Constraints in the current drawing.
  - Show the Constraint Bar Settings, which was previously discussed.
- You can move the Constraint Bars from its default position to any other desired position by clicking and dragging it.

19.6.2 Showing and Hiding

- You can control the visibility of the Constraint Bar. You can also control whether to show it or not for all or some objects.
- To issue this set of commands, locate the **Geometric** panel:

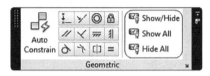

- The commands are:
  - Show/Hide: You can select some of the objects to show the constraints for and some to hide.
  - Show All: This shows all constraints for all objects.
  - Hide All: This hides all constraints for all objects.

19.7 RELAXING AND OVER CONSTRAINING OBJECTS

19.7.1 Relaxing Constraints

- Relaxing an object means removing some of the constraints applied to it. In order to fulfill a command, you can change the object's status. Assume you have a line with a Parallel constraint to a vertical line and you try to rotate it. How does AutoCAD respond to such an action? You will see the following message:

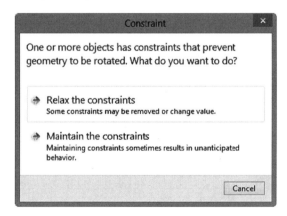

- AutoCAD gives you the choice either to relax the constraints or to maintain them. When you select relaxing the constraint, AutoCAD removes only the associated constraint, which allows the commands to be fulfilled.

19.7.2 Over Constraining an Object

- Assume you applied a Horizontal constraint to a line and then you apply the Perpendicular constraint to an attached line. Applying the Vertical to the second line would mean this object is in over constraint but AutoCAD does not allow this to happen. It shows you the following message:

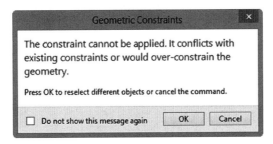

- The message means AutoCAD will not allow the action so you should click OK or Cancel to abort the process.

PRACTICE 19-1

Applying Geometric Constraint

1. Start AutoCAD 2015.
2. Open the file, Practice 19-1.dwg.
3. On the Status bar, make sure that Infer Constraint is off.
4. Start Coincident, select Autoconstraint, and then select the whole shape.
5. Start Concentric and select the arc then the circle.
6. Start Vertical and select the right vertical line.
7. Start Perpendicular then select the right vertical line and the upper horizontal line.

8. Repeat the same with the lower horizontal line.

9. Start Parallel, select the right vertical line, and then the left vertical line.

10. Start Horizontal and select the lower horizontal line. What is the message that AutoCAD produced? _____

11. Click on the Infer Constraint in the Status bar.

12. Using the Fillet command, set Radius = 2, and fillet the upper-right corner of the shape.

13. Draw a circle with Radius = 1, using the center of the arc you just added using the Fillet command.

14. Select the whole shape using grips. Click the midpoint of the right vertical line, and move it. Notice how the whole shape is responding to the movement.

15. Switch off the Infer Constraint button.

16. Pan to the shape at the right.

17. Start the Auto Constraint command, select Settings, and then click the Select All button to select all constraints.

18. Click OK and select the whole shape. How many constraints were added? _____ How many objects? _____

19. Check Equal constraints for both circles and arcs.

20. Select to Hide all constraints then select to show them only for some objects.

21. Save and close the file.

19.8 USING DIMENSIONAL CONSTRAINTS

- Geometric constraints are not enough. You need to set Dimensional constraints to apply the constraint concept to its full. Dimensional constraints specify a length for a line, an angle between two lines, a radius, or a diameter for an arc or a circle. It even links these different dimensional constraints using formulas.

- To reach all the Dimensional constraints commands, select the **Parametric** tab and then locate the **Dimensional** panel:

- The following discusses all types of Dimensional constraints.

19.8.1 Using Linear, Horizontal, and Vertical Constraints

- These are three different constraints, which deal with distances between points. If you click the button, you see the following options:

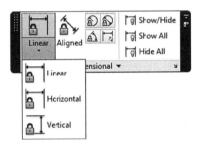

- According to AutoCAD, Linear will constrain either horizontal, or vertical between two points (even if the two points form an angle other than 0, 90, 180, or 270). Horizontal commands will constraint the X-axis distance between two points and the Vertical will constraint the Y-axis distance between two points. You then see the following prompt:

```
Specify first constraint point or [Object] <Object>:
Specify second constraint point:
Specify dimension line location:
Dimension text = 6.6615
```

- Prompts for the three commands are the same. If you select the first point (a red cross with a circle will appear), so then select the second point. You will be asked to specify dimension line location (just as you did in dimensioning) then whether to accept the real measured distance by pressing [Enter] or to input your own value. Of course, the length will change according to the new value. At the first prompt, if you selected Object option, you will see the following prompts:

```
Select object:
Specify dimension line location:
Dimension text = 9.7586
```

- The following illustration shows the shape of the dimensional constraint:

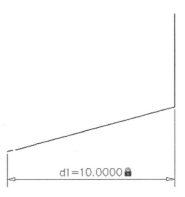

19.8.2 Using Aligned Constraint

- To issue this command, locate the **Dimensional** panel and then select the **Aligned** button:

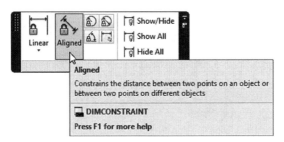

- According to AutoCAD, this will constraint the distance between two points whether on the same objects or on different objects. You will see the following prompts:

```
Specify first constraint point or [Object/Point &
line/2Lines] <Object>:
```

- If you specify the first constraint point, you are asked to specify the second point. If you use Object option, you are asked to select an object. If you choose the Point & line, option you will see the following prompts:

```
Specify constraint point or [Line] <Line>:
Select line:
Specify dimension line location:
Dimension text = 2.2466
```

- Select a point and then a line (while selecting a line, make sure you are selecting the right point). If you select 2 Lines option, you will see the following prompts:

```
Select first line:
Select second line to make parallel:
Specify dimension line location:
Dimension text = 2.9135
```

- Select two line objects. The second line will be made parallel to the first line. Aligned constraint controls the distance between the two lines.
- The following illustration shows Aligned constraint:

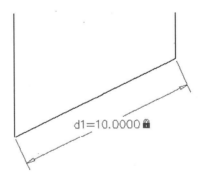

19.8.3 Using Radial and Diameter Constraints

- To issue this command, locate the **Dimensional** panel and then select the **Radial** or **Diameter** button:

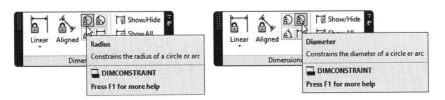

- According to AutoCAD, these will constraint the radius or the diameter of circle or arc. You will see the following prompts:

```
Select arc or circle:
Dimension text = 3.4359
Specify dimension line location:
```

- The following illustration shows the radial and diameter constraints:

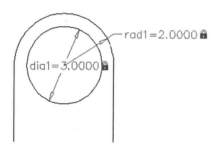

19.8.4 Using Angular Constraints

- To issue this command, locate the **Dimensional** panel and then select the **Angular** button:

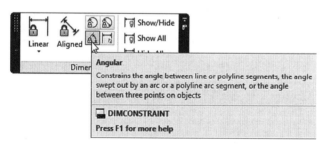

- According to AutoCAD, this will constraint the angle between two lines or polyline objects, in an arc or using three points. You will see the following prompts:

```
Select first line or arc or [3Point] <3Point>:
Select second line:
Specify dimension line location:
Dimension text = 45
```

- Select two lines (or polyline segments) or an arc. If you select the 3 Points option, you will see the following prompts:

```
Specify angle vertex:
Specify first angle constraint point:
Specify second angle constraint point:
```

- Select the vertex point first then the two other points to form an angle. The following illustration shows the angular constraint:

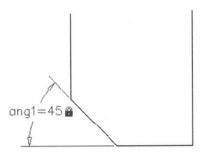

19.8.5 Using the Convert Command

- To issue this command, locate the **Dimensional** panel and then select the **Convert** button:

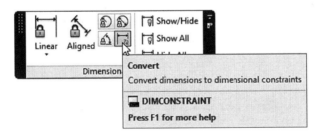

- This command converts normal dimensions to dimensional constraints. You will see the following prompts:

```
Select associative dimensions to convert:
Select associative dimensions to convert:
```

- Simply select the desired dimensions to be converted to constraint. The following illustration shows the process:

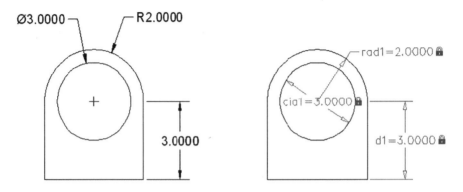

19.9 CONTROLLING DIMENSIONAL CONSTRAINTS

- This section discusses how to control Dimensional constraints by:
  - Using the Constraint Settings dialog box
  - Deleting Dimensional constraints
  - Showing and hiding Dimensional constraints

19.9.1 Constraint Settings Dialog Box

- In this dialog box, you can control the appearance of the dimensional constraint on the screen. To issue this command, select the **Parametric** tab, locate the **Dimensional** panel, then select the **Constraint Settings** and **Dimensional** button:

- The following dialog box comes up:

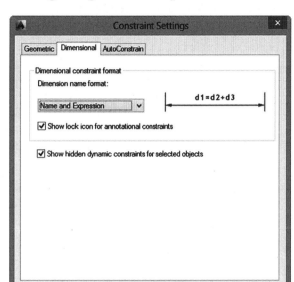

- This dialog box controls what to appear at the dimensional constraint. There are three choices:
  - Name
  - Value
  - Name and Expression
- You can select whether to show or hide the lock symbol, which appears near the measured value. Finally, you can select whether to show the dimensional constraint for any hidden dimensional constraint when you select this object. See the following example:

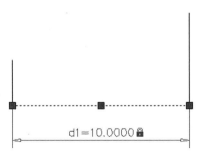

- In the preceding, the Dimensional constraint is hidden but when you select the object, the Dimensional constraint appears.

19.9.2 Deleting Constraints

- You can use the normal [Del] key on the keyboard to get rid of any Dimensional or Geometric constraint. Another way is to locate the **Manage** panel and then select the **Delete Constraints** button:

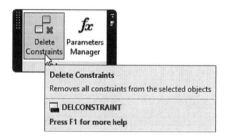

- The following prompt appears:

```
All constraints will be removed from selected objects…
Select objects:
```

- Keep selecting the undesired constraint, then press [Enter] to end the command.

19.9.3 Showing and Hiding Dimensional Constraints

- These commands are similar to what you learned about in terms of the Geometric constraint. To issue this command, locate the **Dimensional** panel, then select one of the following three buttons:

- The Show/Hide button shows a hidden constraint or hides a visible constraint by selecting. You will see the following prompts:

```
Select objects:
Select objects:
Enter an option [Show/Hide]<Show>:
```

- The Show All button shows all dimensional constraints.
- The Hide All button hides all dimensional constraints.

19.10 USING THE PARAMETERS MANAGER

- The Parameters Manager is the place to create equations involving dimensional constraint, linking all dimensional constraints together. Using this method, any change in one of the dimensional constraints means that the other will be affected as well. To create the right equations, AutoCAD allows you to create user-defined parameters. To issue this command, locate the **Manage** panel and then select the **Parameters Manager** button:

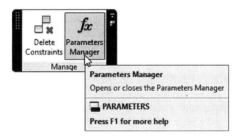

- You will see something like the following palette:

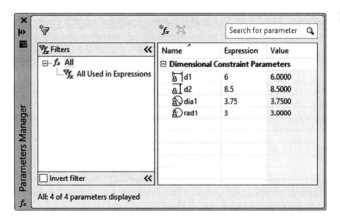

- The Parameters Manager palette is displayed showing all the existing Dimensional Constraint Parameters. As you can see, AutoCAD is using the default names but you have the ability to rename these parameters. Simply click the name once and you can input your own name:

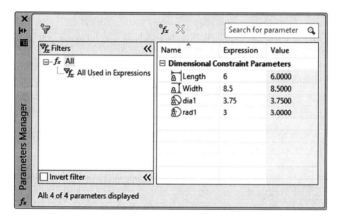

- You can create an equation including one or more of the parameters as in the following example:

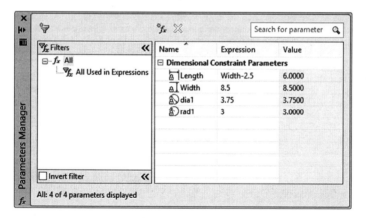

- The means whenever you change Width then the Length will change as well. You can create your own parameters by creating user-defined parameters, using the following button:

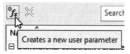

- Input the name of the user-defined parameters and an expression (if valid); if not, input the current value. Later, you can on include an expression involving dimensional constraint parameters. See the following example:

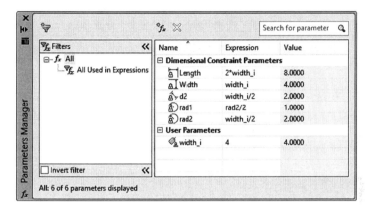

- The previous example shows that both Length and Width are linked with the user parameter called width_i.
- AutoCAD provides filters to categorize your parameters in the category you made. To create a new filter, click the following button:

- Once you click this button, a new filter is added. Simply type the name of the filter, then press [Enter]. To fill it with parameters, drag parameters from the right and drop it at the name of the filter. You will see something like the following:

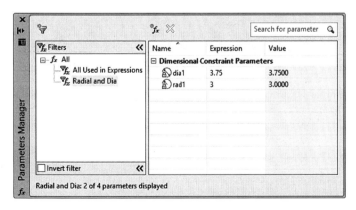

- You can differentiate between normal parameters and equations. Equations starts with **fx:**. User can overwrite the equation by double-clicking it and inputting a value.

19.11 WHAT IS ANNOTATIONAL CONSTRAINT MODE?

- The blocks displayed after you add a dimensional constraint are not printed and zoom in and zoom out commands will not affect its displayed size. This means you do two tasks instead of doing only one; you will add normal dimensions and dimensional constraints. However, AutoCAD can help you cut the workload and do it only once. You have the choice to convert all of the dimensional constraints to Annotational constraints. Annotational constraints can be printed and use the current dimension style.
- There are two ways to do this:
  - Before you start, you can change the mode to an Annotational constraint. With this method, you combine the two actions into a single action.
  - For the dimensional constraints already placed, you can convert them using the Properties palette.

19.11.1 Annotational Constraint Mode

- Before you start placing dimensional constraints, locate the **Dimensional** panel and click the arrow at the bottom to show more buttons. Check that the **Annotational Constraint Mode** is selected:

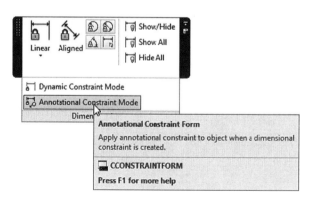

- Add a dimensional constraint using the methods previously discussed. You will see something like the following:

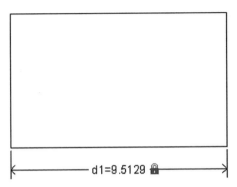

- Using the Dimensional Constraint Settings dialog box, you can choose to show/hide the lock symbol along with the showing the name.

19.11.2 Converting Dimensional Constraints to Annotational

- If you already placed dimensional constraints, you can choose to convert them to Annotational constraints using the Properties palette. Simply select the desired dimensional constraint to be converted, right-click, and select Properties. You then see the following palette. Click the **Constraint Form** and select **Annotational** instead of **Dynamic**:

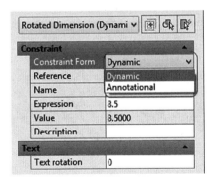

19.12 USING DIMENSIONAL GRIPS

- You can manipulate dimensional constraints using grips. The following image include three types of dimensional constraint, Linear, Aligned, and Angular:

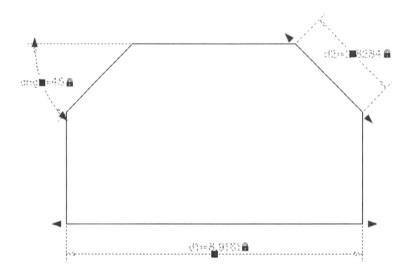

- As you can see, there are grips at the middle of each dimensional constraint that will move the whole block back and forth. The other two arrows will stretch the dimensional constraint, stretching the object by itself. Dimensional parameters have their own grips depending on the type.
- In the following example, there are radius and diameter constraints. The grip helps you relocate the dimensional constraint block. While radius has one arrow to change the value of the radius, the diameter has two arrows to change the value in either direction. Arcs and circles will be affected by this change:

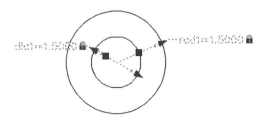

PRACTICE 19-2

Applying Dimensional Constraint

1. Start AutoCAD 2015.
2. Open the file, Practice 19-2.dwg.
3. Using different types of dimensional constraints, add the following constraints using the same names:

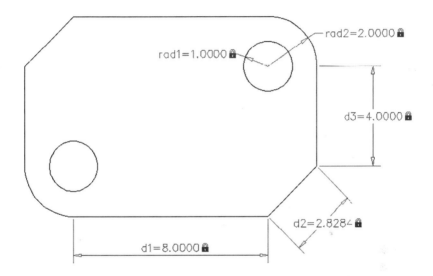

4. Using Parameters Manager do the following:
    a. Rename d1 to become Length.
    b. Rename d3 to become Width.
    c. Create a user-defined parameter width_i with default value of 4.
    d. Change Width = width_i.
    e. Change Length = 2 * width_i.
    f. rad2 = width_i / 2
    g. d2 = width_i / 2
    h. rad1 = rad2/2

5. Using the Parameters Manager change the value of width_i to the following values: 2, 3, 5, 6. How did the other objects react to the change? _____

6. Change width_i to 4 and close the Parameters Manager.

7. Change one of the dimensional constraints to be Annotational. Using Zoom in and Zoom out how did this block respond in comparison to the other blocks?

8. Using the Dimensional Constraint Settings, choose to show only the values without the lock symbol.

9. The lock symbol is still there because all the constraints are equations.

10. Save and close the file.

NOTES:

CHAPTER REVIEW

1. One of the following is not related to dimensional constraints:

    a. It will not be affected by zooming.

    b. Linear

    c. Equal

    d. Angular

2. Parallel, Fix, and Coincident are all _____ constraints.

3. Infer Constraints can be applied while:

    a. Adding line command

    b. Filleting two lines

    c. Stretching objects

    d. Chamfering two lines

4. You can add a single block works as a normal dimension and as a constraint.

    a. True

    b. False

5. There is an option called Autoconstraint in the Collinear command.

    a. True

    b. False

6. You have to select a spline as a first object for _____ constraint.

7. In _____, you can create equations linking different dimensional constraints together.

CHAPTER REVIEW ANSWERS

1. c

3. c

5. b

7. Parameters Manager

CHAPTER 20

DYNAMIC BLOCKS

In This Chapter
- What are Dynamic Blocks?
- Parameters and Actions
- Dynamic Blocks and Constraints

20.1 INTRODUCTION TO DYNAMIC BLOCKS

- We know that AutoCAD can provide the facility to create and insert a block with one shape and dimension that can be scaled larger or smaller. But AutoCAD also provides the possibility to create blocks with multiple shapes, sizes, and views. This type of block is called a Dynamic Block and can be created in the Block Editor.
- Dynamic blocks use Parameters and Actions. Actions are very similar to AutoCAD modifying commands such as Stretch, Scale, and Array. Each action needs a parameter to work on. Another method is to create dynamic blocks using Geometric and Dimensional constraints as discussed in Chapter 19.

- To differentiate between the normal blocks and the dynamic blocks, dynamic blocks will always have a lightning symbol to its right, whether you are at the Insert command dialog box or tool palettes. Here are some examples:

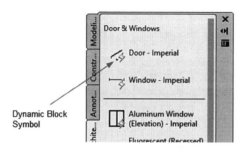

- Dynamic blocks are very handy tools to help you finish producing your drawing faster, with less mistakes, and they reduce the number of blocks in your drawing.

20.2 METHODS TO CREATE A DYNAMIC BLOCK

- Dynamic blocks are made in the Block Editor. You can reach the Block Editor several ways:
  - Using the **Open in block editor** checkbox in Block Definition dialog box.
  - By double-clicking an existing block, then selecting the name of the block.
  - By issuing the **Block Editor** command.

20.2.1 Using the Block Definition Dialog Box

- In general, when you create a normal block, you will use Block Definition dialog box. Simply click Open in block Editor checkbox on and then click OK. AutoCAD takes you directly to the block editor where you can add some dynamic features to your normal block:

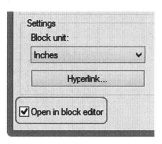

20.2.2 Double-Clicking an Existing Block

- You can convert an existing normal block to be dynamic by double-clicking it. You will see the following dialog box:

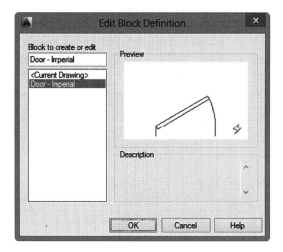

- AutoCAD shows all the blocks in the current drawing. Select the block you want to change and then click OK to go directly to the block editor to add features.

20.2.3 Block Editor Command

- To issue this command, go to the **Insert** tab and **Block Definition** panel, then select the **Block Editor** button:

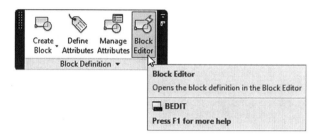

- The following dialog box comes up:

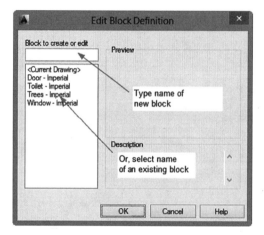

- You can do two things in this dialog box:
  - Type a name of a new block to be created in the block editor.
  - Or select the name of one of the existing blocks to convert (or edit).

20.3 INSIDE BLOCK EDITOR

- Using one of the previous methods, you will be inside the block editor. Once there, three things will happen:
  - The background changes to a different color (most likely it will be dark grey).
  - A new contextual tab called **Block Editor** appears, which includes lots of panels.

- The **Block Authoring** Palette appears, which contains four tabs called Parameters, Actions, Parameter Sets, and Constraints:

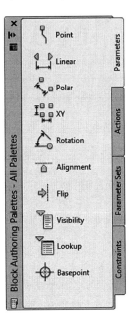

- You can use both the context tab "Block Editor" panels and the Block Authoring palettes to add dynamic features to your block.

20.4 WHAT ARE PARAMETERS AND ACTIONS?

- There is a very simple rule for creating dynamic blocks. First, add a parameter to your block and then assign an action to work on it. Parameters are geometric features of the objects forming the block such as linear, polar, visibility, lookup, etc. Actions are simulation of AutoCAD modifying commands such as Move, Stretch, Scale, etc.
- You will find that some parameters do not need actions and some actions do not need parameters to work on. The rule-of-thumb is: parameter first, then action second. In order to help expedite your work, AutoCAD offers Parameter Sets, which include a set of parameters with actions.
- Both parameter and action have properties, which control the outcome of the dynamic block.
- Chapter 19 discussed geometric and dimensional constraints and how they affect the normal objects. AutoCAD gives the ability to enhance these constraints inside a block.

- The following lists the available parameters:
  - Point parameters add a point definition so the Stretch and Move actions can use it.
  - The Linear parameter measures a distance, which can have any angle. But once applied, you cannot change it. These actions can act on these parameters, Move, Stretch, Scale, and Array.
  - Polar parameters, similar to Linear parameters, have the ability to change the angle. These actions can act on these parameters, Move, Stretch, Polar Stretch, Scale, and Array.
  - XY parameters help define two associated dimensions in X and Y directions. These actions can act on these parameters, Move, Stretch, Scale, and Array.
  - Rotate parameters help you add a rotation parameter to the block definition. Rotate actions can act on this parameter only.
  - Alignment parameters help you add the ability for the block to align itself with an existing object. These parameters do not need an action.
  - Flip parameters flip the existing block to its mirror image using mirror lines. Flip actions can act on this parameter only.
  - Visibility parameters help you create visibility states for some of the objects to be visible and others to be invisible. This parameter does not need an action.
  - Lookup parameters help create tables of possible values of existing parameters. Lookup actions can act on this parameter only.
  - Basepoint parameters define a new base point for the block. This parameter does not need an action
- The following is a list of available actions:
  - Move actions move the selected objects from one position to another. You will see the following prompts:

```
Select parameter:
Specify selection set for action
Select objects:
```

  - Scale action scales the selected objects. You will see the following prompts:

```
Select parameter:
Specify selection set for action
Select objects:
```

- Stretch action stretches selected objects using the Crossing concept (just like in the normal Stretch command). You will see the following prompts:

```
Select parameter:
Specify parameter point to associate with action or
enter [sTart point/Second point] <Second>:
Specify first corner of stretch frame or [CPolygon]:
Specify objects to stretch
Select objects:
```

- Polar Stretch action is similar to Stretch action, but it also has an angle option.
- Rotate action rotates objects using the Rotate parameter. You will see the following prompts:

```
Select parameter:
Specify selection set for action
Select objects:
```

- Flip action works on the Flip parameter only and will create a mirror image using mirror line created in the Flip parameter. You will see the following prompts:

```
Select parameter:
Specify selection set for action
Select objects:
```

- Array action simulates the rectangular array command (the old array prior to the 2012 version). You will see the following prompts:

```
Select parameter:
Specify selection set for action
Select objects:
Enter the distance between columns (|||):
```

- Lookup action only works on the Lookup parameter in order to create a table of existing parameters; it shows them as a list for you to pick one of the available choices.
- The Block Properties Table is discussed after geometric and dimensional constraints and how to include them in a dynamic block are covered.

- Parameter Sets reduce steps needed to complete a drawing and include both parameters and actions, allowing you to work on both at once. There are two exceptions for this:
  - The concept of Pairs means there are two actions acting on a single parameter. There are several pairs such as Linear Stretch Pair, Linear Stretch Pair, etc.
  - The concept of Box Set means there are four actions on a single parameter. There are multiple box sets such as XY Stretch Box Set, XY Array box set, etc.
- You will find the same set of parameters and actions using the context tab Block Editor, located in the **Action Parameters** panel, which looks like the following:

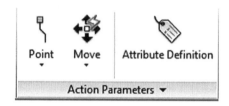

- After you finish adding parameters and actions, you will see something like the following:

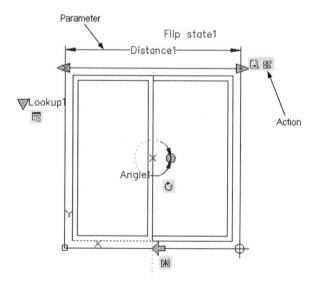

- If you hover over one of the actions, all of the parameters acting on it will be highlighted along with the objects selected:

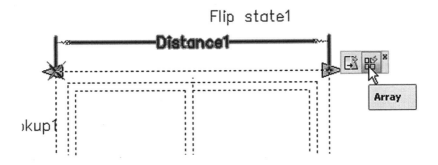

- If you right-click any action, you will see the following menu:

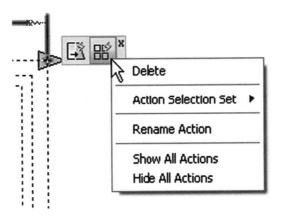

- Using this menu, you can delete the action, change the selection set, rename the actions, and show/hide all actions in the current file. This menu may show different options for different actions.

20.5 CONTROLLING PARAMETER PROPERTIES

- You can control the parameter properties to control the behavior of the block once users start to use it. To see the properties palette, select the parameter, right-click, then select Properties option. Something like the following will come up:

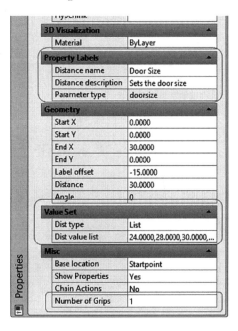

- When you add a new parameter, AutoCAD gives it a temporary name but you can change it to a proper name.
- To do this, locate the **Property Labels** and change the **Distance name** (this example shows the Linear parameter):

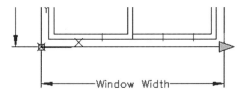

- You can also control size change using the **Value Set**. There are three different types:

- None type: With this type, you can input two values, the Dist minimum and Dist maximum. These two values can be different and can be equal:

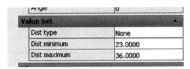

- Increment type: With this type, the value changes using a Dist Increment. You should also input the Dist minimum and Dist maximum:

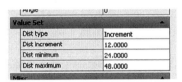

- List type: With this type, the value changes using a list of values which can be input using the small button at the right with the three dots inside it. When clicked, you will see the next dialog box:

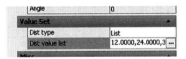

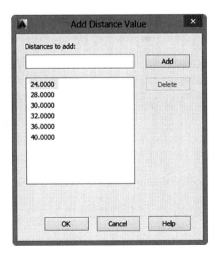

- Finally, you can control the number of grips to appear in the parameter. The default number of parameters in linear are two and there are four in XY. In order to prevent you from scaling or stretching your block from points you do not want, simply reduce or remove the grips. See the following illustration:

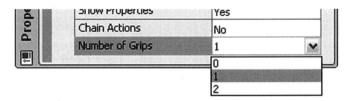

20.6 CONTROLLING THE VISIBILITY PARAMETER

- One of the features you can add to a dynamic block is visibility. The technique is very simple; you have several objects (blocks) and you create a state, which shows some of these objects and hide others. You will create several states. To use this block, select a state from the list beside the block. You have to then follow this procedure:
  - Add a visibility parameter (this parameter does not need an action).
  - Once you add it, the **Visibility** panel in the Block Editor context tab will come to life:

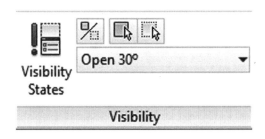

- As you can see, AutoCAD created a new visibility state and gave it the temporary name VisibilityState0. Click the Visibility States button to rename the current visibility state and you will see the following dialog box:

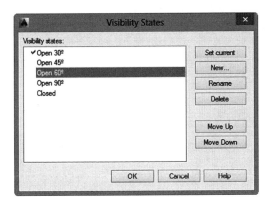

- Click the Rename button and input a new name, then click OK.
- Select the object(s) you want to make visible, then click the **Make Visible** button in the Visibility panel. Next, select the object(s) you want to make invisible, then click the **Make Invisible** button in the Visibility panel. See the following two buttons:

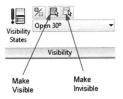

- The default is to hide 100% of the invisible objects. But you can choose to show them with transparency degree, by using the **Visibility Mode** button:

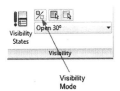

- Create another Visibility State by clicking on the Visibility State button, then clicking the New button. You will see the following dialog box:

- Type the name of the new state and then select one of the choices available. You can select to Hide/Show all of the objects or keep objects with their current visibility settings. Click OK then change the visibility of the objects as you wish.
- Create another state and so on...
- Test your settings by using the list in the panel, as in the following:

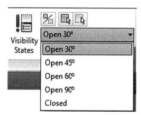

20.7 USING LOOKUP PARAMETER AND ACTION

- Lookup parameter and action are used to display all different sizes of a block in a simple list. With this approach, you can fully control the output of the block because you will limit choices to those listed. You should add the parameters then set the grip and value conditions. The last step is adding the Lookup parameter and Lookup action, or you can add both

at once using Parameter Sets. After adding the parameter and action, do the following steps.
- Right-click the action icon to show the following menu:

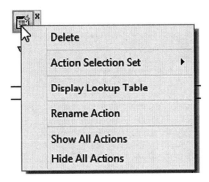

- Select the **Display Lookup Table** option and you will see the following dialog box:

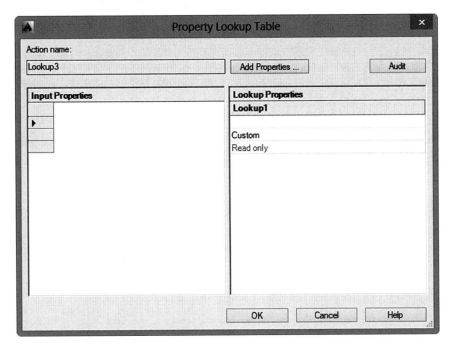

- To add the parameters you want to see in the list click the **Add Properties** button and you will see the following dialog box:

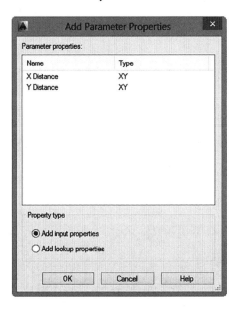

- Select the desired parameters (you can use the [Ctrl] key to select multiple parameters), then click OK to end the adding process.
- The left part of the dialog box then changes to the following:

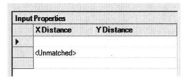

- Start selecting the different values for the Window Width and the corresponding Window Height value. In the right table, input the text that should appear such as "Size 1", "Size 2", etc. This is what you will have:

Input Properties		Lookup Properties
X Distance	Y Distance	Lookup
3'-0"	1'-6"	Window 3' x 1'-6"
3'-6"	1'-9"	Window 3'-6" x 1'-9"
4'-6"	2'-3"	Window 4'-6" x 2'-3"

- The final product should look similar to the following:

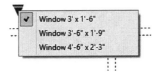

20.8 FINAL STEPS

- Once you are done with adding your features, go to the outmost panel on the left, then the outmost panel on the right. The panel on the left is the Open / Save panel, which looks like the following:

- This panel allows you to:
  - Save the block with the current addition.
  - Test the block without going out of the Block Editor.
  - Edit another block.
  - If you expand the panel, you will see a button to Save As the block under another name.
- The second panel is on the rightmost panel and looks like the following:

- To finish the Block Editor command, you should click the only button in the panel, which is the **Close Block Editor** button. If you did not save changes, AutoCAD shows a warning message to save or discard the changes you made.

PRACTICE 20-1

Dynamic Blocks – Creating a Chest of Drawers

1. Start AutoCAD 2015.
2. Open the file, Practice 20-1.dwg.
3. Create a new block calling it **Chest of Drawers**. Set up the upper-right corner as the base point and the block units as Centimeter. Also, be sure that Open in block editor checkbox is on.
4. Add a Linear parameter for the left vertical line (pick the lower point first then the upper point second). Using the Properties palette, change the name to Height and set the number of grips to one.
5. Using Properties, change Dist Type to Increment, with Dist minimum = 35, Dist maximum = 130, and Dist Increment = 25.
6. Add a Stretch action using the following information:
   a. Select Height parameter.
   b. Select the upper-left corner as your parameter point.
   c. Set the stretch frame (and objects to select) as in the illustration then press [Enter]:

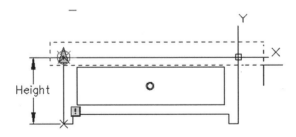

7. Add an Array action selecting the Height parameter and setting the distance to 25.
8. Without leaving block editor, test the block. If everything is OK, save it and test it normally.
9. Save and close the file.

Dynamic Blocks – Door Control

PRACTICE 20-2

1. Start AutoCAD 2015.
2. Open the file, Practice 20-2.dwg.
3. There is a block representing a door. Double-click the block and a dialog box appears. Select Interior Door (it is already selected) and click OK. You are now in block editor.
4. Add a Flip parameter by selecting the midpoint of the two jambs as the reflecting line, specifying the label location beneath the line.
5. Add a Flip action selecting the parameter and all objects representing the door.
6. Test the block.
7. Add a Flip parameter. To specify the reflecting line, use [Shift] + right-click and select the Mid between Two Points option. Select the lower-left corner of the left jamb and the lower right corner of the right jamb. AutoCAD selects the point between these two points. Then using Polar Tracking, select any point upward or downward. Select the label location to be at the middle of the door opening.
8. Add a Flip action selecting the parameter and all objects representing the door.
9. Test the block.
10. Draw a rectangle (using the Rectangle command or Line command) to represent that the door is closed, as in the following:

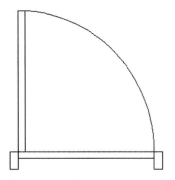

11. Add a Visibility parameter, selecting the upper-left part of the door as the location of the parameter. Note how the Visibility panel was turned on.
12. Using the Visibility panel to create a new state and call it Open, by renaming the existing state.
13. Select the rectangle (or lines) you just added then click the Make Invisible button.
14. Click the Visibility Mode button to see the Invisible objects (they will be transparent).
15. Create a new state calling it Closed.
16. Select the arc and the vertical lines then click the Make Invisible button. Select the rectangle (or lines) you just added and click the Make Visible button.
17. Test the block.
18. Save the block. Test it outside the block editor.
19. Save and close the file.

PRACTICE 20-3

Dynamic Blocks – Wide Flange Beams

1. Start AutoCAD 2015.
2. Open the file, Practice 20-3.dwg.
3. Create a block and name it Wide Flange Beam. Select the lower-left corner as the base point and the block unit as Inch. Also, ensure Open in block editor is on.
4. Add four Linear parameters and name them as follows:

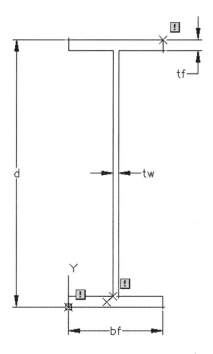

5. Select all the linear parameters and set Grips to 0.

6. Add a Scale action and select d parameter, and all objects.

7. Using the following table, set the Value Set to List and input the following values (the first value is already set so no need to reinput it):

d (in)	tw (in)	bf (in)	tf (in)
44.02	1.02	15.95	1.77
34.21	1.63	15.865	2.95
27.63	0.61	10.01	1.10

8. Going to the Parameter Sets tab, select Lookup Set and add to the drawing. Select the proper location for it.

9. Right-click the Lookup action icon and select the Display Lookup Table option. A dialog box opens so click the Add Properties button. You will see a list of all parameters added previously. Select them all and click OK.

10. Using the preceding table, select all the matched values together and name the first W 44x335, the second W 30x477, and the third W 27X129.

11. Test the block.

12. Save and test the block outside block editor.

13. Save and close the file.

20.9 USING CONSTRAINTS FOR DYNAMIC BLOCKS

- In Block Editor, you can replace using parameters and actions with using geometric and dimensional constraints. This simplifies the steps and cuts into the time it takes to produce dynamic blocks. Using geometric and dimensional constraints while creating dynamic blocks is the same as using it in ordinary objects. Often, the perception is that geometric and dimensional constraints are meant for blocks rather than ordinary objects. Overall, you will see more power in using and controlling geometric and dimensional constraints but parameters and actions are still relevant as well.
- It is important to know that some parameters like Visibility and some actions like Array cannot be replaced in geometric and dimensional constraints.
- One of the things you can do in dimensional constraints is set the Value set like linear parameters using increment or list.
- The interface is similar to what we learned in Chapter 14, with only three new buttons. They are:
  - In the Dimensional panel in block editor, there is a button called Block Table.
  - In the Manage panel in block editor, there is a button called Construction.
  - In the Manage panel as well in block editor, there is a button called Constraint Status.

20.9.1 Block Table Button

- This button is a replica of Lookup parameters and actions. It works with dimensional constraints and in the same way. Why was this feature added? The answer is simple. You cannot use Lookup parameters and actions on dimensional constraints, so you need this feature to produce a pop-up list to select sizes, dimensions, etc. To use this feature, while you are in block editor, locate the **Dimensional** panel and click the **Block Table** button:

Dynamic Blocks • 627

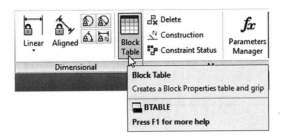

- The following prompts come up:

```
Specify parameter location or [Palette]:
Enter number of grips [0/1] <1>:
```

- You are asked to specify any proper location for the list to appear then to specify the number of grips to include.
- The following dialog box appears:

- To add dimensional constraints to the list, click the following button:

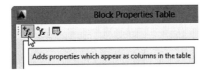

- The following dialog box comes up:

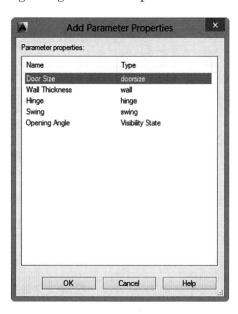

- Select the desired dimensional constraints and then click OK. You will see the following:

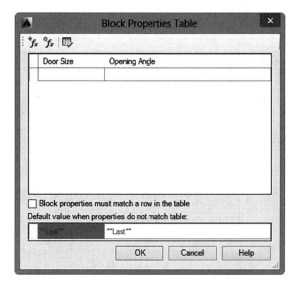

- Pick the first field in the first row and select one of the available values:

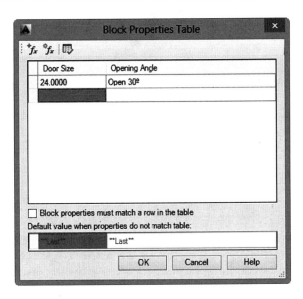

- Keep inputting values until you get something like the following:

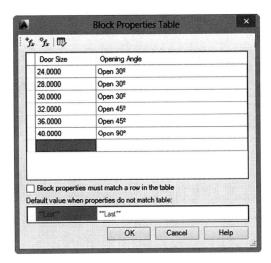

- In order for these values to appear, you should go to the Properties palette of each dimensional constraint and add a list or increment.

- The other two buttons at the top left of the dialog box are to create user parameters. You will see the following dialog box:

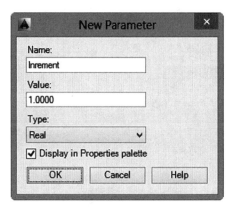

- The third button is to audit the block table. If it does not contains any errors and everything is fine, you will see this message:

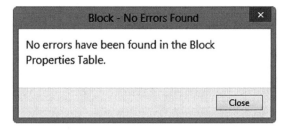

- But if AutoCAD finds a mistake (like one of the fields is empty), you see something similar to the following message:

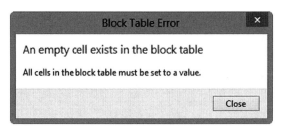

20.9.2 Construction Button

- The Construction object is an object that allows you to define your dynamic block. Inside block editor, locate the **Manage** panel, and click the **Construction** button:

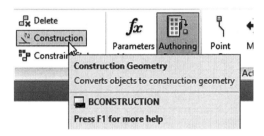

- You will see the following prompt:

```
Select objects or [Show all/Hide all]:
```

- Select the desired object(s).

20.9.3 Constraint Status Button

- This button allows AutoCAD to check your current geometric constraint status. There two options: either blue, which means partial constraints, or magenta, which means full constraints. This is an on/off button. It is preferable to keep it switched on to monitor your block. If you add all of your constraints and you still get blue, you need to add Fix constraints to one of the points, which is an AutoCAD requirement. To switch on/off this button while you are inside the block editor, locate the **Manage** panel, then select the **Constraint Status** button:

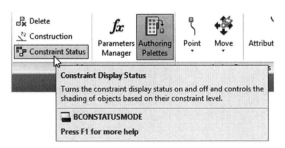

PRACTICE 20-4

Dynamic Blocks Using Constraints – Wide Flange Beams

1. Start AutoCAD 2015.
2. Open the file, Practice 20-4.dwg.
3. Redo Practice 20-3, using geometric and dimensional constraints instead of parameters and actions.
4. Save and close the file.

PRACTICE 20-5

Dynamic Blocks Using Constraints – Creating a Window

1. Start AutoCAD 2015.
2. Open the file, Practice 20-5.dwg.
3. Create a new block from the existing shape, naming it **Window Elevation**. Use the lower-left corner as the base point, Units as Inch, and that Open in block editor is on.
4. Locate the Geometric panel, select the Auto Constrain button, and then select all objects to define the geometric constraints. Next, hide all of them.
5. Using dimensional constraints, set the following linear constraints keeping the naming as follows:

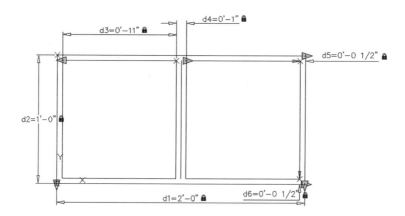

6. Using the **Parameters Manager**, make the following changes to the names: d1=Width, d2=Height, and d3=Glazing.
7. Set the following equations:
    a. Width = Height * 2
    b. Glazing = (Width/2)-d4
8. Select all dimensional constraints except Height and set Grips = 0.
9. Set the Value Set for Height to be Increment, set the min = 1', max = 2', and increment = 6".
10. Set the line between the two glazings to be the Construction object.
11. Test the block.
12. Save the block and test it outside the block editor.
13. Save and close the file.

NOTES:

CHAPTER REVIEW

1. You can add an action to dimensional constraints.

    a. True

    b. False

2. The name of palette in the block editor is _____.

3. One of the following parameters does not need an action.

    a. Linear

    b. Visibility

    c. Lookup

    d. Rotation

4. Rotate action works only with Rotation parameters.

    a. True

    b. False

5. While you are in the Properties palette of a linear parameter, you can set the value set to:

    a. None, Increment

    b. None, List

    c. Increment, List

    d. None, increment, and List

6. All actions need parameters.

    a. True

    b. False

7. You can test your block inside the block editor before you save the changes made.

    a. True

    b. False

CHAPTER REVIEW ANSWERS

1. b
3. b
5. d
7. a

CHAPTER 21

BLOCK ATTRIBUTES

In This Chapter
- What are Block Attributes?
- How to create a block attribute
- How to edit block attributes and value
- How to extract attribute values from drawings

21.1 WHAT ARE BLOCK ATTRIBUTES?

- The previous chapter discussed how to add dynamic features to the block to become more versatile. But, the block still does not include any data. Using Block Attributes you can include information inside a block. For this feature in AutoCAD, it is helpful to ask yourself whether you are the creator of the block or the person using the block. This will help you understand the roles:
    - If you create the block that contains attributes, you will define the attribute definitions, including Color, Cost, Height, and Width.
    - If you are inserting the block, you will input values (e.g., Red, 12.99, 3'-6", and 2'-6").
- The procedure is very simple:
    - Draw the graphical shape of the block.
    - Using attribute definition command define the attributes you want to collect data for.

- Using the Block command (the normal command to create blocks) define the block which contains the attributes.
- Insert the block and fill in the values.
- Set the visibility of the attributes.
- Edit the attributes either as values or definitions.
- Extract data to a table inside the drawing, or as an Excel or database file.

21.2 HOW TO DEFINE ATTRIBUTES?

- This is the first step in creating the attribute definitions. To issue this command, go to the **Insert** tab, locate the **Block Definition** panel, then select the **Define Attributes** button:

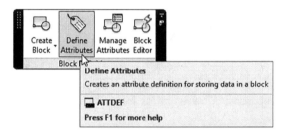

- The following dialog box then comes up:

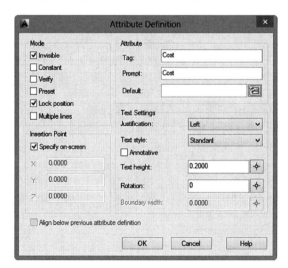

- There are four parts in this dialog box. They are as follows:

21.2.1 Attribute Part

- This part contains three different settings:
  - Tag: the attribute name, you can select any desired name.
  - Prompt: the message that appears to you asking for a piece of information.
  - Default: If there is a value that is repeated more than once, this is the place to input it. If this value is connected to a field, then click the button at the right and input this field.

21.2.2 Mode Part

- There are six modes for the attribute definitions:
  - Invisible: allows you to create an invisible attribute that does not appear in the drawing.
  - Constant: allows you to create a constant value for all of the insertions.
  - Preset: allows you to create an attribute that will be equalized to the default value.
  - Verify: you will be asked to input the attribute twice to verify the value (applies only for the command window input).
  - Lock position: this mode locks the position of the attribute definition.
  - Multiple lines: allows you to input a multi-line text rather than single line text.

21.2.3 Text Settings Part

- This part controls the appearance of the attribute definition:
  - Justification: allows you to specify text justification.
  - Text style: specifies the used text style (recommended to create text style prior to this step).
  - Whether the text will be annotative or not.
  - Text height: inputs the text height to be used if the text style height is 0.
  - Rotation: inputs the rotation angle of the attribute definition.
  - Boundary width: used only when Multiple lines option is selected. It will the line length for multiple lines.

21.2.4 Insertion Point Part

- This part allows you to specify the position of the attribute definition either by typing the coordinates or by selecting **Specify on-screen** option. If you are inserting the second attribute, you have the choice of aligning the new attribute with the previously inserted attribute. Make sure the following checkbox is on:

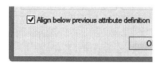

- After defining the attributes, start the Block command and create a block from the graphics and the attributes. While selecting the attributes, select them in the same order you would like them to appear.

21.3 INSERTING BLOCKS WITH ATTRIBUTES

- You should control how the attribute questions appear and system variable ATTDIA will do the job. There are two values to choose from:
  - If **ATTDIA = 0**, questions are shown in the command window.
  - If **ATTDIA = 1**, questions are shown in a dialog box, just like the following:

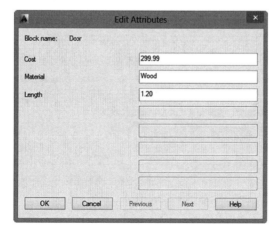

21.4 HOW TO CONTROL ATTRIBUTE VISIBILITY?

- Attributes are either visible or invisible. This section discusses how to change this status temporarily. In order to control attribute visibility, go to the **Insert** tab and locate the **Block** panel. Extend the panel and click the list:

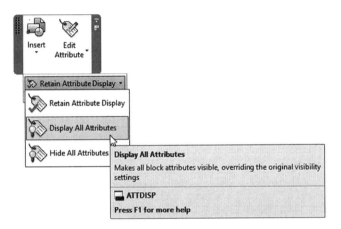

- As previously shown, you have three choices:
  - **Retain Attribute Display** means AutoCAD will show the visible and hide the invisible.
  - **Display All Attributes** means AutoCAD will show all attributes.
  - **Hide All Attributes** means AutoCAD will hide all attributes.

PRACTICE 21-1

Defining and Inserting Blocks with Attributes

1. Start AutoCAD 2015.
2. Open the file, Practice 21-1.dwg.
3. Zoom to the shape at the left.

4. Define the following attributes, inserting them below the door shape:

Tag	Material	Cost	Color
Prompt	Material	Cost	Color
Default	Wood	299.99	Dark
Invisible	✓	✓	✓
Preset	✗	✓	✗
Justification	Left	Align	
Text Style	Arial10	Align	

5. Define a new block naming it "Door", selecting the lower-left corner of the door as the insertion point, selecting all objects except attributes, then selecting attributes one-by-one from top to bottom. To finish, click OK.

6. Make layer A-Door current.

7. Using the ATTDIA command, make sure the value is 1.

8. Insert the new block into the three openings available in the architectural plan, using the following information:

   a. For the one on the left, leave the default values.

   b. For the one in the middle, change the material to be Aluminum.

   c. For the one on the right, change the cost to 399.99 and the color to Light.

9. Change the visibility mode to show all attributes.

10. Save and close the file.

21.5 HOW TO EDIT INDIVIDUAL ATTRIBUTE VALUES?

- This part discusses your ability to edit an individual block attributes. You can edit not only the values, but many other things as well. There are two ways to reach this command:
  - Double-click an existing block with attributes.

- Go to the **Insert** tab, locate the **Block** panel, click **Edit Attribute**, and then select the **Single** button.

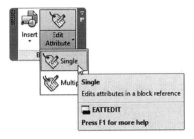

- You will see something like the following:

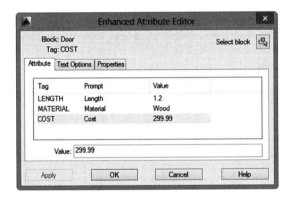

- There are three tabs, Attribute, Text Options, and Properties. You see the Attribute tab when you start editing. In this tab, you can change the value of the attribute for this specific block. Click the **Text Options** tab to see the following:

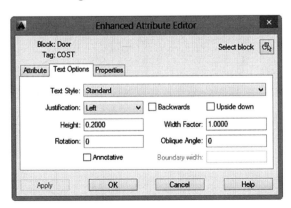

- As you can see, you can change everything related to text style for this specific block and attribute. Select the **Properties** tab to see the following:

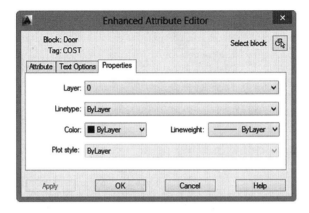

- You can change anything related to properties such as layer, linetype, color, and line weight.

21.6 HOW TO EDIT ATTRIBUTE VALUES GLOBALLY?

- What if you have so many incidences of a block but you want to change one specific value for all incidences? The previous method will be lengthy and tedious. AutoCAD offers a simple command, which everybody knows, as the **Find/Replace** command. To issue this command, go to the **Annotate** tab, locate the **Text** panel, you will find a field with **Find text** inside it:

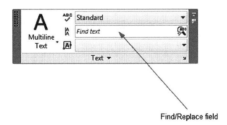

Find/Replace field

- Click inside the field and type the value of attribute you want to change globally, then click the small button at the right. You will see the following dialog box:

- Extend the dialog box, by clicking the small arrow at the lower-left corner of the dialog box to see it in full. You will see something like the following:

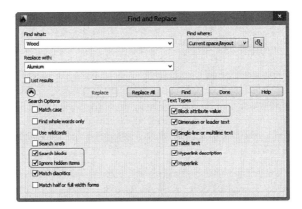

- Confirm the following:
  - Under Text Types, make sure the Block attribute value is on.
  - Under Search Options, make sure that Search blocks is on.
  - Under Search Options, make sure that Ignore hidden items is off.
- Finally, click one of the three execution commands, Find, Replace, or Replace All.

PRACTICE 21-2

Editing Attribute Values Individually and Globally

1. Start AutoCAD 2015.

2. Open the file, Practice 21-2.dwg.

3. There have been some mistakes in inputting the attribute values and you need to unify them. Change the following:

   a. All doors are wood and none are aluminum.

   b. All doors are 299.99 and none are 399.99.

   c. All doors are dark and none are light.

4. Double-click any two doors and make sure that changes took place.
5. Show all attributes.
6. Double-click the main door, then select the Cost attribute.
7. Go to the Text tab and set the Text style to be Standard. Set the Height = 0.15.
8. Go to Properties tab and select Text layer.
9. Click OK and compare this attribute to the others.
10. Retain Attribute Display.
11. Save and close the file.

21.7 HOW TO REDEFINE AND SYNC ATTRIBUTE DEFINITIONS?

- What if you found that there is a missing attribute in your block? You want to remove an unneeded attribute? Or, you want to edit the modes of an existing attribute? You need to redefine the attributes using Block Attribute Manager in each of the preceding examples. To issue this command, go to the **Insert** tab, locate the **Block Definition** panel, then select the **Manager Attributes** button:

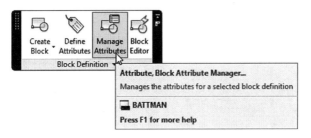

- You will see the following dialog box:

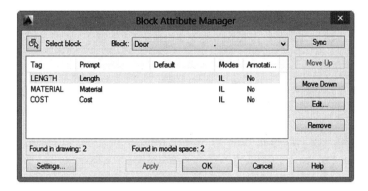

- There are two ways to start working:
  - Click the button at the top right to select a block.
  - Click the list to the right of the button to select the desired block.
- Based on the selected block, you will see a list of attributes for this block. You can do four things with these attributes:
  - Change the position of the attribute in relation to other attributes, by moving the selected attribute one position up.
  - Change the position of the attribute in relation to other attributes, by moving the selected attribute one position down.
  - Remove (delete) the selected attribute.
  - Edit the definition of the selected attribute. You will see the following dialog box:

- In this dialog box, you can edit Tag, Prompt, Default value, Mode, and all other text options and properties as well.
- What about adding a new attribute to an existing block? To answer this question, follow the following steps:
  - Insert the desired block in your drawing.
  - Explode it.
  - The existing attributes will appear.
  - Define the new attribute.
  - Redefine the block using the same name.
- Whether you edit, delete, or add new attributes, this will affect the new insertions of the block. What about the already inserted incidences? You should synchronize to see the changes. In order to issue this command, go to the **Insert** tab, locate the **Block Definition** panel, then select the **Synchronize** button:

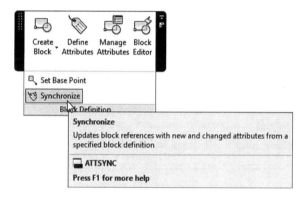

- You will see the following prompt:

```
Enter an option [?/Name/Select] <Select>:
```

- Type the name of the block or choose Select option to select it. You will see the following prompt:

```
ATTSYNC block Window? [Yes/No] <Yes>:
```

- If you input Yes (default), AutoCAD will carry on the synchronize process. The new added attributes will be empty but you can fill them by double-clicking the desired block.

PRACTICE 21-3

Redefining Attribute Definitions

1. Start AutoCAD 2015.
2. Open the file, Practice 21-3.dwg.
3. Using the Block Attribute Manager, do the following:
    a. Remove Color attribute.
    b. Make Cost attribute modes only Invisible.
    c. Change the cost to appear first in the list of questions.
4. Insert block Door in an empty space in the drawing, then explode it.
5. Make layer 0 current.
6. Add a new attribute with the following data:
    a. Tag: SUP
    b. Prompt: Supplier
    c. Default: MYNE Wood
    d. Modes: Invisible
    e. Insert it below the existing two attributes.
7. Redefine the block under the same name, using the same base point (very important).
8. By double-clicking any of the blocks, you will see that the Supplier attribute is not added and the sync command was not issued.
9. Issue the Synchronize command and select one of the doors.
10. Now double-click one of the blocks again, do you see the Supplier attribute? _____
11. Save and close the file.

21.8 HOW TO EXTRACT ATTRIBUTES FROM FILE(S)?

- The final step is extracting the data from drawing(s) and putting it in a table inside the same drawing or another drawing. You can also create an Excel sheet or Access database. Extracting data in AutoCAD is done through a wizard that produces the desired table without hassle. To access this command, go to the **Insert** tab, locate the **Linking & Extraction** panel, then select the **Extract Data** button:

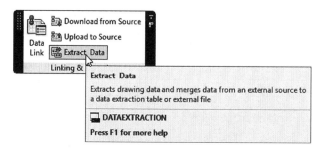

- The following dialog box comes up:

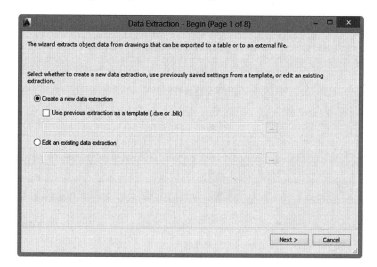

- You can pick one of the three available options:
  - Create a new data extraction.
  - Use the previous extraction as a template.
  - Edit an existing data extraction.

- Let us assume you want to create a new extraction. You should select the first option and click the **Next** button. AutoCAD then asks you to save this extraction (file format is *.dxe) for future use. The following dialog box will appear:

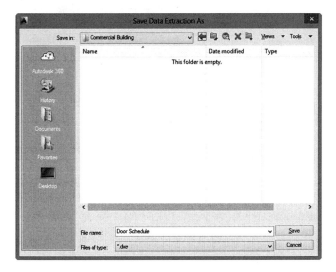

- After saving the dxe file, you will see the following dialog box:

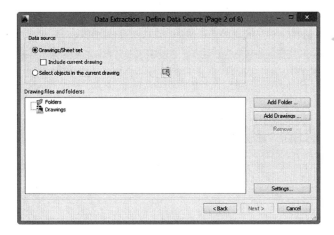

- AutoCAD needs to know where it will find the files to extract data. You have two choices to choose from:
    - Certain files (or sheet sets) and whether to include or to exclude the current file.
    - Select objects in the current drawing.

- Selecting the first choice means you have to specify the folders and draw the files using the two buttons at the right. Clicking the **Add Folder** button, you will see the following dialog box:

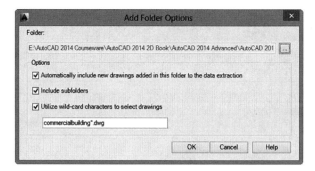

- There are four things to control in this dialog box:
  - Click the small button with three dots to specify the folder that contains the desired files.
  - Select whether or not you want to "Automatically include new drawings added in this folder to the data extraction."
  - Select whether or not you want to include all subfolders of the selected folder.
  - Select whether or not you want to "Utilize wild-card characters to select drawings." This is similar to what you can do in the Windows Search tool. For instance, CommercialBuidling*.dwg means you want to select all drawing files starting with CommercialBuilding.
- If you want to go with the second choice, simply click the small button at the right to select the desired objects you want to extract data from. If you make a wild card, you may see something like the following:

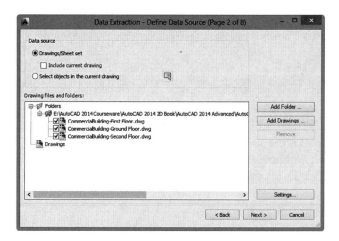

- When finished, click the **Next** button and you will see the following dialog box:

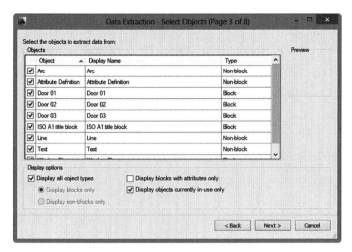

- The list is showing objects and blocks. In order to refine this list, use the checkboxes beneath. You have three choices:
  - Select whether or not to Display all the object types. If you turn this off, you have to select whether to Display blocks only or to Display non-blocks only.
  - Select whether or not to Display blocks with attributes only.
  - Select whether or not to Display objects currently in-use only.
- The following example shows only blocks with attributes. You should deselect one of the blocks that you do not want to include in the extraction process:

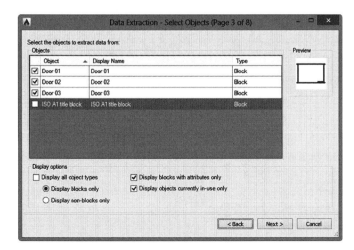

- When done, click the **Next** button and you will see the following dialog box:

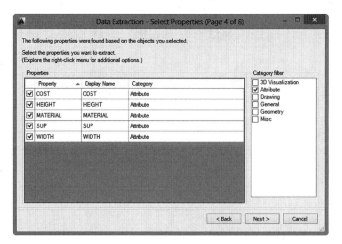

- At the right and under the Category filter, deselect all choices except Attribute. Also, you can further deselect the attributes that you do not want to include in the extraction process. When done, click **Next** to continue to the fifth step in the extraction wizard. You will see the following dialog box:

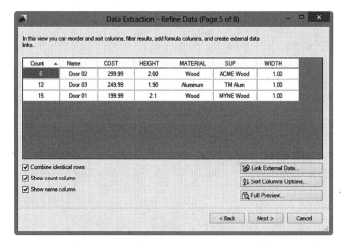

- There are some basic functions you can do with this page:
  - Sort the table ascending or descending by clicking the desired column header.
  - Change the location of any column, by clicking the header and drag and drop.

- Change the column width by dragging the line separating any two columns.
- You can select to Combine identical rows.
- Show count column or not.
- Show name column or not.
■ If you right-click any header, you will see the following menu:

■ Use this menu to do any of the following:
  - Rename a column.
  - Hide and Show columns.
  - Set Column Data Format.
■ You will see the following dialog box:

- Insert a new formula column and you will see the following dialog box. In this dialog box, type the name of the new column, then drag-and-drop the names of the attributes that you want to use, using the arithmetic function between them.

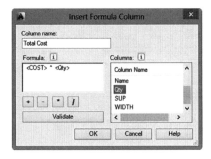

- Edit and Remove Formula Column.
- Insert Totals Footer.
- Create a filter for the selected column.
- If you want to establish a link between this table and an Excel file, click Link External Data button at the right.
- If you want to set advanced sorting criteria then click Sort Columns Options button at the right.
- If you want to establish a data link between these extracted data and Excel sheet.
- The last button (Full Preview) will show a preview of the table
- When done, click Next to go to the next step and you will see the following dialog box:

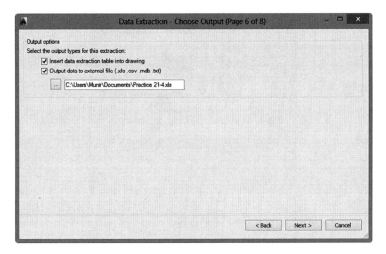

- You can select to insert the data extraction as a table in the current file and produce an external file such as an Excel sheet file or an Access database file.
- When done, click **Next** to go to the seventh step and the following dialog box comes up:

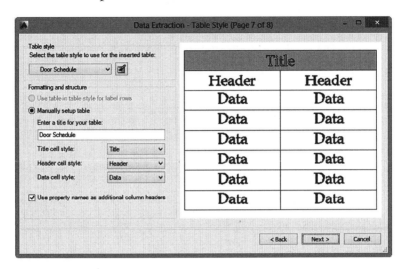

- Select the table style you want to use. When done, click **Next** to go to the last step. The following dialog box comes up:

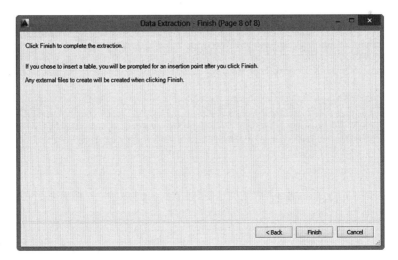

- Click **Finish** to wrap up the extraction process.

PRACTICE 21-4

Extracting Attributes

1. Start AutoCAD 2015.
2. Open the file, Practice 21-4.dwg.
3. Extract data from the following three files which exist in your Chapter 3 folder of practices:

   a. CommercialBuilding-Ground Floor.dwg

   b. CommercialBuilding-First Floor.dwg

   c. CommercialBuilding-Second Floor.dwg

4. This is the final product of the extraction. Do whatever it needs to produce an identical replica:

Door Schedule								
Name	HEIGHT	WIDTH	MATERIAL	Supplier	Qty	COST	Total Cost
Door 01	2.1	1.00	Wood	MYNE Wood	15	199.99	$2,999.85
Door 02	2.00	1.00	Wood	ACME Wood	6	299.99	$1,799.94
Door 03	1.90	1.00	Aluminum	TM Alum	12	249.99	$2,999.88
					33		$7,799.67

5. Save and close the file.

CHAPTER REVIEW

1. While defining a new attribute, you can make the default value equal to a field.

    a. True

    b. False

2. One of the following statements is not true about the extracting process.

    a. While extracting, you can add a formula column.

    b. You can add more than one formula column.

    c. Formula column uses attributes and fields.

    d. You can add a total footer.

3. You can edit attribute values by double-clicking the desired block.

    a. True

    b. False

4. After inserting block containing attributes in your drawing, you can do all but:

    a. Edit the values one-by-one.

    b. Edit the attribute definitions.

    c. Allow AutoCAD to synchronize the values automatically.

    d. Add a new attribute to the block.

5. _____ mode will create an attribute that will be equalized to the default value.

6. Invisible and constant are _____.

7. _____ is a system variable that allows you to control to show or not show a dialog box while inputting values for attributes.

CHAPTER REVIEW ANSWERS

1. a
3. a
5. Preset
7. ATTDIA

CHAPTER **22**

EXTERNAL REFERENCING (XREF)

In This Chapter
- What is an External Reference (Xref)?
- Inserting Xrefs
- Xref Layers
- Editing Xrefs
- Xref Functions
- Xref Clippings
- Xref Special Functions
- eTransmit and Xrefs

22.1 INTRODUCTION TO EXTERNAL REFERENCE

- Today's work environment is collaborative and workers need to share work. External Reference (or Xref) helps engineers/draftspeople to collaborate and coordinate their drawings by allowing users to import their drawing while segregating the content of each drawing to keep it organized and segregated. This is not a new technique in AutoCAD but in the last two or three versions of AutoCAD, only allowed importing file formats such as DWF, DGN, PDF, and any image file format.

- Xref files establish a link with the original file so whenever a change takes place in the original file, AutoCAD notifies you of the update. When you open an Xref file, the latest version of the file is loaded into the drawing.
- External Reference is a better method than using the **Insert** command to bring in contents from DWG files, as this method will add the whole file into your file, you will not be able to update it, and it will mix all coming layers and blocks with your current layers and blocks.

22.2 INSERTING EXTERNAL REFERENCE DIFFERENT FILE FORMATS

- There is a single command that can insert all the five types of files in the current drawing file. To reach this command, go to the **Insert** tab, locate the **References** panel, then select the small arrow at the lower right:

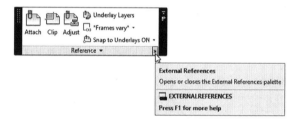

- The **External Reference** palette comes up showing only the name of the current file:

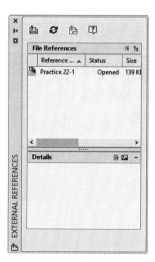

- To attach any file format, simply go to the top left part of the palette where a small list button exists and click it. You will see something like the following:

- As shown, you can select one of the following file formats to bring in your current file:
  - Attach DWG file
  - Attach Image file
  - Attach DWF file
  - Attach DGN file (Microstation V8 file)
  - Attach PDF file
  - Attach Point Cloud

22.2.1 Attach a DWG File

- You can use this command to attach AutoCAD drawing in the current drawing. The Normal Open dialog box will appear and once you click OK, you see the following dialog box:

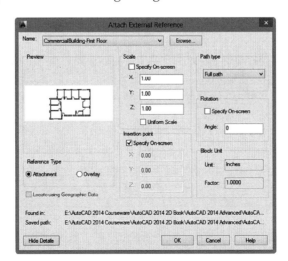

- As the previous illustration shows, four parts of this dialog box are identical to the Insert block command:
  - Scale
  - Insertion point
  - Rotation
  - Block unit
- Path type and Reference type are new.
- As for Path type, you already read that AutoCAD keeps a link to the original file to keep tracking of the changes will take place. In order to do this, AutoCAD saves the current path of the attached DWG file. It will ask you the type of the path and there are three choices in hand.

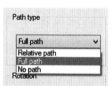

  - **Full path**: AutoCAD saves the exact path of the attached DWG file, remembering the drive and folder(s), which means if the file is moved to a different place, AutoCAD will report an error.
  - **Relative path**: The default option. AutoCAD only save the name of the folder containing the file, ignoring the other superior folders and drive. If you are moving your folder to another place, AutoCAD will be able to locate it.
  - **No path**: AutoCAD does not save the path for the DWG file. As an alternative, it will search the current folder then Search paths mentioned in the **Options** dialog box, the **Files** tab, and both paths mentioned under Project and Support.
- As for Reference Type, this option deals with multiple people using Xrefs to import DWG files. Assume that person B imports a DWG to his/her file from person A. Then, person C imports person B's file. What will person C see in his/her file?
  - Attachment: Person C will see both person A's drawing and person B's drawing.
  - Overlay: Person C will see only person B's drawing.

- As a final note, AutoCAD does not allow a user to reference the same DWG file twice in the same drawing.

22.2.2 Attach an Image File

- You can use this command to attach an image file in the current drawing. You will see a normal Open dialog box to select your desired file. Once you click OK, you see the following dialog box:

- All the parts appearing in this dialog box have been discussed. AutoCAD permits you to reference the same file more than once in the same drawing.

22.2.3 Attach a DWF File

- You can use this command to attach a DWF file in the current drawing. You will see a normal Open dialog box to select your desired file. Once you click OK, you should see the following dialog box:

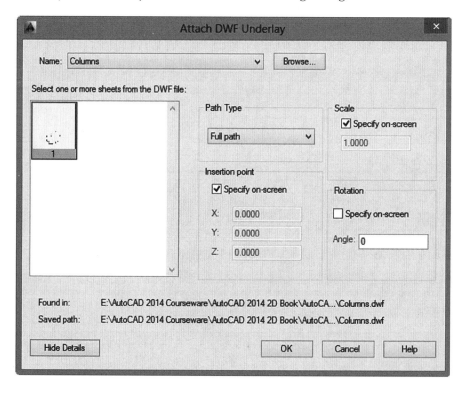

- This dialog box is identical to attaching the DWG file dialog box, except the part to select which sheet of the DWF file you want to attach, that is, if you have a multiple sheet DWF file. AutoCAD allows you to reference the same file more than once in the same drawing.

22.2.4 Attach a DGN File

- You can use this command to attach a DGN (Microstation) file in the current drawing. You will see a normal Open dialog box to select your desired file. Once you click OK, you will see the following dialog box:

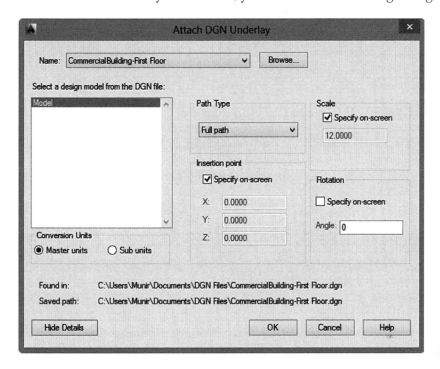

- This dialog box is identical to DWG file except for one part: Conversion Units. If you know that Microstation files deals with two units, Master and Sub, you have to tell AutoCAD the current AutoCAD unit (Master or Sub). AutoCAD allows you to reference the same file more than once in the same drawing.

22.2.5 Attach a PDF File

- You can use this command to attach a PDF file in the current drawing. You will see a normal Open dialog box to select your desired file. Once you click OK, you will see the following dialog box:

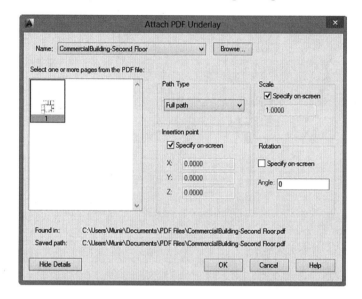

- This dialog box is identical to attaching the DWG file dialog box, except the part to select which sheet of the PDF file you want to attach, that is, if you have a multiple sheet PDF file. AutoCAD allows you to reference the same file twice in the same drawing.
- Note the following:
    - The reference file is always located in the current layer. It is advised to remember the name of the current where you attach the reference file so as not to freeze it in the future.
    - The first time you insert a reference file of any type, you will notice at the right part of the Status bar an icon called **Manage Xrefs**. Clicking it once will show the External Reference palette:

22.3 EXTERNAL REFERENCE PALETTE CONTENTS

- Mastering the different parts of the palette is very helpful to get the desired information. The following sections discuss each part of the palette and what it does for you. The upper part has two views:
  - List view: this shows all the files attached to your file. Each attachment has a different icon to its left to distinguish from the others. Also, you can see to the right of the name information about the file such as status, size, path, etc.

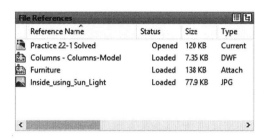

  - Tree view shows the same files but in tree view. This is very handy if your referenced DWG file has reference files by itself. This view has no details.

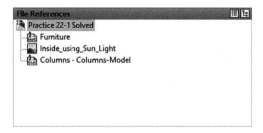

- The lower part of the palette has two views:
  - Either Detail shows something like the following. You can see that the Saved Path is editable and can be changed:

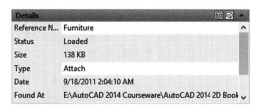

- You will also see a preview of the attached file, similar to the following:

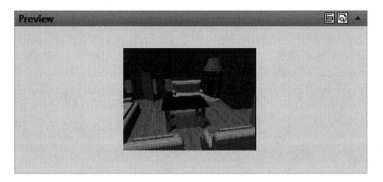

22.4 USING THE ATTACH COMMAND

- AutoCAD also offers the **Attach** command, another way to attach all types of file without using a palette. To access this command, go to the **Insert** tab, locate the **Reference** panel, then select the **Attach** button:

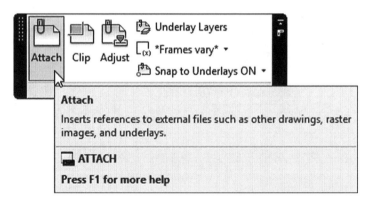

- The following dialog box appears. First select the type of the file (use the list shown), then select the desired file:

22.5 REFERENCE FILE AND LAYERS

- As previously mentioned, Xref segregates coming layers from reference DWG files and the original layers. To view this, you can visit the Layer Properties Manager palette after the attachment and you will see something like the following:

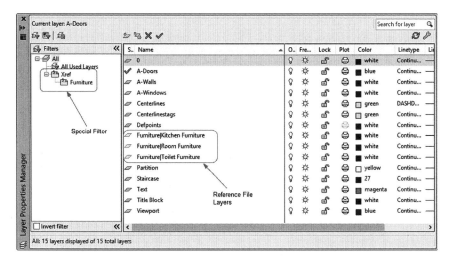

- As you can see from the palette, you can easily distinguish the original layers of the current file and the layers coming from the DWG reference file. At the left pane, you can see an automatic filter called Xref.
- The Layers coming from a DWG reference file always starts with the name of the file, a pipe (|), then the name of the layer: *filename|layername*.
- There are two important facts about layers in a reference file:
  - AutoCAD does not allow you to draw on reference layers and you cannot make any of them current.
  - AutoCAD allows you to change the properties of referenced layers (color, linetype, lineweight, etc.) and to change the status of a layer (on/off, freeze/Thaw, etc.) without affecting the original file. Once you save the file, the changes made will be saved. If you want to change this setting, go to the Application menu and select **Options**. Then go to the **Open and Save** tab and click off the **Retain changes to Xref layers** checkbox (you can do the same using the system variable VISRETAIN, by settings its value to be 0 instead of 1):

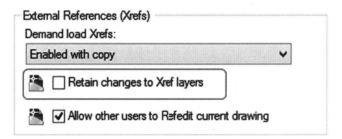

22.6 CONTROLLING FADING OF A REFERENCE FILE

- When you attach a DWG file to your current file, you will notice it is faded. This happens because the default setting for any DWG reference file is 70% as shown in the Reference panel.
- You can do two things regarding this issue:

- Control the fading percentage by increasing or decreasing it using the Reference panel:

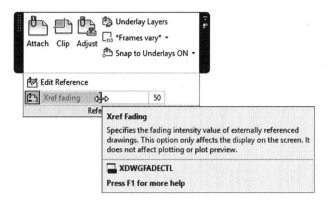

- Turning on/off the fading using the Reference panel:

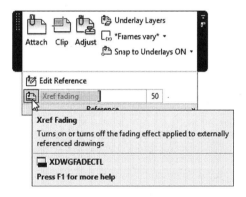

PRACTICE 22-1

Attaching and Controlling Reference Files

1. Start AutoCAD 2015.
2. Open the file, Practice 22-1.dwg.
3. Make sure that layer 0 is the current layer.
4. Using the External Reference, attach Furniture.dwg file using 0,0.
5. Using the External Reference, attach Columns.dwf using 0,0.

6. Go to D-Size Architectural Plan layout.
7. Using the External Reference, attach Inside_using_Sun_Light.jpg using your base point and scale. Go to the frame and insert it in the upper part of the legend using the proper scale to fit within.
8. Close the External Reference palette.
9. Start the Layer Properties Manager.
10. How many layers are added with the Furniture.dwg? _____
11. What are their names? _____, _____, _____
12. Change the color of the Toilet Furniture to Red.
13. Freeze layer Kitchen Furniture.
14. Go to Options dialog box and select Open and save tab. Turn off the checkbox and Retain changes to Xref layers.
15. Save the file and close it.
16. Reopen the file. What happens to the reference file layers? _____
   _____
   _____ Why? _____
   __ _____
17. Turn off the fading from reference DWG file, then turn it on to see the difference.
18. Click the Manage Xrefs button on the Status bar to show the External Reference palette then examine the four parts of the palette.
19. Save and close the file.

22.7 EDITING AN EXTERNAL REFERENCE DWG FILE

- AutoCAD allows you to edit any external referenced DWG file using two methods.
  - Using the Edit Reference command
  - Using the Open command
- Ethically you should not edit another person's file without his/her permission. The following discusses the possibility of doing so.

22.7.1 Using the Edit Reference Command

- This command can edit the external referenced DWG file. To issue this command, go to the **Insert** tab, locate the **References** panel, then select the **Edit Reference** button:

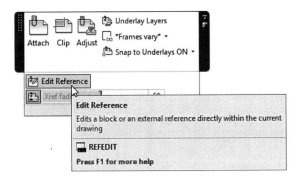

- You then see the following prompt:

```
Select reference:
```

- Click on the desired DWG to edit and you see the following dialog box:

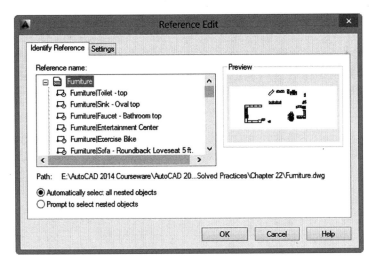

- As you can see, the filename of the file selected appears with all of the blocks coming with it, which proves that this command can edit not only external referenced files but blocks as well. If this is the file you want to edit, click the OK button. Once you do, two things take place:
  - All the drawings dim except the selected file.
  - A new context tab titled Edit Reference appears showing four buttons.
- Meanwhile you can use all the Modifying commands in order to alter the file in its place.
- The context tab contains four buttons:
  - Add to Working Set: allows you to add more objects to the current set of objects to edit.
  - Remove from Working Set: allows you to remove objects from the current set of objects to edit.
  - Save Changes: this is the final step after you finish editing.
  - Discard Changes: this is the final step after you finish editing and decide no more changes need to take place.

22.7.2 Using the Open Command

- This command opens the external referenced DWG file for editing, then saves the file with the new amendments. To do this, click any object in the external referenced DWG file then right-click. A long menu will appear, select the Open Xref option:

- The file opens ready for editing. Make the changes you need, then save the file and close it.
- There two things worth noting about this issue:

- There is no way you can prevent anyone who referenced your file from opening it and making changes. But you can prevent them from editing. To do this, go to the **Options** dialog box, then select **Open and Save** tab, and click off "Allow other users to Refedit current drawing" option.

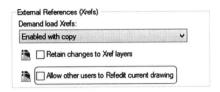

- Regardless of the method used, if any changes were saved to the external referenced file, you see a balloon stating that changes took place to the file. You need to reload it, just like the following:

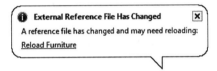

22.8 EXTERNAL REFERENCE FILE RELATED FUNCTIONS

- When you are working with the top part of External Reference, you can right-click the name of the file to get the following menu:

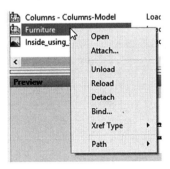

- As shown, there are six commands:
  - Open (already discussed)
  - Attach (shows you the Attachment dialog box to edit the settings of attachment)
  - Unload
  - Reload
  - Detach
  - Bind
  - Xref Type (Attach or Overlay)
  - Path (Make Absolute, Make Relative, Remove Path)
- The following sections discuss these six new commands:

22.8.1 Unload Command

- Unloading an attached file means the file will disappear from the display, but the link will stay. This is (in principle) similar to the freeze function in layers.
- The unloaded file will look like the following in the list:

Reference Name	Status	Size	Type	Date
Practice 22-1 Solved*	Opened	120 KB	Current	9/16/201˙
Columns - Columns-Model	Loaded	7.35 KB	DWF	9/16/201˙
Furniture	Unloa...		Attach	
Inside_using_Sun_Light	Loaded	77.9 KB	JPG	5/23/201˙

22.8.2 Reload Command

- The reload command can be used to reload an unloaded file. Or, you can use it to refresh all of your current links if needed.

22.8.3 Detach Command

- This command allows you to make the link disappear just like the Unload command, except you lose the link as well. Therefore, you will not see the file in the list. If you need the file again, you have to attach it again.

22.8.4 Bind Command

- If for any reason, you opt to keep the latest copy of an external referenced DWG file in your current file and cut the link between the two files, then you should use the Bind command. Bind commands allow you to bind the contents of the desired file inside your current drawing and remove the link. Bind command offers two methods for binding, as the following dialog box illustrates:

- As shown, you have two choices:
  - Bind
  - Insert
- The difference between the two methods is the naming of layers, styles, and blocks after the binding process ends. Assume you have a layer called:

  Furniture|Frame

- If you use bind, the name becomes: Furniture0Frame (the name of the layer still has the name of where the file comes from and | was converted to 0)
- On the other hand, if you use Insert, it becomes Frame.
- This applies to styles (text style, dimension style, etc.) and blocks.

22.8.5 Xref Type

- This option allows you to change the Xref type from Attach to Overlay and vice-versa.

22.8.6 Path

- Upon the attachment of Xref, you should specify whether you wants to make the path Absolute, Relative, or No Path. With this option, you can change this setting to any of the other options.

PRACTICE 22-2

External Reference Editing

1. Start AutoCAD 2015.

2. Open the file, Practice 22-2.dwg.

3. Click on any object in Furniture.dwg, then right-click, and select Open Xref option.

4. Erase the table and chairs in the kitchen and save the file.

5. A bubble appears to reload the file. Reload the file to get the latest version of it.

6. Using the External Reference palette, unload Furniture.dwg.

7. Reload the Furniture.dwg.

8. Open Layer Properties Manager palette and take a look at the three layers coming from Furniture.dwg.

9. Using the External Reference palette, bind Furniture.dwg, using the Bind option.

10. List the things that take place after binding:

    a. _____

    b. _____

    c. _____

11. Save and close the file.

22.9 CLIPPING AN EXTERNAL REFERENCE FILE

- AutoCAD allows you to clip an external referenced file (of any type) to show portions of the file. To reach to this command, go to the **Insert** tab, locate the **Reference** panel, then select the **Clip** button:

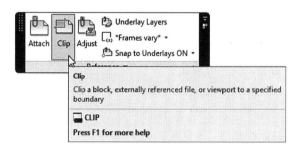

EXTERNAL REFERENCING (XREF) • **681**

- The following prompt comes up:

  ```
  Select Object to clip:
  ```

- You can select any type of external referenced file. For simplicity, the DWG file prompts are shown. These prompts are:

  ```
  Enter clipping option [ON/OFF/Clipdepth/Delete/ generate Polyline/New boundary] <New>:
  ```

- The default option is to create a new boundary to clip the existing referenced file. When you press [Enter], the second prompt comes up:

  ```
  [Select  polyline/Polygonal/Rectangular/Invert  clip] <Rectangular>:
  ```

- Using this prompt, you can:
  - Select a drawn polyline to act as a boundary.
  - Define a new boundary using polygon option and can draw any irregular shape.
  - Define a new boundary using rectangular shape.
  - Invert an existing boundary.
- Based on the option you select, define the desired boundary and you will see some of the referenced file rather than all of it.
- The other options are:
  - Turn the clipping on/off while keeping it.
  - Deleting the clipping boundary.
  - Clip depth (this has to do with the 3D).
- By default, you cannot see the clipping boundary because of the hidden nature of it. To show this and manipulate it, go to the **Insert** tab, locate the **Reference** panel, check the second list at the right, and you will see:

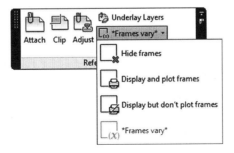

- There are four choices to choose from:
  - Hide frames
  - Display and plot frames
  - Display but do not plot frame
  - Frames vary
- While the frame is shown, you can click the frame itself to show something like the following:

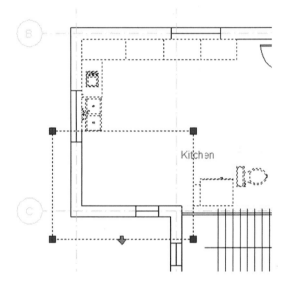

- With the grips displayed, you can:
  - Resize the frame.
  - Invert the frame.
  - Delete the frame (deleting the frame using the [Del] button mean deleting the external reference file).

22.10 CLICKING AND RIGHT-CLICKING A REFERENCE FILE

- There are two actions that you can make while you are dealing with an Xref file:
  - When you click it, a context tab appears with related panels for editing, clipping, and other stuff.
  - When you click then right-click, similar specific functions appear with the menu relating to the type of referenced file selected.

22.10.1 Clicking and Right-Clicking a DWG File

- Clicking (one click) a DWG referenced file opens the External Reference context tab, which includes three panels as shown:

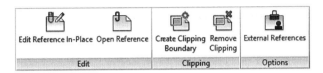

- The Edit panel contains the Edit Reference In-Place and Open Reference buttons (both discussed). The Clipping panel contains the Create Clipping Boundary and Remove Clipping buttons (both discussed), and finally the Options panel contains the External References button, which shows the External Reference palette.
- Right-clicking a DWG referenced file shows the following menu:

- These options are identical to the buttons in the context tab.

22.10.2 Clicking and Right-Clicking an Image File

- Clicking (one click) an image referenced file opens the Image context tab, which includes three panels as shown:

- The Adjust panel contains Brightness, Contrast, and Fade slider to control the image colors. The Clipping panel contains the Create Clipping Boundary and Remove Clipping buttons. The Options panel contains the Show Image, Background Transparency (whether to show objects behind the image), and External References buttons.

- Right-clicking an image referenced file shows the following menu:

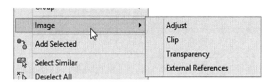

- These options are identical to the buttons in the context tab.

22.10.3 Clicking and Right-Clicking a DWF File

- Clicking (one click) a DWF referenced file opens the DWF Underlay context tab, which includes three panels as shown:

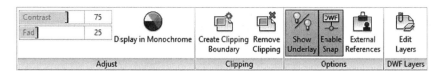

- The Adjust panel contains the Contrast, Fade sliders, and Display in Monochrome button. To control the image colors, the Clipping panel contains the Create Clipping Boundary and Remove Clipping buttons. The Options panel contains the Show Underlay, Enable Snap, and External References buttons. The DWF layers panel includes Edit Layers, which shows the following dialog box:

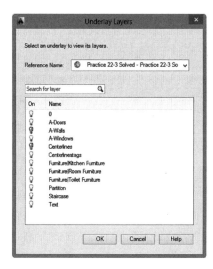

- Right-clicking a DWF referenced file shows the following menu:

- These options are identical to the buttons in the context tab.
- The DGN and PDF tabs are identical to DWF so you do not need to repeat them.
- Overall, you will notice that there are two buttons in the Reference panel that were not discussed. They are:
  - Underlay Layers button: shows a dialog box of Edit Layers previously mentioned for all DWF, DGN, and PDF attached to your current drawing.
  - Snap to Underlays button: acquires a point on DWF, DGN, and PDF files, as shown in the following list:

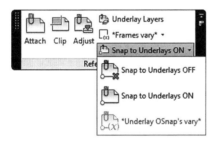

PRACTICE 22-3

External Reference File Clipping and Controlling

1. Start AutoCAD 2015.

2. Open the file, Practice 22-3.dwg.

3. Using the Insert tab and Reference panel, click Clip, click one of the pieces of furniture, select a rectangular shape, and create a rectangular around the kitchen and toilet.

4. What happened to the other pieces of furniture? _____

5. Select to show the frame of the clip.

6. Click the frame and shrink it to include only the kitchen.

7. Invert the clipping frame.

8. Using the clip command delete the clipping frame.

9. Go to the D-Size Architectural Plan layout and zoom to the image at the top-right of the title block.

10. Click the frame of the image. What happens to it? _____

11. Change the Fade % to 30%.

12. Using the context tab, clip the image to a smaller size.

13. Go back to Model space, select one of the columns, then right-click, and set DWF snap to off.

14. Hide frames of clipping.

15. Save and close the file.

22.11 USING THE eTRANSMIT COMMAND WITH EXTERNAL REFERENCE FILES

- If you attached several DWG files to your current file, along with image and DWF files, and you need to send your file to someone via e-mail, you need to use the eTransmit Command. If you do not use this command, your file is dependent on other files to look complete, hence AutoCAD will show the recipient error message "Can't locate xxxx.dwg attached" or "Can't locate xxxx.jpg attached."
- Using the eTransmit command (using a simple click), you can collect all related files to your current file, zip them, and prepare them to be sent either using e-mail or other means.

- Using the Application Menu, select **Publish/eTransmit**. If you did not save your file, you will see the following message:

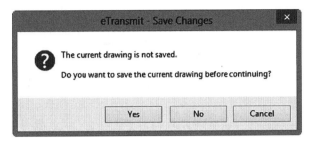

- Save your file by clicking Yes, and the following dialog box comes up:

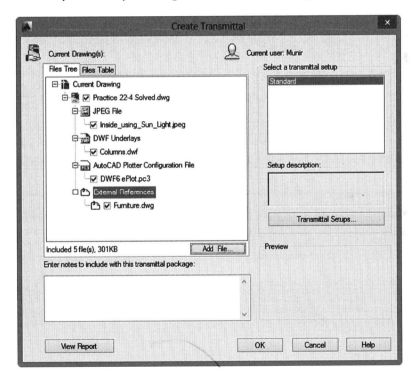

- At the top-left, you see two tabs; the default is Files Tree and the other is Files Table, as in the following:

- You can include/exclude any undesired file. Using the button labeled **Add File**, you can add other files to the existing list.
- Use the Transmittal Setups button to create your own parameters. Click the Transmittal Setups button, then click the New button, and you will see the following dialog box:

- Type the name of the new setup, then click the Continue button to see the following dialog box:

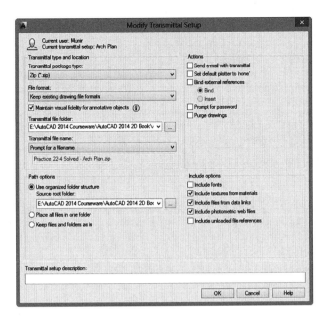

- All the options in this dialog box are very simple and easy to understand.
- Make the necessary changes and then click OK to end this dialog box.
- As the final step, click the button at the lower-left labeled View Report to check everything and you will see something like the following:

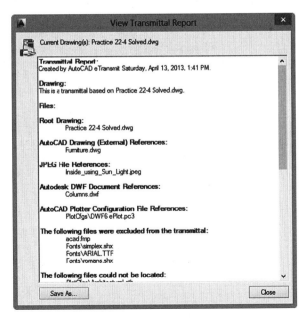

PRACTICE 22-4

Using the eTransmit Command

1. Start AutoCAD 2015.
2. Open the file, Practice 22-4.dwg.
3. Start the command eTransmit.
4. See the Files Tree and Files Table. Is the AutoCAD Plotter Configuration File listed? _____ If yes, what is it? _____
5. Create a new Transmittal Setup and name it Architectural Drawing. Make sure of the following:

    a. Package type = Zip file

    b. File format = AutoCAD 2007

    c. Include fonts = yes

6. As you noticed, fonts files are added.
7. View the report.
8. Save the file as Architectural Drawing.zip.
9. Save and close the file.

NOTES:

CHAPTER REVIEW

1. You can attach DOC and XLS files to your AutoCAD drawing using the External Reference palette.

    a. True

    b. False

2. There are only one type of reference type called "Overlay."

    a. True

    b. False

3. If you attach a DWF file, all of the following statements are correct, except one.

    a. You can control DWF layers.

    b. You can Unload and Reload.

    c. You can Bind.

    d. You can Detach.

4. You can attach the same image file more than once in the current file.

    a. True

    b. False

5. One of the following statements is not true:

    a. Once you attach one file, an icon appears at the right part of the Status bar.,

    b. The current layer is always the carrier of the reference file.

    c. Layer 0 is always the carrier of the reference file.

    d. Relative path is better than Full path.

6. If you have a layer in a reference DWG file named Furniture|Chair, after using the Bind command and Bind option, it will be _____.

7. _____ collects all related files to your current file, zips them, and prepares them to be sent either using e-mail or other means.

CHAPTER REVIEW ANSWERS

1. b
3. c
5. c
7. eTransmit

CHAPTER 23

SHEET SETS

In This Chapter
- What are sheet sets?
- Types of sheet sets
- Understanding Sheet Set Manager
- Example sheet set creation steps
- Existing drawing sheet set creation steps
- eTransmit, Archive, and Publish sheet sets
- Label blocks and Callout blocks

23.1 INTRODUCTION TO SHEET SETS

- The ultimate goal for any architectural or engineering construction firms is to print out the final design in a sheet set. However, printing individual drawings then gathering them into a set, labeling them, and putting them in a certain sequence can be time consuming. International bodies such as the U.S. National CAD Standard (NCS, http://www.nationalcadstandard.org) have a CAD standard, which addresses the sheet sets, specifying discipline designator, sheet type designator, and sheet sequence number. Companies that want to adopt NCS standard expect AutoCAD to adhere to these standards

and print fast and accurately. Since 2005, AutoCAD has provided this feature and for Sheet Set creation and manipulation.
- AutoCAD will create a sheet set then save it as a template to be used several times or as needed, which helps you avoid having to repeat any work.
- There are two methods to create a sheet set:
  - **Example sheet set**: Example Sheet set is a premade sheet set to provide the organizational structure of the sheet set depending on the discipline. There will be no sheets attached to the sheet set, hence you should create sheets, add views, and scale them.
  - **Existing drawings**: This method uses the layouts in the drawings as sheets and folders as subsets.
- There are pros and cons for each method. The first method is lengthy but gives you full control over all the elements. The second method is also tedious because you have to create the layouts twice, one for the layout and the second for sheet sets, but it also offers a certain amount of control.

23.2 DEALING WITH SHEET SET MANAGER PALETTE

- This section, discusses:
  - How to open and close an existing sheet set
  - How to deal with Sheet Set Manager palette

23.2.1 How to Open and Close an Existing Sheet Set

- To show the Sheet Set Manager palette, go to the **View** tab, locate the **Palettes** panel, then select the **Sheet Set Manager** button:

- If there are no opened files, you can click the same icon in the Quick Access menu. Either way, you will see the following palette:

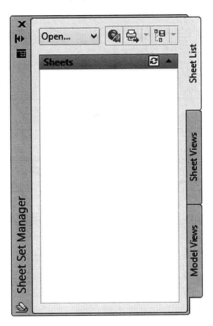

- From the list at the top, select Open:

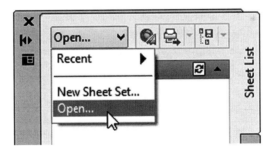

- The Normal Open dialog box will appear. Select your desired *.DST file to open it. There are several sample DST files in the following path: \Autodesk\AutoCAD 2015\Sample\Sheet Sets\ You will find three folders: Architectural, Civil, and Manufacturing.

- Something similar to the following appears:

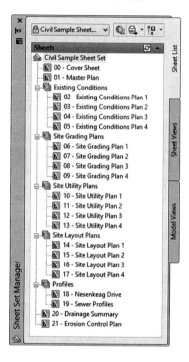

- To close an existing sheet set click the name of the sheet set in the Sheet Set Manager, then right-click. A menu will appear and you should select the first option **Close Sheet Set**:

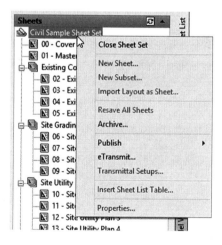

23.2.2 Working with the Sheet Set Manager Palette

- As previously shown, the Sheet Set manager palette has three tabs:
  - Sheet List tab
  - Sheet Views tab
  - Model View tab
- When you are in the Sheet List tab (default tab), you see the following:

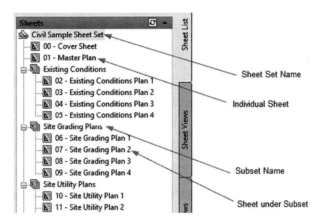

- In the Sheet List tab, you see the following:
  - Sheet Set name
  - Individual sheets (not listed beneath Subsets)
  - Subsets (method to group sheets)
  - Sheets under Subsets
  - Sheets are sorted from top to bottom
  - For each sheet, there is a number and a title (for example, 01 is the number for Master Plan sheet).
- This section discusses the Model View tab followed by the Sheet Views tab. When you click, you see something like the following:

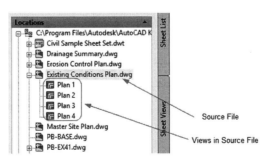

- The Model View tab contains the AutoCAD files named as source files. They contain the views that are used in the first method of sheet sets (you should be using drag-and-drop method to insert these views as an external reference to the current file).
- Clicking on the Sheet Views shows you something similar to the following:

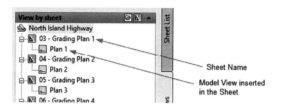

- Using the Model Views tab, if you drag and drop a model view, then the Sheet View tab shows the name of the sheet and the name of the view inserted in it.
- While you are in any of the three tabs, you can always see a preview of the file, sheet, or view. See the following two examples:
  - In the Sheet List tab, if you point (without clicking) on any of the sheets, you will see something like the following:

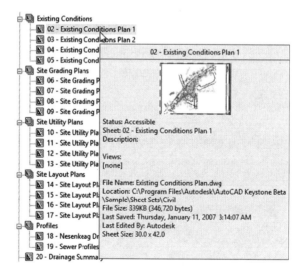

- While you are at Model View tab, if you point to one of the views, you will see something like the following:

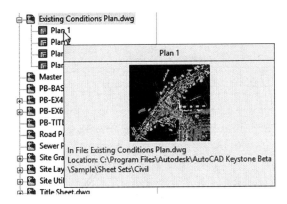

23.3 SHEET SET FILE SETUP

- As discussed, AutoCAD will create a DST file that holds the association and information that defines the sheet set. The DST file does not contain the files and views but rather the subsets and links to views (or layouts) of the source files. When the sheet set is created, AutoCAD will ask you about the location of the DST file along with template files. AutoCAD will also create for each sheet a DWG file, containing the sheet and the model views inserted in it. If you go to the folder that contains your DST file, you will see something like the following:

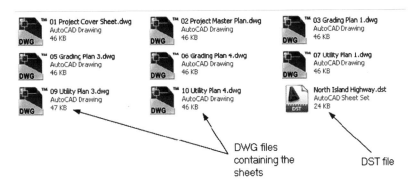

PRACTICE 23-1

Opening, Manipulating, and Closing Sheet Set

1. Start AutoCAD 2015.
2. Open Sheet Set Manager palette (regardless of whether a file is opened or not).
3. Use the Open command to open the following DST file (depending on the installation of AutoCAD in your machine): Program Files\Autodesk\AutoCAD 2015\Sample\Sheet Sets\Civil\Civil Sample Sheet Set.dst
4. Double-click sheet 00-Cover Sheet to open the sheet.
5. Close it without saving.
6. Point to sheet 06, which shows more information about it. What is the Sheet size? _____
7. Go to the Model Views tab.
8. Click the plus sign to expand the list.
9. Click the Existing Conditions Plan.dwg. How many views are there? _____
10. Point to several files (or views) to view a preview of each one of them.
11. Double-click Erosion Control Plan.dwg. Did you open Model Space or layout? _____
12. Close the file without saving.
13. Go to the Sheet List tab and close the Civil Sample Sheet Set.

23.4 SHEET SET USING AN EXAMPLE

- This is the first of two methods to create a new sheet set. The merit of this method is the fact that you will have full control over it but this method is lengthy and may take a long time to develop.
- In order to create a new sheet set, make sure that at least one file (even if it was empty) is opened.

- Open the Sheet Set Manager palette, and from the menu at the top select the New Sheet Set option. You will see the following dialog box:

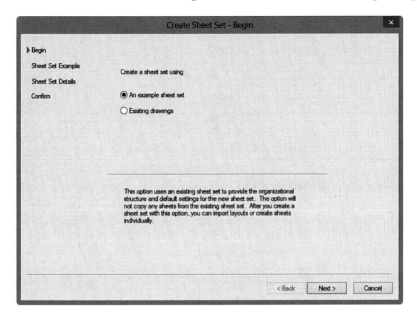

- The default option is an example sheet set. Click Next to go to the second step and you will see the following dialog box:

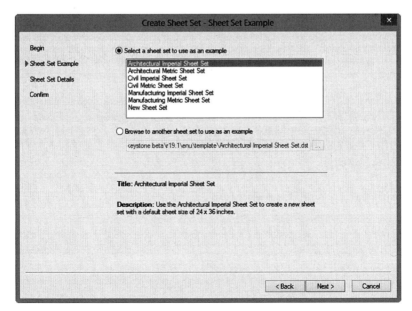

- There are six example: two for architectural (metric and imperial), two for civil (metric and imperial), and manufacturing (metric and imperial). These examples contain only subsets using templates that come with AutoCAD. Another option is to use a previously user-created sheet set (which can be company's template). When done, click the Next button to go to the third step and you will see the following dialog box:

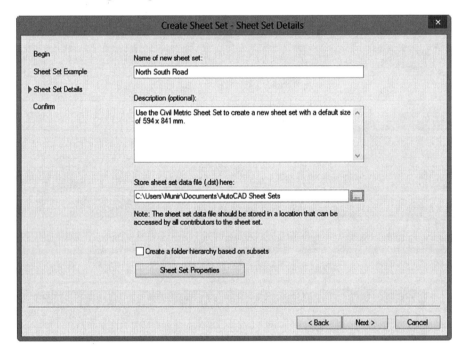

- You should do the following:
  - Input the sheet set title.
  - Input the sheet set description.
  - Specify the folder to store DST file.
  - Specify whether or not to Create a folder hierarchy based on subsets.

- Click the Sheet Set Properties button and you will see the following dialog box:

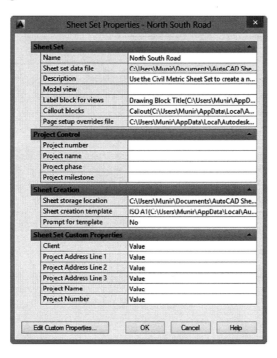

- Almost all of the data is already filled using default values. Project Control information are fields related to sheet set whereas Sheet Set Custom Properties is information created by AutoCAD. You can add, delete, or modify them. Click the Edit Custom Properties button to see the following dialog box:

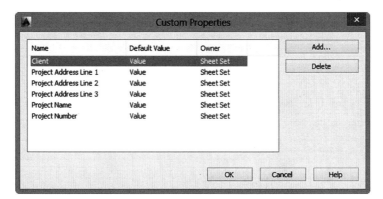

- There are two different pieces of information you should consider controlling:
  - Sheet storage location: this information allows you to specify the location of the sheets you generate. It is preferable to make it the same folder for the DST file.
  - Sheet creation template: this information allows you to specify the location of the DWT file which sheets will be created using it
- When done, click the OK button to end Sheet Set Properties and then click Next to go to the fourth step.
- You will see the following dialog box:

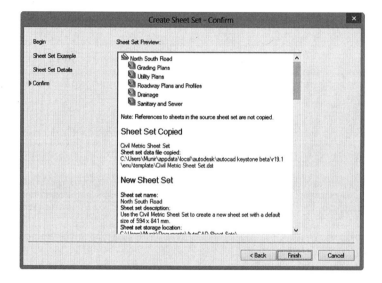

- This is the confirmation page, where you can review your settings. After, click Finish to end the creation process and you go back to the Sheet Set Manager palette. You should see something similar to the following:

23.4.1 Adding Sheets in Subsets

- To add sheet in subsets, simply right-click the name of the subset and you will see the following menu:

- In this menu, you can do the following:
  - Create a new Sheet in the current subset. When you see the following dialog box, fill in the sheet number and title, then click OK, and this will create a new sheet using the template file designated in the creation process:

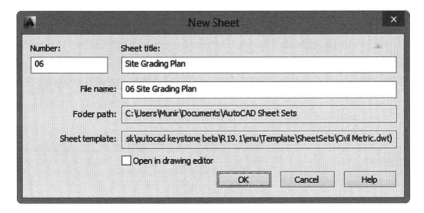

- Create a new subset inside the current subset and you should see the following dialog box. Fill in the subset name, sheet location, and template:

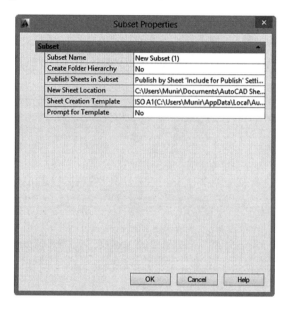

- Import Layout as Sheet
- Rename Subset
- Remove Subset
- Publish
- Properties (which is the same you got once you create it)

23.4.2 Sheet Control

- If you right-click a sheet, you will see the following menu:

- Using this menu, you can:
  - Open the sheet (double-clicking the sheet will lead to the same result).
  - Open the sheet as read-only.
  - Create a new sheet in the same subset.
  - Rename and Renumber a sheet.
  - Remove Sheet
  - Publish the current sheet.
  - eTransmit
  - Properties of the current sheet

PRACTICE 23-2

Creating a Sheet Set Using an Example Sheet Set

1. Start AutoCAD 2015.
2. Make sure a new file is open.
3. Start the Sheet Set Manager and create a new sheet set using an example sheet set.
4. Select the Civil Imperial Sheet Set.
5. Name the new sheet set "North Island Highway."
6. Save the DST file in: Practices\Chapter 5\Sheet Sets\Practice 23-1,23-2
7. Click the Sheet Set Properties and input the following:

    a. Project Number = HW-03

    b. Project Name = Designing North Island Highway

    c. Project Phase = 03

    d. Client = North Island Governorate

8. Finish the creation process.
9. Click the name of the sheet set, then right-click and add the following two sheets:

    a. Sheet Number 01 – Title = Project Cover Sheet

    b. Sheet Number 02 – Title = Project Master Plan

10. Move the two new sheets to be at the top of the list (using the drag and drop technique).
11. Remove the following subsets: Drainage, Sanitary, and Sewer.
12. Under the Grading Plans subset create the following sheets:
    a. 03 – Grading Plan 1
    b. 04 – Grading Plan 2
    c. 05 – Grading Plan 3
    d. 06 – Grading Plan 4
13. Under the Utility Plans subset, create the following sheets:
    a. 07 – Utility Plan 1
    b. 08 – Utility Plan 2
    c. 09 – Utility Plan 3
    d. 10 – Utility Plan 4
14. Using My Computer, go to the folder where you the saved DST file and look at the files created by sheet set. How do their naming and numbering compare to the sheets? _____
    _____
15. Leave the Sheet Set Manager open.

23.5 ADDING AND SCALING MODEL VIEWS

- The next step is to fill the sheets with model views from source files. Model views should be prepared prior to this step and has the four following steps:
  - Specify folder(s) that contain the files holding the desired model views.
  - Open the desired sheet.
  - Drag-and-drop the views in the sheets.
  - Scale views.

- Model views can be considered as external reference in the sheet. Any change in the original file updates the model views as well.
- To accomplish the first step, select the Model Views tab and you will see the following:

- Double-click the Add New Location button to open a normal dialog box and specify the desired folder. You can repeat this step as many times as needed to add more folders. Once you are done, you will get something like the following:

- Some of the files have a plus sign to their left; click it to expand and see the related model views, just like the following:

- Open the desired sheet by double-clicking it, then drag the desired view inside it. Before you decide the exact insertion point, right-click and you will see a list of scales to use:

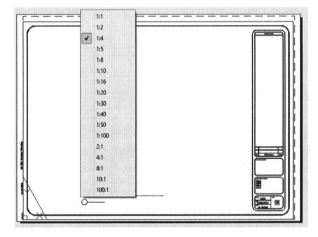

- Select the desired scale and then specify the insertion point. You should see something like the following:

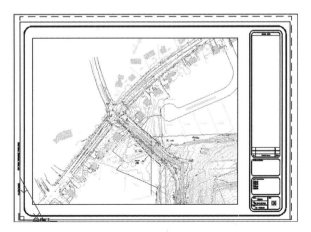

- The view you add in the sheet is considered a sheet view. To see this, select the Sheet Views tab and you see something like the following:

PRACTICE 23-3

Adding and Scaling Model Views

1. Start AutoCAD 2015.
2. Make sure that the Sheet Set Manager and North Island Highway sheet set are open.
3. Double-click sheet number 03, named Grading Plan 1, and notice how a lock is displayed beside it to indicate that this sheet is in-use.
4. Go to the Model Views tab, double-click the Add New Location, and go to \Program Files\Autodesk\AutoCAD 2015\Samples\Sheet Sets\Civil.
5. A set of files will be listed, click the plus sign beside Site Grading Plan. dwg file and you will see four model views. Drag Plan view into the current sheet and scale it 3/16"=1'
6. Close and save the file.
7. Do the same for the following:
    a. Plan 2 view inside sheet # 04 (same scale)
    b. Plan 3 view inside sheet # 05 (same scale)
    c. Plan 4 view inside sheet # 06 (same scale)

    d. Plan 1 view in Site Utility Plan.dwg inside sheet # 07 (same scale)

    e. Plan 2 view in Site Utility Plan.dwg inside sheet # 08 (same scale)

    f. Plan 3 view in Site Utility Plan.dwg inside sheet # 09 (same scale)

    g. Plan 4 view in Site Utility Plan.dwg inside sheet # 10 (same scale)

8. Close the sheet set.

23.6 SHEET SETS USING EXISTING DRAWINGS

- If you do not want to create layouts twice, you can let AutoCAD take all of your layouts and create from them sheets. Also, you can ask AutoCAD to take the folder structure you created to be your subset structure. With this method, whatever in the layout will be considered part of the sheet, but the viewport will not be external reference
- While the Sheet Set Manager palette is open, choose the New Sheet Set option and the following dialog box will appear:

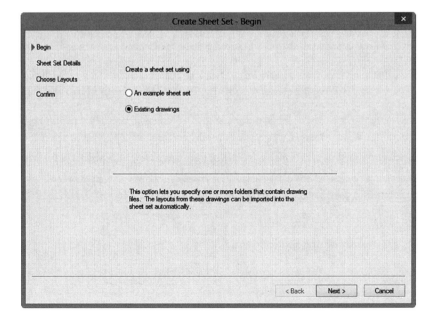

- Select the Existing drawings option, click **Next**, and you will see the following dialog box:

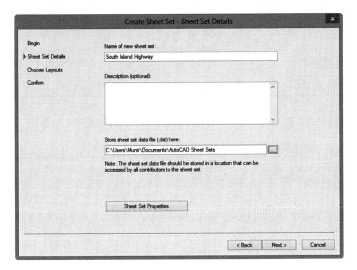

- You should do the following:
  - Input the sheet set title.
  - Input the sheet set description.
  - Specify the folder to store DST file.
  - Click the Sheet Set Properties button and you will see the following dialog box:

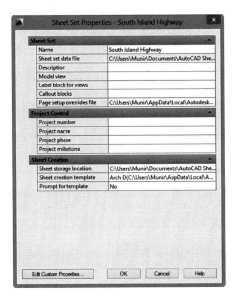

- These are the same as the properties discussed in the first method. Click OK, then click Next, and you will see the following dialog box:

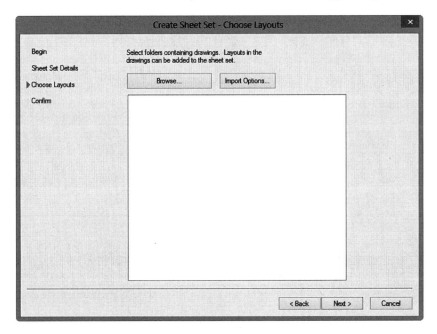

- To control importing of layouts, click the Import Options button to see the following dialog box:

- You can Control any of the following:
  - Prefix sheet titles with file name
  - Create subsets based on folder structure
  - Ignore top level folder

- When you are done, click OK, then click the Browse button to select the folder containing the desired files with the desired layout. You are able to pick and choose which files/layouts will be included in the importing process. You should see something similar to the following:

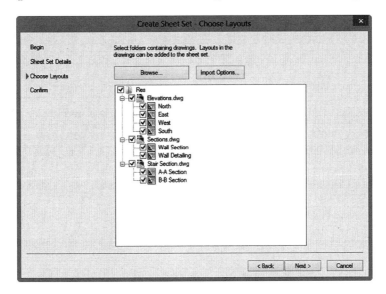

- When done, click Next to move to the final step.
- You will see the Confirm page. Click Finish button to finish the creation process:

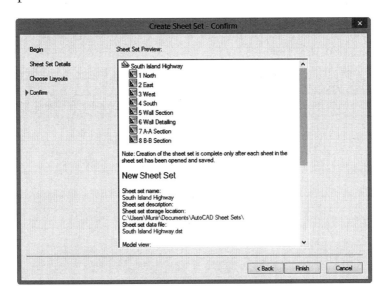

- The Sheet Set Manager looks similar to the following:

- As you can see, only sheets were brought so you need to create subsets to arrange them using the drag-and-drop technique. See the following illustration:

- Another helpful tool that can be used with both the first and the second method is importing layouts as sheets. Go to the desired subset, then right-click, and you will see the following menu. Pick Import Layout as the Sheet option:

- Click the Browse for Drawings button and select the desired file. You then see something like the following:

- In the previous example, you selected two files, each containing one layout; you can import both layouts or just one of them. Click the Import Checked button to end the importing process.

PRACTICE 23-4

Creating a Sheet Set Using Existing Drawings

1. Start AutoCAD 2015.
2. Make sure at least one empty file is open.
3. Create a new sheet set titled "Mira House" using the existing drawing. Saving it in Practice\Chapter 5\Sheet Sets\Practice 23-4.
4. You will find the drawings that contain the desired layouts in Practice\Chapter 5\Sheet Sets\Practice 23-4\Res.
5. Uncheck the Stair Section.dwg file, taking with you only layouts from Elevations.dwg and Sections.dwg.

6. Create three subsets: Elevations, Sections, Stair Sections.

7. Using the drag-and-drop technique, move North, East, West, and South to Elevations subset. Then, Wall Section and Wall Detailing to Sections.

8. Right-click Stair Sections subset and select Import Layout as Sheet.

9. Browse for the following folder: Practice\Chapter 5\Sheet Sets\Practice 23-4\Res, and select Stair Section.dwg.

10. Import both layouts.

11. Renumber and Rename the two imported sheets to be 7 – A-A Section, and 8-B-B Section.

12. Close the sheet set.

23.7 PUBLISHING SHEET SETS

- Using the Sheet Set Manager, if you right-click the name of the sheet set, a menu appears with the Publish option. You will see the following sub-menu:

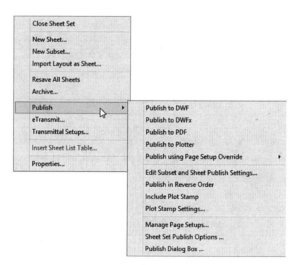

- The first four options were discussed previously. They are: Publish to DWF, Publish to DWFx, Publish to PDF, and finally Publish to Plotter. The fifth is new, which says Publish using Page Setup Override; this option helps you to override the default page setup. In order for this option to be available, you should set the Page setup Override in the Properties dialog box, as in the following illustration:

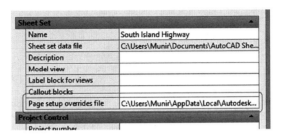

- If this option is properly set, you will see the following:

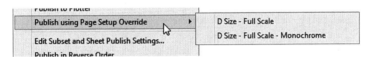

- The following option is the Edit Subset and Sheet Publish Settings, as shown in this dialog box:

- It allows you to specify the subset/sheet to be included in the Publish command.
- The rest of the options have been previously discussed or are easy to understand.

23.8 USING eTRANSMIT AND ARCHIVE COMMANDS

- AutoCAD provides two methods to exchange and archive sheet sets, using the eTransmit and Archive commands. Both commands allow you to group all the related files and pack them in a single file.

23.8.1 Using the eTransmit Command

- To create a package of all files of the sheet set, simply right-click the name of the sheet set and you will see the following menu:

- You then see the following dialog box:

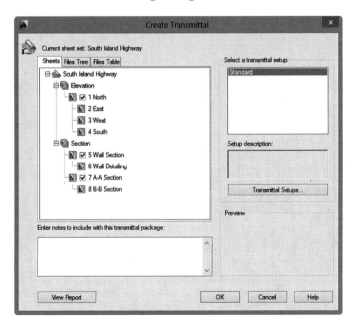

- This dialog box is identical to the one discussed in Chapter 22, but with only one change, which is the Sheet tab at the upper-left part of the dialog box.

23.8.2 Using the Archive Command

- To issue this command, right-click the name of the sheet set and a menu will appear. Select the Archive command and you will see the following dialog box:

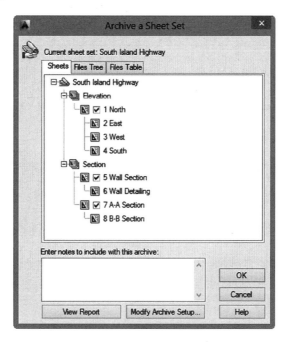

- As you can see from the preceding dialog box, this one is the same as the eTransmit dialog box. However, the Archive command does not save a setup like the eTransmit command. Instead, you have to Modify Archive Setup each time you issue it.

PRACTICE 23-5

Publishing and eTransmitting a Sheet Set

1. Start AutoCAD 2015.
2. Open the Mira House sheet set, which you created in the previous practice.
3. Right-click the sheet set name, select Publish, and then Publish to DWF.
4. Name the file Mira House.dwf and save it at the same folder you saved the DST file in.
5. You can open it using the Autodesk Design Review software to view it.
6. Right-click the sheet set name again and select eTransmit.
7. Make sure that all sheets are included.
8. Go to the Files Table tab and ensure that the Mira House.dst file is included.
9. View the report.
10. Produce and eTransmit the package and save it in the same folder you saved the DST file in.
11. Close the sheet set.

NOTES:

CHAPTER REVIEW

1. How many methods create a sheet set?

    a. One method

    b. Three methods

    c. Two methods

    d. Four methods

2. _____ is a method that helps you to bring in layouts from other drawings and consider them as sheets.

3. Using an Example Sheet Set, you will have viewports ready for you.

    a. True

    b. False

4. Using an Example Sheet Set:

    a. AutoCAD creates sheets.

    b. AutoCAD creates subsets.

    c. AutoCAD creates both subsets and sheets.

    d. AutoCAD creates both subsets, sheets, and then add views.

5. Using the Existing drawings method, AutoCAD will consider folders as subsets and layouts as sheets.

    a. True

    b. False

6. In the Sheet List tab, if you point (without clicking) on any of the sheets, you will see _____ of the sheet.

CHAPTER REVIEW ANSWERS

1. c
3. b
5. a

CHAPTER 24

CAD STANDARDS AND ADVANCED LAYERS

In This Chapter

- What is CAD Standard?
- How to create and configure DWS files
- What are Layer Filters?
- Using the Layer States Manager
- Advanced functions for controlling layers

24.1 WHY DO WE NEED CAD STANDARDS?

- Projects are getting larger and more complex so companies using AutoCAD (or any other software package) often have a need for CAD Managers. CAD Managers work to produce standard dimensions, text, and layering systems. AutoCAD offers a tool to help CAD Managers produce CAD Standard files, link them to DWG files, and make sure that no standards are violated.
- The process of creating a standard file is lengthy and encompasses the following steps:
  - Find an international standard or create your own.
  - Sit with all the parties whom will be affected by the new standard (stakeholders) and get their inputs.
  - Put the standard on paper first and get the approval of all stakeholders.
  - Create a standard file using AutoCAD.

- Test it and get feedback from users.
- Make any corrective steps needed.
- Put the standard in use and monitor it.
■ Before this tool was introduced in AutoCAD, CAD Managers used to check the adherence to the standard file manually using a random check method. Now, CAD Managers have the ability to check several files at the same time.
■ Another tool is the Layer Translator, which allows you to map layers in your current file, compare it to another file, and find the match and mismatch. It also allows you to correct any layer names.

24.2 HOW TO CREATE A CAD STANDARD FILE

■ To create a CAD Standard file (*.DWS), do the following steps:
  - Open a DWG file containing all the desired layers, dimension styles, text styles, and linetypes. Or, you can create a new file using a template file.
  - Go to the Application menu, select Save As, and then the AutoCAD Drawing Standards option:

- You will see the following dialog box:

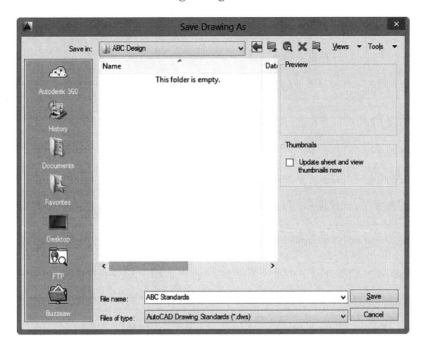

- AutoCAD gives the file the extension of DWS automatically. Input the name of the new standard file then click the **Save** button. This process takes place only once.

24.3 HOW TO LINK DWS TO DWG FILES AND CHECK THEM

- After you create the DWS file, you should perform two more steps:
  - Configure (link) it to DWG file.
  - Perform the checking process to find any violation to the standard file.

24.3.1 Configuring (Linking) DWS to DWG

- To link, do the following steps:
  - Open the desired DWG file to be checked.

- Go to the Manage Using **Manage** tab, locate the **CAD Standards** panel, and select the **Configure** button

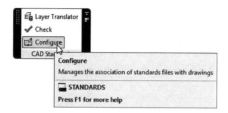

- You will see the following dialog box:

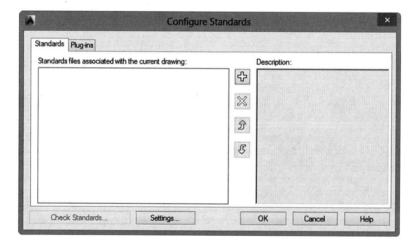

- Click the Plus sign button to link the current DWG with the desired DWS:

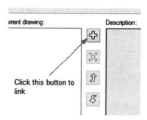

- You will see a normal file dialog box. Select the DWS file and then click OK.

- To control what to check for, click the Plug-ins tab, and the following appears:

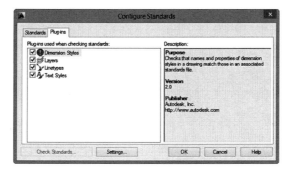

- As mentioned, AutoCAD can check only for four things: Dimension, Text styles, Layers, and Linetypes. To have full control over the process of checking, click the Settings button and the following dialog box appears:

- Control the Notification settings by selecting one of the following:
  - Disable standards notifications.
  - Display alert upon standards violation (this is used when you are still working with the DWG file).
  - Display standards status bar icon (default):

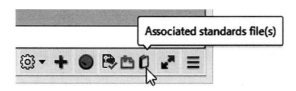

- The Control Check Standards settings can change any of the following:
  - Automatically non-standard properties
  - Show ignored problems
  - Choose your preferred DWS file

24.3.2 Checking DWG Files

- The first step should be to open the desired DWG file you want to check for Standard compliance. Then go to **Manage**, locate the **CAD Standards** panel, and select the **Check** button:

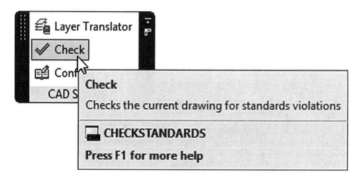

- You will see the following dialog box:

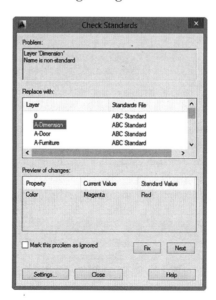

- You can see that the dialog box is divided into three parts:
  - The Problem section is where AutoCAD shows the problem that violates the standard. In the following example, the name of the layer in your DWG file is not in the standard file:

  - The Replace section is where AutoCAD proposes to replace the mistake with something already in the standard file. In the following example, the Dimension layer is replaced with the A-Dimension layer. In this case, AutoCAD removes the old layer and transfers all the objects in it to the new layer:

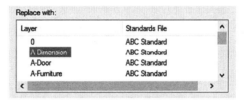

  - The Preview of changes section is where AutoCAD shows changes between the old and the new. In the following example, the name and the color change.

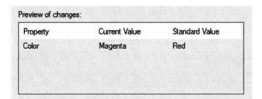

- AutoCAD offers three solutions to any problem:
  - Fix: To fix the problem using the Replace with option.
  - Next: Skip the problem and go to the next problem.
  - Ignore: Ignore this problem and go to the next problem.

- When this command finishes, it shows the following message:

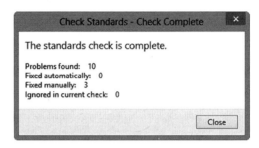

- You can choose to go with another approach, which is to link the DWS file with an incomplete DWG. This means when you make any changes to dimension, text styles, or layers, an instant message at the lower-right appears similar to the following:

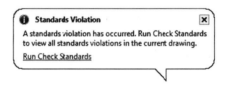

- If you click the Run Check Standards hyperlink at the end of the bubble, this takes you directly to the same dialog box previously mentioned.

PRACTICE 24-1

Using CAD Standards Commands

1. Start AutoCAD 2015.
2. Open ABC Standard.dwg and create from it ABC Standard.dws, then close the file.
3. Open the file, Practice 24-1.dwg.
4. Using Configure link Practice 24-1.dwg to ABC Standard.dws.
5. Go to the Plug-ins tab and make sure that all four available options are turned on, then click OK to end the command.
6. Before you start checking the compliance, check the color of layer Dimension. What is the color? _____

7. Start Check command and make the following changes:

Problem	Action
Dim Style: Outside Walls	Replace it with Outside
Dim Style: Inside	Replace it with Inside
Layer: Dimension	Replace it with A-Dimension
Layer: Furniture	Replace it with A-Furniture
Layer: Text	Replace it with A-Text
Layer: Title Block	Replace it with Title Block
Layer: TobeHidden	Replace it with 0
Text Style: Room Titles	Replace it with Room Titles
Text Style: TitleBlock_Bold_Mine	Replace it with TitleBlock_Bold
Text Style: TitleBlock_Regular	Replace it with TitleBlock_Regular
Text Style: Arial_09	Replace it with Standard from DWS file

8. How many problems did AutoCAD find? _____

9. Go to the Layer Properties Manager palette and create a new layer called "Test." What is the reaction of AutoCAD? _____

10. Run Check Standard and replace Test with 0.

11. Save and close the file.

24.4 USING THE LAYER TRANSLATOR

- The Layer Translator is a tool that allows you to translate a file's layering system to your layering system. You can use the following steps:
  - Open the desired DWG file.
  - Link it with a file (DWG, DWS, or DWT) holding your layering system.
  - Use the Layer Translator to translate layers (naming, colors, linetype, etc.).

- Save the translation in a file, so you can use it for other files.
- To issue this command, go to the **Manage** tab, locate the **CAD Standards** panel, then select the **Layer Translator** button:

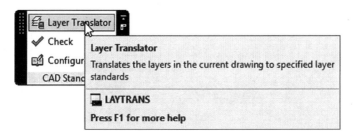

- You will see the following dialog box:

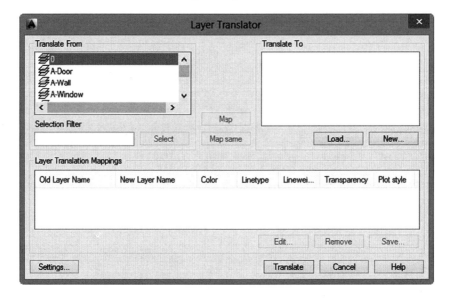

- At the top-right part, you can see the Translate From list, which contains the list of layers that need to be translated to your layering system. The Translate To list is empty due to the lack of your file. To load your file, click the **Load** button; you can load DWG, DWT, and DWS files.
- This is what you see after you load your file:

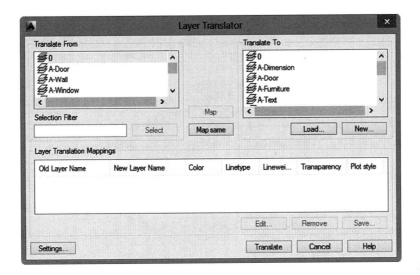

- To find the common layers between the two files, click the **Map same** button in the middle and you will get something like the following:

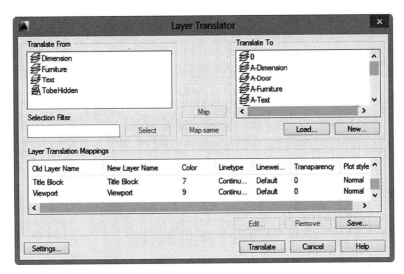

- Whatever is left in **Translate From** part are the layers that do not match your file. Select one layer from Translate From and select one layer from Translate To, then click the **Map** button. Keep doing this until you finish all the layers. You can also leave some layers without matching, which means these layers will be left in the file without changing.

- Use the **Save** button to save the translation for future use.
- To manipulate the process, click the Settings button and you will see the following dialog box:

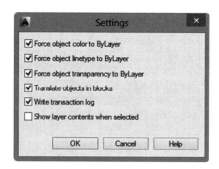

- All the settings are self-explanatory. To finalize the command, click the Translate button to perform the translation process.

PRACTICE 24-2

Using the Layer Translator

1. Start AutoCAD 2015.
2. Open the file, Practice 24-2-A.dwg.
3. Start the Layer Translator command and load ABC Standard.dws.
4. Using the Map Same button, see if there are any common layers between the two files.
5. Using the Map button, map the following layers:
    a. Dimension to A-Dimension
    b. Furniture to A-Furniture
    c. Text to A-Text
    d. TobeHidden to 0
6. Save the translation as the same name of the file with extension DWS.
7. Close Practice 24-2-A.dwg and open Practice-2-B.dwg

8. Load the DWS file you saved in step 6.

9. See how all the layers are already translated.

10. Save and close the file.

24.5 DEALING WITH LAYER PROPERTIES MANAGER

- AutoCAD provides you all the necessary tools to control the Layer Properties Manager palette. To issue the command, go to the **Home** tab, locate the **Layers** panel, then select the **Layer Properties** button:

- You will see the following palette:

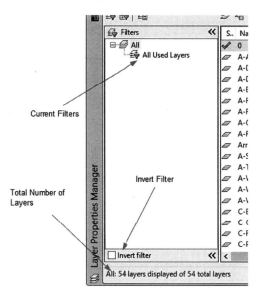

- Note the three arrows that identify the three part of the palette:
  - Total number of layers: This shows the used and unused layers in the current drawing.
  - Current filters: By default, you will see a single filter which is "All Used Layers" and shows only used layers.
  - Invert filter: This shows the inverse of your current filter.
- While the palette is open, you can do any of the following:
  - Hide unneeded column. Simply right-click one of the columns and a menu will appear. All displayed columns has (ü) at its left. To hide any of the columns, click the name:

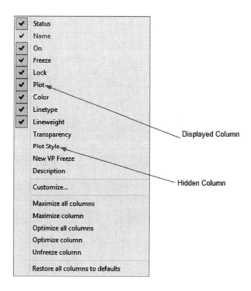

  - Sort layer list based on column either ascending or descending. Simply click the title of the column and an arrow appears. If it is pointing upward, this means ascending but if it is pointing downward this means descending.

- Relocate the column position comparing to other columns, using the drag-and-drop technique:

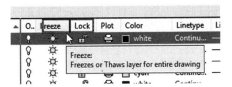

- Change the column width or freeze and unfreeze a column.
- By right-clicking the heading of a column, the following menu appears. At the bottom, there are six commands: Maximize all columns (show the heading or the contents, whichever are larger), Maximize column (the one you right-click), Optimize all columns (show the contents, ignoring the heading), Optimize column, Freeze/Unfreeze column, which means if you horizontally scroll this column is always shown. Finally, you can restore all columns to their defaults:

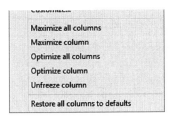

24.6 CREATING PROPERTY FILTERS

- With the complexity of today's drawings, you can expect an enormous number of layers in a single drawing. AutoCAD allows you to create a filter to show some of the layers based on a single property or more. Start the Layer Properties Manager and then click the New Property Filter button:

- You will see the following dialog box. Do the following steps:
  - Type the name of the filter.
  - Under Filter definition, use the fields below to input one criteria or more. In the following example, two criteria are used, the name (A-*) then color (color = Cyan).

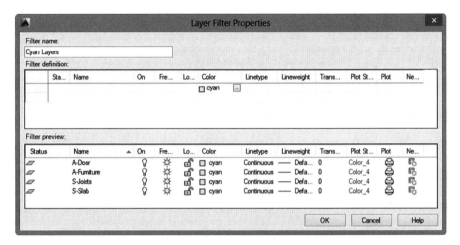

- Automatically AutoCAD shows layer(s) that satisfy the criteria (in this example there are only four layers).
- When you click OK and go back to the Layer Properties Manager, you will see on the left side that the new property filter is added, as in the following:

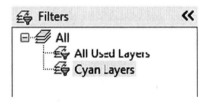

- At the lower-left corner, you will see the following:

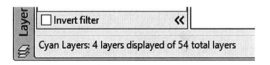

- At the same time, the layer list in Layers panel does not show only the layers of the filter:

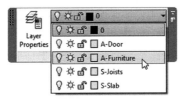

24.7 CREATING GROUP FILTERS

- Group filter is a filter to group layers, which have nothing in common between them but for a reason or another you want to put them together.
- While the Layer Properties Manager is open, click the New Group Filter button:

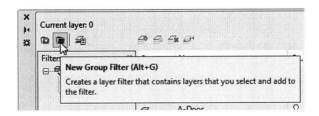

- When the new filter is added, type your desired name. Initially this filter is empty. Click All filter, then using the drag-and-drop technique add the desired layer:

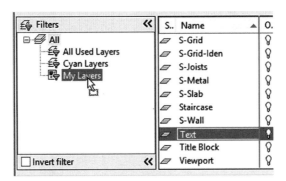

24.8 THINGS YOU CAN DO WITH FILTERS

- One advantage of filters is to show some of the layers and to hide some, but there are more benefits than just a single advantage. AutoCAD allows you to do collective actions on all the layers in the filter in one shot.
- When you right-click the filter, a special menu appears for each type.

24.8.1 Property Filter Menu

- Right-click the property filter and you will see the following menu:

- You can do any of the following:
  - Change the visibility of the layers. You can turn them on or off, freeze them, or thaw them.
  - Lock or unlock the layers.
  - If you are in a layout, you can choose to freeze or thaw layers in the current viewport.
  - If you are in a layout, you can choose to isolate layers in the all viewports or active viewport only.
  - Create a new property or group filter.
  - Convert the property filter to group filter.
  - Rename and delete property filter.
  - Change the criteria the filter uses by selecting Properties. This option will show you the same dialog box you create the property filter with.

24.8.2 Group Filter Menu

- Right-click group filter and you will see the following menu:

- Almost all of the options were discussed in the property filter menu, except instead of having the Properties option, you have the Select Layers option. You will see the following sub-menu:

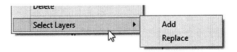

- This option will add/replace more layers to the group filter using normal selecting methods. You will see the following prompt:

```
Add layers of selected objects to filter...:
```

- Click the desired object so the layer containing this object is added to the group filter.

PRACTICE 24-3

Layer Advanced Features and Filters

1. Start AutoCAD 2015.
2. Open the file, Practice 24-3.dwg.

3. Open Layer Properties Manager.

4. Layers are sorted by name (ascending). Sort them by color descending. What is the name of the first layer in the list? _____

5. Maximize all column widths.

6. Show Description column and hide Plot Style column.

7. Put Freeze column to the left of On column.

8. Unfreeze Name column.

9. Create a new property filter, name it "Architectural Layers with Cyan color," and set the proper criteria for such a filter.

10. Create a new group filter and call it My Layers. Add Text, Title Block, Frame, and Viewport.

11. Right-click the group filter and choose the Select Layers/Add option, then select the arrow on the stairs. Check the filter, is there any new layer? What is the name? _____

12. Right-click the property filter and freeze the layers in it.

13. Save and close the file.

24.9 CREATING LAYER STATES

- A Layer State is used to save and retrieve a set of layers with their current state from color, linetype, lineweight, on/off, freeze/thaw, etc. If you saved a certain state, retrieving it will lead you to retrieve all the states of the layer at one time, hence saving lots of time. There are multiple ways to use this feature:
  - Make sure the Layer States Manager is displayed then click Layer States Manager button:

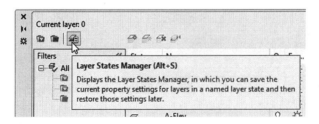

- You can also use the Layers panel. Use the first list and click New Layer State button:

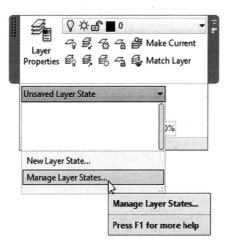

- Either way, the following dialog box will appear:

- To create a new layer state, click the New button and you will see the following dialog box:

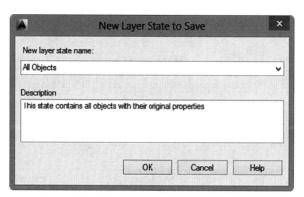

- You should input the name of the new layer state and a good description for its contents. When you click the OK button, the current state of layers will be saved. To see what properties are saved, click the arrow at the lower-right corner of the dialog box and you will see something like the following:

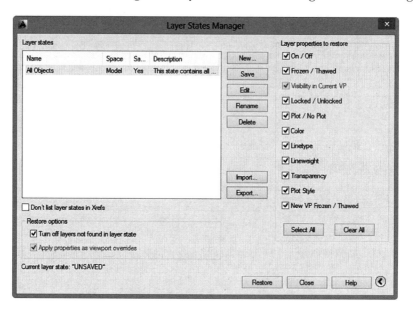

- You can see that AutoCAD saves all of layer properties. If you click the Edit button, you have the choice to alter the state of the layer saved in layer state. You will see something like the following:

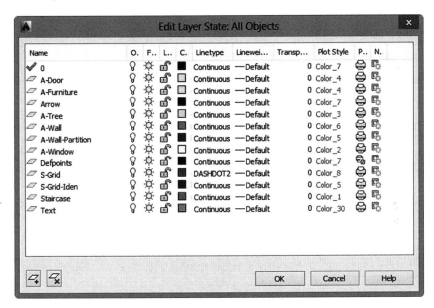

- The rest of the keys are self-explanatory
- To retrieve layer states, do the following:
  - You can use the Restore button at the Layer States Manager.
  - You can also use the first list in the Layers panel.

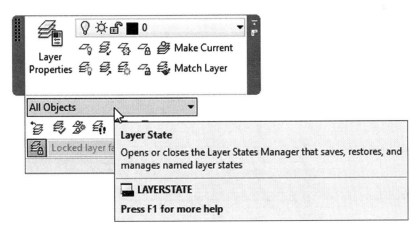

PRACTICE 24-4

Using Layer States

1. Start AutoCAD 2015.
2. Open the file, Practice 24-4.dwg.
3. Start the Layer States Manager, create a new state from the current state, and call it "All Objects." Type the following in the description: "This state contains all layers with their original properties."
4. Freeze layer A-Tree, A-Furniture, and change the A-Window layer's color to blue.
5. Save this state under the name "No Trees and Furniture," typing in the following description: "All Objects except Trees and Furniture."
6. Restore All Objects state.
7. Using the Edit button, edit No Tree and Furniture by freezing the layer Arrow and changing the color of A-Walls-Partition to Cyan.
8. Switch between the two states.
9. If you have time, create your own state.
10. Save and close the file.

24.10 SETTINGS DIALOG BOX

- The Settings dialog box is used to control three things related to layers: what to do when new layer are added, what are the settings for isolated layers, and some dialog settings. To display this dialog box, make sure that the Layer Properties Manager is displayed and then click the Settings button:

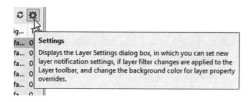

- You will see the following dialog box:

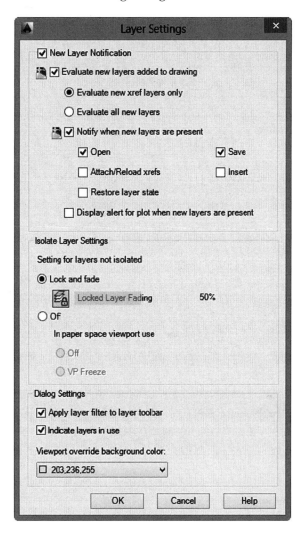

- As you can see, there are three parts in this dialog box:
  - New Layer Notification
  - Isolate Layer Settings
  - Dialog Settings

24.10.1 New Layer Notification

- In this part, AutoCAD wants to know whether to inform you when a new layer is added. See the following:

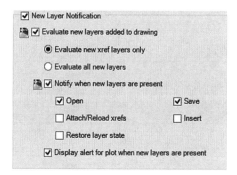

- You can control any of the following settings:
  - If you want to be notified of new layers.
  - Should AutoCAD evaluate all layers or only Xref layers?
  - When the notification should take place: when opening the file, saving the file, attaching/reloading the file, inserting the file in other files, or when you restore layer state.
  - Should AutoCAD display an alert for plot when new layers are present?

24.10.2 Isolate Layer Settings

- When you issue the Isolate command for layers, how should AutoCAD treat layers not isolated? There are two choices, either Lock and Fade (user should control the fading percentage) or the not isolated layers to be turned off, and, if so, what to do in layout's viewport (turn them off or use Viewport freeze). See the following:

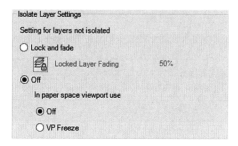

24.10.3 Dialog Settings

- In this part, you should tell AutoCAD how to deal with several issues related to layers. They are:

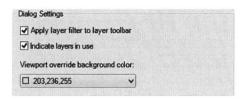

- Control the following:
  - Apply layer filter to layer toolbar.
  - Indicate layers in use.

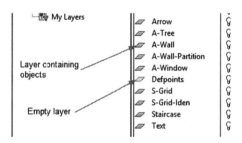

  - If a layer has an override setting in viewport, how AutoCAD will distinguish it from other layers? By giving it a background color in Layer Properties Manager? If so, set the color.
- Now what will happen if a new layer is introduced in a drawing? AutoCAD will show you a bubble just like the following, telling you that a new layer is found and you need to reconcile it:

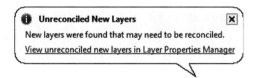

- In the Layer Properties Manager, in the Filters part, you will see the following:

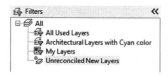

- A new filter named Unreconciled New Layers appears, which contains all the new layers. You can right-click it and choose the **Reconcile Layer** option to accept these new layers or Delete Layer to reject these layers:

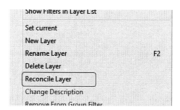

PRACTICE 24-5

Using Settings Dialog Box

1. Start AutoCAD 2015.
2. Open the file, Practice 24-5.dwg.
3. Start Settings dialog box.
4. Turn on the New Layer Notification checkbox and make the following settings:

   a. Evaluate new layers added to drawing = on

   b. Evaluate all new layers = on

   c. Notify when new layers are present = on

   d. Open & Save = on

5. Settings for Layer not isolated = Lock and fade, and fading percentage = 70%

6. Start the Isolate command and select one line from the outer wall, then press [Enter]. What happened to other layers? _____.
   Get closer to one of the pieces of furniture and you will notice all objects are locked.

7. Add a new layer, call it "Test," and then save the file. AutoCAD should inform you that the new layer was added and you need to reconcile it.

8. Go to the Layer Properties Manager and you will find a new filter called Unreconciled New Layers. Right-click it and choose to reconcile it.

9. Save and close the file.

NOTES:

CHAPTER REVIEW

1. On of the following statements is wrong:
   a. You can use Layer Translator instead of to CAD Standards.
   b. The CAD Standard will check for Dimension Style.
   c. The CAD Standard will check for Text Style.
   d. The File extension for Standard file is DWS.

2. If you right-click _____ filter, you have an option to add layers by selecting them.

3. Property filter will group layers that have the same properties.
   a. True
   b. False

4. In Layer Translator, using the _____ button will find common layers between two files.
   a. Map
   b. Map similar
   c. Map same
   d. Map all

5. The Settings dialog box contains the method to notify you if a new layer is already added.
   a. True
   b. False

6. _____ is to save and retrieve a set of layers with their current state from color, linetype, lineweight, on/off, freeze/thaw, etc.

CHAPTER REVIEW ANSWERS

1. a
3. a
5. a

CHAPTER 25

DRAWING REVIEW

In This Chapter
- Review of Publish command
- Using Autodesk Design Review
- How to read DWF files in AutoCAD and republish them
- Checking and Comparing DWF files

25.1 INTRODUCTION TO DRAWING REVIEW

- Normally, design drawings are checked and modified several times before the final design. Checking may happen in any design stage by the design engineer, chief engineer, senior manager, or owner, etc. The draftsperson or engineer who create and own the drawing as well as all the involved parties should keep track of modifications such who asked for a revision and when.
- Previously, this cycle was done manually using hard copy printouts of each drawing, which led to waste of paper, ink, and electricity plus the heavy usage of plotters. Plus the confusion resulting from huge paper stacks laying on your desk with the multiple people checking the drawing.
- AutoCAD offers a new fully electronic way to complete the revision cycle without even leaving your desk. You can keep track of all the changes and

who asked for it and when, which allows management to know who is responsible for what action. This cycle is based on the following items:
- DWF file produced from AutoCAD drawing.
- Autodesk Design Review software. This software comes with AutoCAD, or if not, you can download it from autodesk.com for free.
- Markup Set Manager, which is a tool in AutoCAD.

- The process is very simple:
  - Produce DWF (single sheet or multiple sheet) and send it to who checks (checker) it via e-mail.
  - Using Autodesk Design Review, the checker opens your DWF and puts all of the comments and remarks. The Autodesk Design Review register all the actions took place such as when and by whom. The checker will send it back to you via e-mail.
  - Using the Markup Set Manager, you are able to open both DWF and DWG, making all the changes required. You are able to republish the DWF again with the first revisions. You can send it back to the checker via e-mail.
  - The checker will open both files to the original DWF and the first revision, then compare them to make sure that all modifications asked for were completed. The checker may then ask for more changes before sending it back to you
  - And so on…

25.2 FIRST STEP – CREATING A DWF FILE

- Chapter 10 covered how to do this part. There are three ways to create a DWF from file the current DWG file. They are:
  - Using the **Plot** command and *DWF6 ePlot.pc3* printer. This method produces for each layout along with the Model space a separate DWF.
  - Using the **Batch Plot** (**Publish**) command. This command produces a single DWF file containing all layouts and Model Space for a file or more.
  - Using the **EXPORTDWF** command. This command produces a single DWF file containing all layouts of the current file.

25.3 SECOND STEP – USING AUTODESK DESIGN REVIEW

- When you create a DWF using methods 2 and 3, Autodesk Design Review will open automatically by default. If not, you will find a shortcut at your desktop. Simply double-click it then use the normal open file procedure. You will see something like the following:

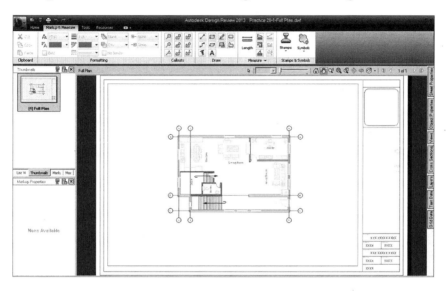

NOTE *Autodesk may opt not to update Autodesk Design Review 2013, with a promise to keep in mind any future need for updating with new features. Therefore, using Autodesk Design Review 2013 (free software) that can be downloaded from the Autodesk website is still a valid thing to do.*

- The Autodesk Design Review interface looks like the AutoCAD interface using Ribbons, Application Menu, and the Quick Access Toolbar. At the right, there are seven tabs that expand when you hover your mouse over them:
  - Sheet Properties include full information about the sheets.
  - Views are a list of the named views in the sheets.
  - Layers show the AutoCAD layers used.
  - Text Data

- Grid Data
- Cross Sections
- Object Properties
■ Below the ribbon you will see a canvas containing all the viewing commands:

■ All are similar commands to what was learned in AutoCAD. The Home icon would always reset the view to the original view.
■ At the left there are multiple windows:
  - Models show the Model Space if you include it in the DWF file.
  - List Views show the sheets in a list view.
  - Thumbnails show the sheets as thumbnails.
  - Markups
  - Markup Properties
■ You can combine all windows in one window using the drag-and-drop technique. To do this, hold the title of the window and drag it to the list of the tabs at the bottom of the existing window. Also, you can create a separate window by using the drag-and-drop technique. To do this, simply hold the name of the tab at the bottom of the window and drag it outside.

25.4 MARKUP AND MEASURE TAB

■ This tab contains all the tools you need to markup and measure the DWF file. It contains six panels:
  - Clipboard
  - Formatting
  - Callouts
  - Draw
  - Measure
  - Stamps and Symbols

25.4.1 Clipboard Panel

- This panel looks like the following:

- Any markup or measure objects you add to the DWF can be selected so you can Cut, Copy, or Paste.

25.4.2 Formatting Panel

- Once you start adding markup and measure objects, and just before adding, you should format it first. You will use the following panel:

- In this panel, there are three parts: the first to format the text (one of text tools should be selected) and the second to format lines (one of the other markup and measure should be selected to activate this set of tools). Finally, the third is to format areas. Here are the text formatting tools:

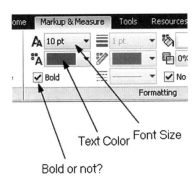

- Here is the line formatting part:

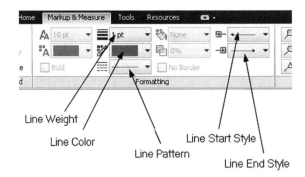

- Finally, here is the area formatting part:

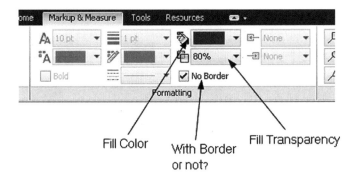

25.4.3 Callouts Panel

- This panel offers you nine different shapes for callouts. Six of the nine have a cloud to help you highlight the desired area. Formatting text, line, and area are suggested before inserting the callout. You will see the following panel:

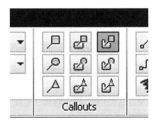

- Here is an example of a callout:

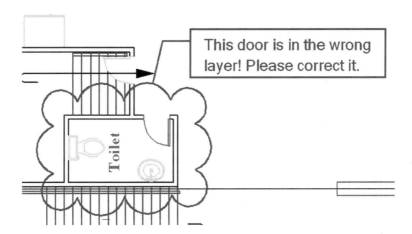

25.4.4 Draw Panel

- The panel looks like the following:

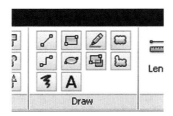

- This panel contains ten different drawing tools. For the respective tool, use the suitable formatting tool:
  - Line
  - Polyline
  - Freehand
  - Rectangle
  - Ellipse
  - Text
  - Freehand Highlighter
  - Rectangle Highlighter
  - Rectangular Cloud
  - Polycloud

25.4.5 Measure Panel

- The panel looks like the following:

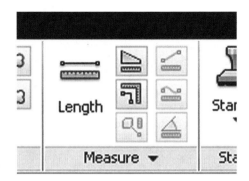

- There are three options to measure 2D DWF files; the rest is for 3D DWF. These three are:
  - Length means you need to specify two points in any angle and you will see a dimension displayed. By default, Autodesk Design Review snaps to the object's end effectively. See the following illustration:

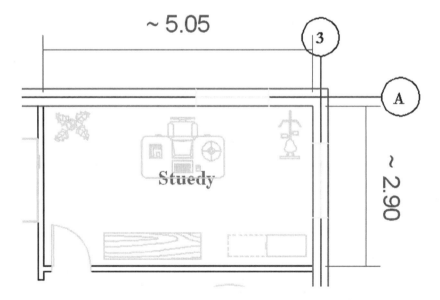

- Area means you need to specify as many points as desired to get a closed shape to measure its area. See the following illustration:

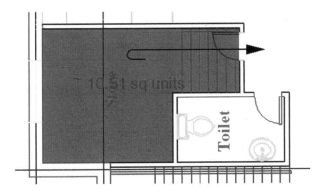

- Polyline measures the total length of group of points (you may measure perimeter of a shape). See the following illustration:

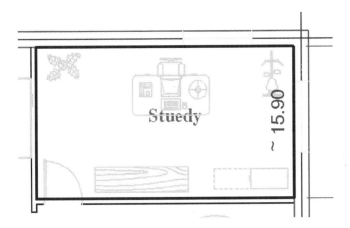

25.4.6 Stamps and Symbols Panel

- The panel looks like the following:

- You can insert seven types of stamps on the DWF file such as Approved, For Review, etc.
- Symbols button will create catalog of symbols from the current DWF or from other DWF files.

25.5 HOW TO EDIT MARKUP OBJECTS

- To edit an existing markup object, do the following steps:
  - From the upper canvas, choose the Select tool.

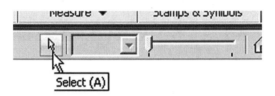

- Click the objects to be edited.
- You will see small yellow dots appear at certain places, depending on the object selected. You should see something like the following:

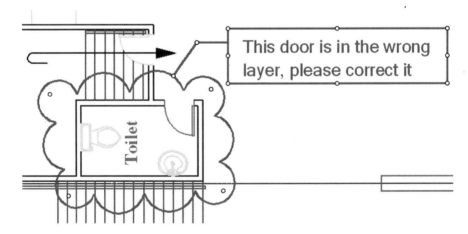

- In this mode, you can simply hit the [Del] key to delete the selected markup object, move it, or resize it (in the previous example, you can resize the callout alone and the cloud alone). You can also edit the text.

25.6 CONTROLLING MARKUPS

- As shared at the beginning of this chapter, there will be two windows at the left. The first one is Markups window and the second Markup Properties window.

25.6.1 Markup Window

- You will see something like the following:

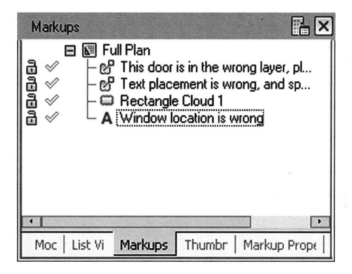

- At the top you, will see the name of the sheet containing the markups (in this example, it is Full Plan), then you will see a list of the markups added in the current DWF. In the list, you can see the type of the markup (rectangle callout, cloud, or text) and the attached text. The list is sorted using the time of insertion. Selecting one of the markups in the list will lead to selecting the markup object and zoom to it.

25.6.2 Markup Properties Window

- If you select any markup object whether graphically or using the Markups window, go to the Markup Properties window and you will see something like the following:

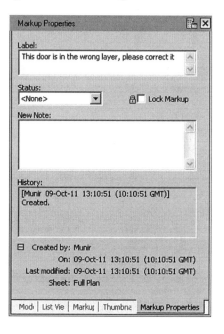

- In this window, you see full information of the selected markup such as who created the markup, at what time, and on what date. The state of the markup is also shown as Done or for Review, etc.

PRACTICE 25-1

Creating DWF and Using Markup Tools

1. Start AutoCAD 2015.
2. Open the file, Practice 25-1.dwg.
3. Create from the Full Plan layout only a DWF keeping the default name.
4. Open the created file using the Autodesk Design Review software.

5. Measure the area of the store. What is the value? _____

6. Measure the width and length of Study. What are the two values? Width = _____, Length = _____

7. From the Callout panel, choose the Rectangle Callout with Rectangle Cloud tool and put a callout around the toilet. Typing the following: "This door is not in the right layer."

8. From the Draw panel, put a rectangle cloud around the title of the room called Study, then add text from the same panel stating the following: "Wrong Spelling and wrong position."

9. From the Callouts panel, select the Rectangle Callout tool, pointing to the upper window of the Study room. Type the following statement: "Window location is wrong."

10. Save the DWF under the same name and close it.

25.7 USING MARKUP SET MANAGER IN AUTOCAD

- After mark up, the DWF file should go back to the creator of the DWG file to make the necessary corrections based on the DWF file. In order to do this, you should take the following steps after receiving the DWF file:
  - Start AutoCAD.
  - Open your original DWG file and go to Model space.
  - Go to the **View** tab, locate the **Palettes** panel, and then select the **Markup Set Manager** button:

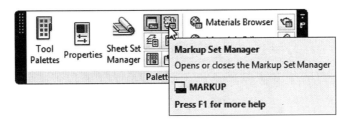

- In the Markup Set Manager, you will see the following palette:

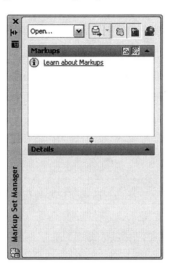

- From the upper pop-up list, select the Open option, which shows a normal open file to select your desired DWF file. The palette will change to something like the following:

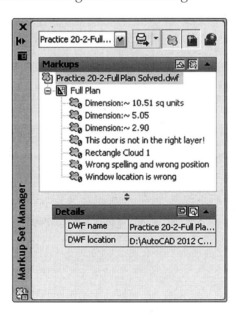

- You will see the name of the DWF file, name of the layout, and all of the markups listed. Selecting one the markups will show details below the list of the markups, just like the following:

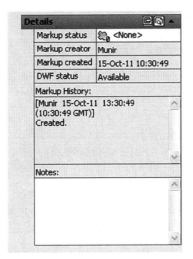

- As you can see, you will see the Markup status, Markup creator, Markup created date and time, DWF status, and Markup history and Notes.
- In order to see the markups on the DWG drawing, simply double-click the markup. This takes you to the layout and then you can zoom to the part of the layout that was marked.
- Go to the Model space (or double-click) the viewport in order to correct the necessary changes.
- After finishing the correction, go to the Markup Set Manager palette and change the status of the markup to be done, as in the following:

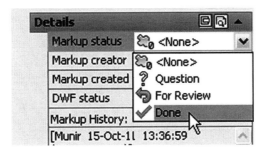

- While both DWG and DWF objects are shown, at the top-right side of the palette you will see the following:

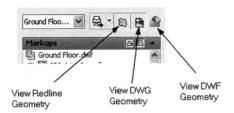

- The two buttons at the right are View DWG Geometry and View DWF Geometry; if the first is on, the other is off.
- The third button will turn on/off the markup objects.
- The final step should be to republish the DWF file again with the new changes. Click the Republish Markup DWF button and you will see the following selections.

- Select one of the following two choices: Republish All Sheets or Republish Markup Sheets only. You can also use the Batch Plot command to produce the newly revised DWF file.
- When you republish, you should give the DWF a new name to indicate the revision has been made along with a number to make the revision number meaningful. The name should be similar to, name_rev_01.dwf, so you can make comparisons later on.

PRACTICE 25-2

Creating a DWF and Using Markup Tools

1. Start AutoCAD 2015.

2. Open the file, Practice 25-1.dwg.

3. Open the Markup Set Manager palette.
4. Open DWF file you created from the previous exercise.
5. You will see seven markups, but only three need to be fixed. The other four are either measurements or clouds.
6. Double-click. The markup starts with: "This door ..." This will take you to the Full Plan layout and you can zoom to the door. Go to the Model space and move the door to A-Door layer.
7. Do the same for the other two remarks, correcting the misspelled word, moving the text downward a bit, then moving the window to the left by 1 unit.
8. Change the status of all markup objects to "Done."
9. Using the Batch Plot command create a DWF from the Full Plan layout and name it Practice 25-1-Full Plan_Revision_001.dwf.
10. Using the Markup Set Manager, close the DWF file and then close the palette.
11. Save the DWG file and close it.

25.8 COMPARING DWF FILES

- To fully complete the first round of reviewing a file, you should check whether all the remarks were met.
- You need to open the old (first) file, which contains the markups using Autodesk Design Review software, and then open the new one to make the comparisons. After opening the old file, go to the **Tools** tab, locate the **Canvas** panel, and then select **Compare Sheets**:

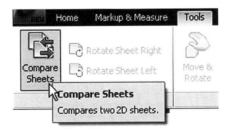

- You will see the following dialog box:

- Click the **Browse** button and select your new file. You will get something like the following:

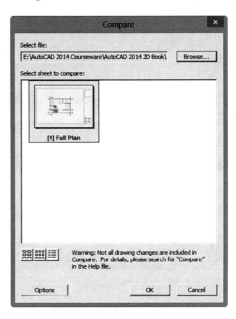

- When you click the **Options** button, the following dialog box comes up:

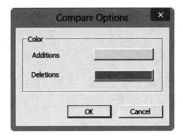

- Additions are displayed in Green and deletions are displayed in Red.
- Accordingly, you will see the additions and deletions as in the following:

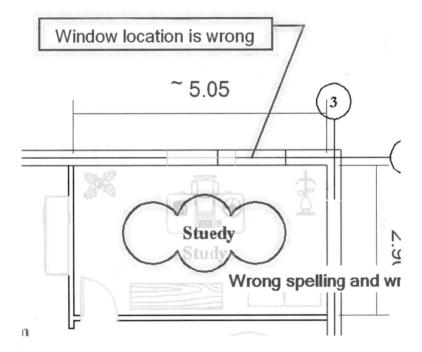

PRACTICE 25-3

Comparing DWF Files

1. Start the Autodesk Design Review software.
2. Open the file, Practice 25-1-Full Plan.dwf.
3. Compare it with Practice 25-1-Full Plan_Revision_001.dwf.
4. Check whether the three objects need to be corrected and that all corrections were rightly made _____.
5. Close the Autodesk Design Review.

CHAPTER REVIEW

1. The name of the software that can open and mark DWF is Autodesk Design Review.

    a. True

    b. False

2. You can _____ two DWF files using Autodesk Design Review software.

3. One of the following cannot be done in Autodesk Design Review software.

    a. Calculate an area

    b. Input a dimension of a line using two points

    c. Put some reline objects

    d. Import DWG and DWF file in the same session

4. While you are using Autodesk Design Review software, you can:

    a. Switch the layers on/off.

    b. Check the markup window to see the markup objects included in the current file.

    c. Change the layer of a selected object.

    d. Check the properties of the markup.

5. Using Markup Set Manager, you can open the DWF while in AutoCAD.

    a. True

    b. False

CHAPTER REVIEW ANSWERS

   1. a

   3. d

   5. a

INDEX

2D objects, 313
    Boundary command, 349–350
    construction lines, 333–334
    converting polylines to lines and arcs, 315–316
    Divide command, 339, 340
    Donut command, 323
    Edit Polyline command, using. *See* Edit Polyline command
    Ellipse command, 346–348
    joining lines and arcs to form polyline, 316
    Measure command, 339–340
    Point command, 338
    Point Style command, 337–338
    Polygon command, drawing using, 320–322
    Polyline command, 314–315
    rays, 334–335
    Rectangle command. *See* Rectangle command, drawing using
    Region command, 351–352
    Revision Cloud command, 324–325
    Spline command. *See* Spline command

A

A-Door row, 279
Action Parameters panel, 612
Actions, 605, 609–613
Add area, 164
Add button, 276
Add/Delete Scales option, 531
Add Leader button, 510
Add Scale button, 531
Add Selected command, 398–399
Add Text and Add Separator options, 432–433
Add Vertex, 93
Add Watched Folder button, 391
Adjust Space button, 496
Align button, 125, 511
Aligned button, 230–231
Aligned constraint, 586–587
Alignment parameters, 610
All Drawings, 18
All mode, 63
Allow Docking, 429–430
Allow press and drag on object, 65
Alternate Units tab, 488–489
Ang option, 334
Angle button, 162–163
Angle Type, 287
Angular button, 231–232
Angular constraints, 588–589
Annotation Visibility, 530
Annotational Constraint Mode, 596–597
Annotative feature, 528–533
Annotative text style, 445
Application Menu, 3, 4–5, 18
Apply to Layout button, 282
Arc command, 39–41
Arc Length button, 233, 324
arc length dimension, insertion of, 233–234
architectural project
    imperial, 288–295
    metric, 295–302
Archive command, 723
area, measuring, 163
    complex area, 164–165
    simple area, 163–164
Area button, 163
Area field, 196
Area option, 319–320
Array action, 611

Array command, 102
Array Creation context tab, 117–118, 120
arraying objects, 116
    path array, 122–128
    polar array, 129–132
    rectangular array, 116–120
Attach command, 670–671
Attribute values, global editing of, 644–646
attribute visibility, controlling, 641
attributes extraction from file(s), 650–658
Auto Constrain command, 578–580
AutoCAD 2015 interface, 3
    Application Menu, 3, 4–5
    Command Window, 3, 10
    Graphical Area, 3, 10
    InfoCenter, 9–10
    Quick Access Toolbar, 3, 5, 14, 16
    Ribbons, 3, 6–9
    status bar, 3, 10–11
AutoCAD defaults, 12–13
AutoCAD drawing
    creating new, 14–16
    opening an existing, 16
AutoCAD LT, 1
AutoCAPS button, 212
Autodesk, Inc., 1
Autodesk 360, 2, 3
Autodesk Content Explorer, 389–394
Autodesk Design Review, 539, 762, 763–764
Autodesk Exchange window, 9
Autodesk Online Services, 10
automatic scaling feature, 417–418

B

Background Color, 182
Background Mask, 205
Base point, 173
Baseline button, 243
Baseline command, 242–243
Baseline spacing, 473
Basepoint parameters, 610
Batch Plot (Publish) command, 537–539, 762
BIM (Building Information Modeling), 1
Bind command, 678–679
Bisect option, 334
Block attributes, 637–638
    attribute values, global editing of, 644–646
    attribute visibility, controlling, 641
    attributes extraction from file(s), 650–658

Define Attributes button. *See* Define
    Attributes button
    individual attribute values, editing, 642–644
    inserting blocks with attributes, 640, 641–642
    redefining attribute definitions, 646–649
Block Authoring, 609
Block Definition dialog box, 607
Block Editor button, 436
Block Editor command, 608–609, 626
Block Table button, 626–630
block tools and block editing, 417
    automatic scaling feature, 417–418
    design center, 419–422
    editing blocks, 436–438
    hatch and tool palette, 427–429
    tool palettes, 422–427, 429–436
Block unit, 174
Blocks, 171
    converting blocks to files, 178–179
    creating, 172–174
    exploding, 178
    using (inserting), 174–177
Boolean operation on regions, 352
Borders tab, 453
Both-Automatic detection, 65
Boundaries panel, 294, 301
Boundary button, 349
Boundary command, 349–350
Break at Point button, 134
Break command, 102, 133–134
Browse button, 175
Bullets and Numbering, 208

C

CAD Standard, 729–730, 736–737
    CAD Standard file, creation of, 730–731
    checking DWG files, 734–736
    group filters, 745, 747
    layer advanced features and filters, 747–748
    Layer Properties Manager, 741–743
    linking DWS to DWG files, 731–734
    property filter menu, 746
    property filters, creating, 743–745
    using Layer Translator, 737–741
Callouts panel, 766–767
Cell Format panel, 462–463
Cell Styles panel, 460–462
Center Mark button, 499
Chamfer command, 101

Chamfer Option, 318
chamfering objects
    using Distance and Angle, 109–111
    using Distance Option, 109
Check Spelling button, 219
Chest of Drawers, creating, 622
Circle button, 39
Circle command, drawing circles using, 37–39
circles, drawing, using Circle command, 37–39
circular arcs, drawing, using Arc
    command, 39–41
Clip button, 273, 680–682
Clipboard option, 362
clipboard panel, 765
Close Block Editor button, 437
Close button, 18–19
Close Hatch Creation, 181
Close option, 26
Close panel, 213
Coincident constraint, 564–566
Collect button, 512
Collinear constraint, 566–567
Color-dependent plot style table, 517–522, 527
Color field, 144
Column panel, 117–118
Column Settings dialog box, 209
Column width, 209
Columns button, 208
Columns panel, 459
Command Window, 3, 10, 20
commands, modifying, 59, 101
    arraying objects. *See* arraying objects
    Break command, 133–134
    chamfering objects. *See* chamfering objects
    Copy command, 72–73
    Erase command, 66–67
    extending objects, 114–115
    filleting objects, 106–107
    grips and dynamic input, 91–93
    grips and perpendicular and tangent
        OSNAPs, 93–94
    joining objects, 85–86
    lengthening objects, 84–85
    Mirror command, 78–79
    Move command, 70–71
    objects, selecting. *See* objects, selecting
    offsetting objects. *See* offsetting objects
    Rotate command, 74–75
    Scale command, 76–77
    Selection Cycling, 65–66

Stretch command, 80–81
    trimming objects, 111–113
    using grips to edit objects, 87–91
complex area, calculating, 164–165
Concentric constraint, 567
Configure filtering button, 392
Configure Settings button, 390
Constraint Bar, 580–581
Constraint Settings dialog box, 590–592
Constraint Status button, 631
Construction button, 631
Construction Lines button, 333–334
Content Explorer, use of, 389–394
Context tab, editing using, 120
Continue button, 241
Continue command, 241–242
Control Vertices method, 343–344
Convert command, 589–590
Copy command, 72–73
copying objects, 362
    to New Layer Command, 402
Create Block button, 172
CREATE option, 2
Cross Hairs, 3
Crossing mode, 61, 64
Crossing Polygon mode, 62, 64
Cumulative Area, 196
Current Drawing, 18
Current Layer command, 402
Current layout, 262
Custom Model Views, 378
Custom option, 276
Customization User interface (CUI), 547–548
Customize Palettes, 433–436
Customize Workspace, 555

D

Data Link button, 367, 368
Data panel, 464–465
data sharing
    from MS Excel, 366–367
    from MS Word, 364–366
Decurve option, 328
Define Attributes button
    Attribute part, 639
    Insertion Point part, 640
    Mode part, 639
    synchronization, 646–648
    Text Settings part, 639
Delete command, 404

Delete Constraints, 592
Delete Layer button, 150
Delta method, 84
design center, 419–422
Design Center blocks, palette creation from, 425
design drawings, 761
Design Web Format (DWF) file, 534
    exporting, 534–536
    viewing, 539–540
detach command, 678
DGN file, attaching, 667
Diameter button, 236
Diameter constraint, 587–588
diameter dimension, insertion of, 236
dimension style
    alternate units tab, 488–489
    fit tab, 484–485
    functions, 495–503
        Adjust Space, 496–498
        Break, 495–496
        Center Mark, 499
        Jog line, 498
        Justify, 501
        Oblique, 499–500
        Override, 501–502
        Text Angle, 500–501
    lines tab, 472–474
    primary units tab, 486–488
    sub dimension style, creating, 492–494
    symbols and arrows tab, 475–477
    text tab, 478–484
    Tolerances tab, 489–491
Dimensional constraints, 584, 599–600
    Aligned constraint, 586–587
    Angular constraints, 588–589
    controlling, 590–592
    Convert command, 589–590
    converting to annotational constraints, 597
    Linear, Horizontal, and Vertical Constraints, 585–586
    Radial and Diameter constraints, 587–588
    showing and hiding, 592–593
Dimensional grips, 598
dimensioning, 225–226, 469–470
dimensions, 225
    aligned dimension, insertion of, 230–231
    angular dimension, insertion of, 231–232
    arc length dimension, insertion of, 233–234
    diameter dimension, insertion of, 236
    editing a dimension block
        using grips, 248–251
        using quick properties and properties, 253–254
        using right-click menu, 252
    jogged dimension, insertion of, 238–239
    linear dimension, insertion of, 229–230
    ordinate dimension, insertion of, 239–240
    Quick Dimension command, using, 245–247
    radius dimension, insertion of, 234–235
    series of dimensions, insertion of
        using baseline command, 242–243
        using continue command, 241–242
    types, 226–228
Dimensions Option, 320
Direction button, 287
Display Boundary Objects, 193
Display when creating a new layout, 263
distance, measuring, 160
Distance and Angle, chamfering using, 109–111
Distance button, 160
Distance option, chamfering using, 109
Divide button, 339
Divide command, 339, 340
Divide option, 125
Donut button, 323
Donut command, drawing using, 323
door control, 623–624
Double option, 183
Download from Source button, 465
drafting, in AutoCAD 2015, 25
    Arc command, drawing circular arcs using, 39–41
    Circle command, drawing circles using, 37–39
    Dynamic Input, 27–28
    Line command, drawing lines using, 26
    Object Snap (OSNAP), 33–36
    Object Snaps related to circle and arc, 41
    Object Snap Tracking, 41–43
    ortho versus polar tracking, 29–32
    Polar Snap, using, 53
    polyline
        conversion to lines and arcs, 49
        joining lines and arcs to form, 50
    Polyline command, drawing lines and arcs using, 47–49
    priorities, 25
    Snap and Grid, 51–53
Drafting Settings dialog box, 52, 65

Drag-and-Drop method, 363
Draw panel, 26, 39, 47, 767
Drawing, table insertion in, 454–457
Drawing orientation, 264
Drawing review, 761–762
    controlling markups, 771–772
    creating DWF and using markup tools, 772–773, 776–777
    DWF file, 762, 777–780
    markup and measure tab. *See* markup and measure tab
    markup objects, editing, 770–771
    Markup Properties window, 772
    Markup window, 771
    using Autodesk Design Review, 763–764
    using Markup Set Manager in AutoCAD, 773–776
drawing units, 13–14
Drawing Utilities/Units, 285, 373
Drawing1.dwg, 15
DST file, 701
DWF file
    attaching, 666
    clicking and right-clicking, 684–685
    comparing, 777–780
    creating, 762, 772–773, 776–777
DWFX file
    exporting, 534–536
    viewing, 539
.dwg extension, 15
DWG file
    attaching, 663–665
    checking, 734–736
    clicking and right-clicking, 683
    linking DWS files to, 731–734
Dynamic blocks, 605–606
    Block Editor, 608–609
    chest of drawers, creating, 622
    constraints for, 626, 632–633
    creation methods, 606
    door control, 623–624
    final steps, 621
    Lookup parameter and action, 618–621
    parameter properties, controlling, 614–616
    Parameters and Actions, 605, 609–613
    Visibility parameter, controlling, 616–618
    wide flange beams, 624–626, 632
Dynamic Columns, 209
Dynamic Input, 27–28
Dynamic method, 85

E

Edge option, 359
Edit Dictionaries button, 211
Edit option, 369
Edit Polyline button, 50, 316, 326
Edit Polyline command, 326
    Decurve option, 328
    Edit Vertex option, 327–328
    Fit option, 328
    Join option, 327
    Ltype gen option, 329
    Multiple option, 330–331
    Open and Close options, 327
    Reverse option, 329–330
    Spline option, 328
    Width option, 327
Edit Reference command, 675–676
Edit Vertex option, 327–328
editing
    External Reference, 679–680
    – of dimension block
        using grips, 248–251
        using quick properties and properties, 253–254
        using right-click menu, 252
    Template file, 546–547
editing blocks, 436–438
editing text. *See* text editing
Elevation option, 318
ellipse, drawing
    using axis points, 347–348
    using the center option, 347
Ellipse button, 346
elliptical arc, drawing, 348
End Object Isolation, 399
Equal constraint, 574–575
Erase command, 66–67
Erase Source option, 358
eTransmit command, 722–723, 724
    and External Reference files, 686–690
Excel, pasting a linked table from, 367–371
Excel Data Link1, 368
existing drawing sheet set creation, 714–720
Explode button, 49, 179, 315
Explode command, 49
Explore button, 389
EXPORTDWF command, 762
Extend command, 102
extending objects, 114–115

External Reference (Xref), 661–662
   Attach command, 670–671
   attaching and controlling, 673–674
   clicking and right-clicking, 682
      DWF file, 684–685
      DWG file, 683
      image file, 683–684
   clipping, 680–682
   DWG file, editing, 674–677
   editing, 679–680
   eTransmit Command with External Reference files, 686–690
   fading, control of, 672–673
   file clipping and controlling, 685–686
   functions, 677
      bind command, 678–679
      detach command, 678
      path, 679
      reload command, 678
      unload command, 678
      Xref type, 679
   insertion of, 662
      DGN file, 667
      DWF file, 666
      DWG file, 663–665
      image file, 665
      PDF file, 668
   palette contents, 669–670
   reference file and layers, 671–672
Extract Data button, 650

F

Fence mode, 63, 64
fields button, 406–411
File tab, 3, 14, 17–18
Fillet command, 101
Fillet Option, 318–319
filleting objects, 106–107
Find and Replace command, 212, 219–220
Find text field, 220
Fit option, 72, 328
Fit Points method, 342–343
fit tab, 484–485
Fix constraint, 568
Flip action, 611
Flip parameters, 610
Format/Drawing Limits, 287
Formatting panel, 206, 765–766
Freeze and Off commands, 401
Fuzz distance, 331

G

Gap Tolerance, 188
General tab, 451–452
Generate Boundary, 197
geometric constraints, 564, 583–584
   Auto Constrain, 578–580
   Coincident constraint, 564–566
   Collinear constraint, 566–567
   Concentric constraint, 567
   Constraint Bar, 580–581
   Equal constraint, 574–575
   Fix constraint, 568
   Horizontal constraint, 570–571
   Infer constraint, 576–578
   Parallel constraint, 568–569
   Perpendicular constraint, 569–570
   settings, 575–576
   showing and hiding, 582
   Smooth constraint, 572–573
   Symmetric constraint, 573–574
   Tangent constraint, 572
   Vertical constraint, 571
Graphical Area, 3, 8–9, 10, 18
Grid Snap, 53
Grid spacing, 52
Grid tool, 52
grips
   and dynamic input, 91–93
   editing a dimension block using, 248–251
   editing using, 118–120
   and perpendicular and tangent OSNAPs, 93–94
   using, to edit objects, 87–91
Group filters, 747
   creating, 745
Gutter distance, 209

H

hatch, editing, 195–197
hatch and tool palette, 427–429
Hatch Angle, 182
hatch boundary, 192–194
Hatch Color, 181
Hatch command, 180–181
Hatch Creation context tab, 180, 185
Hatch Edit, 197
Hatch Editor tab, 195
Hatch Layer Override, 183
hatch options, controlling, 186–191
hatch origin, specifying, 185–186
Hatch Pattern Scale, 182

INDEX • 789

hatch properties, controlling, 181–183
Hatch Type, 180
hatching, 180
Hide Objects, 399
Hor option, 333
Horizontal constraint, 570–571
Hyperlink button, 372
hyperlinking AutoCAD objects, 372–373

I

Ignore Island Detection, 191
image file
    attaching, 665
    clicking and right-clicking, 683–684
imaginary circle, using, 321
Import Text button, 212
In All Viewports Except Current option, 278
Increment Angle, 31
Individual attribute values, editing, 642–644
Infer constraint, 576–578
InfoCenter, 3, 9–10
inquiring radius, 161
Inquiry commands, 159
Insert Column Break Alt+Enter option, 209
Insert panel, 208–210, 464
insertion point option, specifying, 455
interface customization, 547–548, 555–556
Isolate command, 401
Isolate Objects, 399
Isolate to Current Viewport command, 403

J

Jog line button, 498
Jogged button, 238
jogged dimension, insertion of, 238–239
Join button, 381, 383
Join option, 50, 316, 327
joining objects, 85–86
Jointype option, 330
Justification of the text, 207

L

Lasso selection, 62
Last mode, 63
Layer advanced features and filters, 747–748
layer controls, 148
    controlling layer visibility, locking, and plotting, 148–150
    deleting and renaming layers, 150–151

Layer Properties Manager, using, 153–154
    making an object's layer current layer, 151–152
    moving objects from one layer to another, 152–153
    undo only layers actions, 152
Layer option, 358
Layer panel, 401
    copying objects to New Layer Command, 402
    Current Layer command, 402
    Delete command, 404
    Freeze and Off commands, 401
    Isolate and Unisolate commands, 401
    Isolate to Current Viewport command, 403
    Layer Walk command, 403
    Lock and Unlock commands, 402
    Merge command, 403–404
    turn all layers on and thaw all layers commands, 402
Layer Previous, 152
layer properties, creating and setting, 143
    color setting for a layer, 144–145
    creating new layer, 143–144
    current layer, setting, 147
    linetype setting for layer(s), 145–146
    lineweight setting for layer(s), 146–147
Layer Properties button, 142
Layer Properties Manager, 142, 143, 144, 145, 146, 147, 150, 741–743
Layer State, creating, 748–752
    settings dialog box, 752–753, 756–757
        Dialog Settings, 755–756
        Isolate Layer Settings, 754
        New Layer Notification, 754
Layer States Manager, 748
Layer Translator, 737–741
Layer Walk command, 403
layer's transparency, 404–406
layers concept in AutoCAD, 141–143
Layout panel, 267
Layout tab, 3, 267, 271
Layout Viewports panel, 277
layouts, 260–265
    creation, using copying, 265–268
Leader Format tab, 505–506
Leader Structure tab, 506–507
LEARN option, 2–3
Length Type, 286
lengthening objects, 84–85

Line button, 26
Line command, drawing lines using, 26
Line Spacing of paragraph, 208
Linear button, 229
linear dimension, insertion of, 229–230
Linear parameter, 610
Linear/Horizontal/Vertical constraints, 585–586
lines and arcs, drawing, using Polyline command, 47–49
lines tab, 472–474
Linetype field, 145
Linetype scale, 304, 307
Lineweight, 146
Load button, 146
locate Origin panel, 185
Lock command, 402
Locking option, 370
Lookup action, 611
Lookup parameters, 610
    and action, 618–621
Ltype gen option, 329

M

Make Current button, 151
Manage Xrefs, 668
Manager Attributes button, 646
markup and measure tab, 764–770
    callouts panel, 766–767
    clipboard panel, 765
    draw panel, 767
    formatting panel, 765–766
    measure panel, 768–769
    stamps and symbols panel, 769–770
markup objects, editing, 770–771
Markup Set Manager in AutoCAD, 762, 773–776
Markups, controlling, 771–772
    Markup Properties window, 772
    Markup window, 771
Match button, 152
match properties, using, 360–361
match properties across files, 363
Match Properties button, 188, 360
Maximize Viewport button, 277
Measure button, 339
Measure command, 339–340
    with Block option, 340
measure panel, 768–769
mechanical project – I (imperial), 306–310
mechanical project – I (metric), 303–306
mechanical project – II (imperial), 311
mechanical project – II (metric), 310–311
Merge command, 403–404
Merge panel, 459–460
Method option, 110
Mirror command, 78–79
Model space, 259, 528
Model tab, 3, 14
Model View tab, 700
Modify button, 263
Modify panel, 6, 49, 50, 64
Modifying commands, 64
Move command, 70–71
moving an object from a layer to another layer, 156
MS Excel, data sharing coming from, 366–367
MS Word, data sharing coming from, 364–366
Mtext, 230
multileader dimension, inserting, 509–514
multileader style, creating, 504–509
    Content tab, 507–509
    Leader Format tab, 505–506
    Leader Structure tab, 506–507
multiline text, 203
    Close panel, 213
    Formatting Panel, 206
    Insert Panel, 208–210
    Options Panel, 212
    Paragraph panel, 206–208
    Spell Check Panel, 210–211
    Style panel, 205
    text editor, 213–214
    Tools Panel, 211–212
Multiline Text button, 203
Multiple option, 103–104, 110, 330–331
Multiple Points button, 338
Multiple polylines, 327
multiple rectangular viewports, 270–271

N

Name field, 144
Named button, 270
Named command, 381–382, 384
Named Plot Style Tables, 522–528
Navigation Bar, 3
New button, 14
New command, creating, 557–559
new dimension style, creating, 470–472
New Layer button, 143
New Layout option, 261

New Panel, creating, 548–552
New Tab, creating, 553
New Workspace, creating, 554–555
New/Delete/Rename Palette, 433
No Islands detection, 191
No trim option, 106, 109
Normal Island Detection, 190
Noun/Verb technique, 64–65

O

Object option, 164
Object Snap (OSNAP), 33
    activating running, 34–36
    OSNAP Override, 36
Object Snap Settings option, 35
Object Snap Tracking button, 41–43
Object Snap Tracking Settings, 43
Object Snaps related to circle and arc, 41
object visibility, 399–400
objects, selecting, 59
    Crossing mode, 61
    Crossing Polygon mode, 62
    Fence mode, 63
    Lasso selection, 62
    Last, Previous, and All modes, 63
    Noun/Verb technique, 64–65
    Window mode, 61
    Window Polygon mode, 61
Oblique button, 499
Offset command, 101, 357
    Erase Source option, 358
    Layer option, 358
    system variable, 358
Offset Distance Option, offsetting using, 102–103
Offset option, 334
offsetting objects
    using Offset Distance Option, 102–103
    using Through Option, 103
OLE (Object Linking & Embedding), 364
OLE Text Size dialog box, 366
Open and Close options, 327
Open button, 15–17
Open command, 676–677
Options Panel, 212
Ordinate button, 240
ordinate dimension, insertion of, 239–240
Ortho function, 29–30
Ortho versus Polar Tracking, 29
    additional angles, 31
    Increment Angle, 31
    polar angle measurement, 32
Outer Island Detection, 190
Over Constraining, 583
Override button, 502
own command, creating, 557–559

P

Page Setup Manager, 262
Paper Size, 263
Paper Space, 259
Paragraph panel, 206–208
Parallel constraint, 568–569
Parameter properties, controlling, 614–616
Parameter Sets, 612
Parameters, 605, 609–613
Parameters Manager, 593–596
Parametric constraints, 563–564
    dimensional. *See* Dimensional constraints
    geometric. *See* Geometric constraints
Partial Load, 413–414
partially opened files, use of, 412–414
Paste as Hyperlink, 373
Paste Special command, 365
pasting objects, 363
path array, 122–128
Pattern panel, 181
PDF file
    attaching, 668
    exporting, 534–536
PEDITACCEPT, 331
Percentage method, 84
Perpendicular constraint, 569–570
Pick Points button, 173, 193, 350
Plot command, 281–282, 762
Plot Offset, 263
Plot options, 264
Plot Scale, 263
Plot style table, 264, 517
    Color-dependent, 517–522
    named, 522–528
plotting, 259
    layouts, 260
        creation, using copying, 265–268
        Model Space and Paper Space, 259
        plot command, 281–282
        steps to create new layout
            from scratch, 260–264
            using a template, 264–265

viewports. *See* viewports
Point command, 338
Point parameters, 610
Point Style button, 337–338
polar angle measurement, 32
polar array, 129–132
Polar command, 43
Polar parameters, 610
Polar Snap button, 53
Polar Stretch action, 611
Polar Tracking dialog box, 30, 53
Polygon button, 320
Polygon command, drawing using, 320
   imaginary circle, using, 321
   using length and angle of one of the edges, 322
Polygonal option, 271, 273–274
polygonal viewport, 271–272
Polyline button, 47, 314
Polyline command, 314–315
   drawing lines and arcs using, 47–49
polylines
   conversion, to lines and arcs, 49, 315–316
   joining lines and arcs to form, 50, 316
Preview button, 282
Previous mode, 63
primary units tab, 486–488
projects, 285
   architectural project (imperial), 288–295
   architectural project (metric), 295–302
   drawing for a new project, 285–288
   mechanical project – I (imperial), 306–310
   mechanical project – I (metric), 303–306
   mechanical project – II (imperial), 311
   mechanical project – II (metric), 310–311
Properties command, 158–159
Properties option, 196, 253
Properties palette, 196
Properties panel, 181
Property filter menu, 746
Property filters, creating, 743–745
Publish Options, 538
Purge command, 373
purging items, 373–375

Q

Quick Access Toolbar, 3, 5, 14, 16, 20
   creating, 553–554
Quick Dimension button, 245
Quick Dimension command, using, 245–247
Quick Properties, 156–158, 195
   editing using, 120
Quick Select command, 394–397

R

Radial constraint, 587–588
Radius button, 161, 235
radius dimension, insertion of, 234–235
Ray button, 334
reading instantaneous information about an object, 156
Recreate button, 193
Rectangle button, 317
Rectangle command, drawing using, 317
   Area Option, 319–320
   Chamfer Option, 318
   Dimensions Option, 320
   Elevation Option, 318
   Fillet Option, 318–319
   Rotation Option, 320
   Thickness Option, 319
   Width Option, 319
rectangular array, 116
   Array Creation context tab, using, 117–118
   editing, using context tab, 120
   editing, using grips, 118–120
   editing, using Quick Properties, 120
   first step, 116–117
Rectangular button, 269
Redo command, 5, 19, 20
Reference Option, 75
References panel, 662
Region button, 351
Region command, using, 351
   performing Boolean operation on regions, 352
Relaxing Constraints, 582–583
reload command, 678
Remove Vertex, 93
Rename option, 261
Retain Boundary pop-up list, 194
Reverse option, 329–330
Revision Cloud command, 324–325
Ribbons, 3, 5, 6–9
Rotate action, 611
Rotate command, 74–75
Rotate parameters, 610
Rotation Option, 320
Rows panel, 118, 459

S

Save Block button, 437
Scale command, 76–77
Select button, 294, 301
Select New Boundary Set button, 194
Select objects button, 173
Select Similar command, 397–398
Selected page setup details, 262–263
Selection Cycling, 65–66
Set Boundary, 197
Set Current button, 147
Set Origin button, 186, 197
Settings dialog box, 752–757
    Dialog Settings, 755–756
    Isolate Layer Settings, 754
    New Layer Notification, 754
Shaded viewport options, 264
Sheet Set Manager, 696, 699, 720
Sheet sets, 695–696
    adding and scaling model views, 710–714
    file setup, 701
    opening, manipulating, and closing, 702
    publishing, 720–721, 724
    Sheet Set Manager palette, 696–701
    using an example, 702–705, 709–710
        adding sheets in subsets, 706–708
        sheet control, 708–709
    using Archive command, 723
    using eTransmit command, 722–723, 724
    using existing drawings, 714–720
Sheet Views, 700
Show Filter Tree, 154
Show Menu Bar, 5
Show/Hide Lineweight button, 147
simple area, calculating, 163–164
Single Line button, 201
Single line text, creating, 447–448
single line text, using, 201–202
single rectangular viewports, 269–270
Smooth constraint, 572–573
Snap and Grid, 51–53
Snap mode, 51
Snap Settings option, 52, 53
Spell Check Panel, 210–211, 219
spline, editing, 344–345
Spline command, 328, 342
    Control Vertices method, 343–344
    editing spline, 344–345
    Fit Points method, 342–343

stamps and symbols panel, 769–770
Standard text style, 445
starting AutoCAD, 1–3
Static Columns, 209
status bar, 3, 10–11
Stretch action, 611
Stretch command, 80–81
Stretch Vertex, 93
Style panel, 205
sub dimension style, creating, 492–494
Subtract area, 164
Symbols, 210
symbols and arrows tab, 475–477
Symmetric constraint, 573–574
Synchronize button, 648

T

table cell functions, using, 458–466
    Cell Format panel, 462–463
    Cell Styles panel, 460–462
    Columns panel, 459
    Data panel, 464–465
    Insert panel, 464
    Merge panel, 459–460
    Rows panel, 459
table cells, formulas in, 457–458
table insertion in drawing, 454–457
    insertion point option, specifying, 455
    window option, specifying, 455–456
table style, creating, 448–453
    Borders tab, 453
    General tab, 451–452
    Text tab, 452–453
tables, creating, 443–444
Tangent constraint, 572
Template file
    creation, 543–545
    editing, 546–547
text, writing. *See* writing text
Text alignment, 483
text and tables, creating, 443–444
Text Angle button, 500
text editing, 216
    double-click text, 216–217
    editing using grips, 218
    quick properties and properties, 217
Text Editor, 205, 213–214
text style, creating, 444–448
Text tab, 452–453

dimension style, 478–484
Thaw All Layers command, 402
Thickness Option, 319
Through Option, offsetting using, 103
Tolerances tab, 489–491
tool palettes, 422–427
    creation from Design Center blocks, 425
    creation from scratch, 424
    filling new palette with content, 424
    tools properties, customizing, 425–427
Tools Panel, 211–212
Total method, 84
Tracking Settings, 31
transparency, 182, 430–431
Trim and Extend, 358
Trim command, 101
Trim option, 107, 109, 110
trimming objects, 111–113
Turn All Layers On command, 402

U

Undo command, 5, 19, 20, 26, 110
Unisolate command, 401
Unload Command, 678
Unlock command, 402
Unlocked option, 370
Update All Data Links option, 368
Upload to Source button, 370
U.S. National CAD Standard (NCS), 695
Use Boundary Set, 194
Utilities panel, 159, 161

V

Ver option, 333
Vertical constraint, 571
View Manager button, 376
View Options, 431–432
ViewCube, 3
Viewport Configuration button, 379–384
Viewport dialog box, 378

Viewport Spacing, 271
viewports, 260, 269
    after creation, 274–275
    creation
        by clipping existing viewports, 273–274
        by converting existing objects, 272–273
    freezing layers in, 278
    layer override in, 279
    multiple rectangular, 270–271
    polygonal, 271–272
    scaling and maximizing, 275–277
    single rectangular, 269–270
    using views in, 378–379
views, creating, 376–378
visibility, of object, 399–400
Visibility parameters, 610
    controlling, 616–618
VP Freeze layer, 278

W

wide flange beams, 624–626, 632
Width option, 48, 314, 319, 327
Window mode, 61, 64
window option, specifying, 455–456
Window Polygon mode, 61, 64
Workspace, 3, 5, 554
Write Block button, 178
writing text, 201
    editing text, 216
        double-click text, 216–217
        editing using grips, 218
        quick properties and properties, 217
    multiline text, using. *See* multiline text
    single line text, using, 201–202
    spell check and find and replace, 219–220

X

Xdatum option, 240
XY parameters, 610